IoT-Based Data Analytics for the Healthcare Industry

IoT-Based Data Analytics for the Healthcare Industry
Techniques and Applications

Edited by

Sanjay Kumar Singh

Department of Computer Science and Engineering, Indian Institute of Technology (BHU), Varanasi, Uttar Pradesh, India

Ravi Shankar Singh

Department of Computer Science and Engineering, Indian Institute of Technology (BHU), Varanasi, Uttar Pradesh, India

Anil Kumar Pandey

Banaras Hindu University (BHU), Varanasi, Uttar Pradesh, India

Sandeep S. Udmale

Department of Computer Engineering and IT, Veermata Jijabai Technological Institute (VJTI), Mumbai, Maharashtra, India

Ankit Chaudhary

Department of Computer Science, The University of Missouri, St. Louis, MO, United states

Series Editor

Fatos Xhafa

Universitat Politècnica de Catalunya, Spain

Academic Press is an imprint of Elsevier
125 London Wall, London EC2Y 5AS, United Kingdom
525 B Street, Suite 1650, San Diego, CA 92101, United States
50 Hampshire Street, 5th Floor, Cambridge, MA 02139, United States
The Boulevard, Langford Lane, Kidlington, Oxford OX5 1GB, United Kingdom

Notices
Knowledge and best practice in this field are constantly changing. As new research and experience broaden our
understanding, changes in research methods, professional practices, or medical treatment may become
necessary.

Practitioners and researchers must always rely on their own experience and knowledge in evaluating and using
any information, methods, compounds, or experiments described herein. In using such information or methods
they should be mindful of their own safety and the safety of others, including parties for whom they have a
professional responsibility.

To the fullest extent of the law, neither the Publisher nor the authors, contributors, or editors, assume any liability
for any injury and/or damage to persons or property as a matter of products liability, negligence or otherwise, or
from any use or operation of any methods, products, instructions, or ideas contained in the material herein.

Library of Congress Cataloging-in-Publication Data
A catalog record for this book is available from the Library of Congress

British Library Cataloguing-in-Publication Data
A catalogue record for this book is available from the British Library

ISBN 978-0-12-821472-5

For information on all Academic Press publications
visit our website at https://www.elsevier.com/books-and-journals

Publisher: Mara Conner
Acquisitions Editor: Sonnini R. Yura
Editorial Project Manager: Rafael G. Trombaco
Production Project Manager: Omer Mukthar
Cover Designer: Victoria Pearson

Typeset by SPi Global, India

Contents

CHAPTER 3   **Characteristics of IoT health data** **31**

Ritesh Sharma, Sandeep S. Udmale, Anil Kumar Pandey, and
Ravi Shankar Singh

CHAPTER 4   **Health data analytics using Internet of things** **47**

Vishakha Singh, Sandeep S. Udmale, Anil Kumar Pandey, and
Sanjay Kumar Singh

SECTION II   IoT services in health industry

Vishakha Singh, Sandeep S. Udmale, Anil Kumar Pandey, and
Sanjay Kumar Singh

Righa Tandon and P.K. Gupta

Sujit Bebortta and Dilip Senapati

SECTION III   Applications of IoT for human

CHAPTER 17   Internet of animal health things (IoAT): A new frontier in animal biometrics and data analytics research 261

Santosh Kumar, Sunil Kumar, Prerna Mishra, and Mithilesh K. Chaube

Contributors

Tanveer Ahmed
Bennett University, Greater Noida, Uttar Pradesh, India

Gaurav Baranwal
Department of Computer Science, Institute of Science, Banaras Hindu University, Varanasi, Uttar Pradesh, India

Sujit Bebortta
Department of Computer Science, Ravenshaw University, Odisha, India

Vandana Bharti
Department of Computer Science and Engineering, Indian Institute of Technology (BHU), Varanasi, Uttar Pradesh, India

Bhaskar Biswas
Department of Computer Science and Engineering, Indian Institute of Technology (BHU), Varanasi, Uttar Pradesh, India

Mithilesh K. Chaube
Mathematical Sciences, Dr. SPM IIIT-Naya Raipur, Atal Nagar, Raipur, Chhattisgarh, India

Ankit Chaudhary
Department of Computer Science, The University of Missouri, St. Louis, MO, United states

P.K. Gupta
Department of Computer Science and Engineering, Jaypee University of Information Technology, Solan, Himachal Pradesh, India

M. Jayashankara
Department of Computer Science and Engineering, P.E.S. College of Engineering, Mandya, Karnataka, India

Abhinav Kumar
Department of Computer Science and Engineering, Indian Institute of Technology (BHU), Varanasi, Uttar Pradesh, India

Dinesh Kumar
Department of Computer Science & Engineering, Motilal Nehru National Institute of Technology, Allahabad, Uttar Pradesh, India

Santosh Kumar
Department of Computer Science and Engineering, Dr. SPM IIIT-Naya Raipur, Atal Nagar, Raipur, Chhattisgarh, India

Sunil Kumar
Marine Engineering Research Institute (MERI), Kolkata, West Bengal, India

Kanak Manjari
Department of Computer Science Engineering, Bennett University, Greater Noida, India

Ashish Kumar Maurya
Department of Computer Science & Engineering, Motilal Nehru National Institute of Technology, Allahabad, Uttar Pradesh, India

Prerna Mishra
Department of Computer Science and Engineering, Dr. SPM IIIT-Naya Raipur, Atal Nagar, Raipur, Chhattisgarh, India

Saniksha Murria
CT Institute of Management & Information Technology, Jalandhar, Punjab, India

Anil Kumar Pandey
Banaras Hindu University (BHU), Varanasi, Uttar Pradesh, India

Jagdish Lal Raheja
Control and Automation Unit, CSIR-CEERI, Pilani, Rajasthan, India

Rajinder Sandhu
Department of Computer Science and Engineering, Jaypee University of Information Technology, Solan, Himachal Pradesh, India

Sonal Saxena
Division of Veterinary Biotechnology, ICAR-Indian Veterinary Research Institute, Bareilly, Uttar Pradesh, India

Dilip Senapati
Department of Computer Science, Ravenshaw University, Odisha, India

Anil Sharma
Lovely Professional University, Phagwara, Punjab, India

Anshul Sharma
Department of Computer Science and Engineering, Indian Institute of Technology (BHU), Varanasi, Uttar Pradesh, India

Ritesh Sharma
Department of Computer Science and Engineering, Indian Institute of Technology (BHU), Varanasi, Uttar Pradesh, India

Sameer Shrivastava
Division of Veterinary Biotechnology, ICAR-Indian Veterinary Research Institute, Bareilly, Uttar Pradesh, India

Kaushal Kumar Shukla
Department of Computer Science and Engineering, Indian Institute of Technology (BHU), Varanasi, Uttar Pradesh, India

Gaurav Singal
Department of Computer Science Engineering, Bennett University, Greater Noida, India

Ravi Shankar Singh
Department of Computer Science and Engineering, Indian Institute of Technology (BHU), Varanasi, Uttar Pradesh, India

Rishav Singh
NIT-Delhi, Delhi, India

Ritika Singh
CSIR-CSIO, Chandigarh, India

Sanjay Kumar Singh
Department of Computer Science and Engineering, Indian Institute of Technology (BHU), Varanasi, Uttar Pradesh, India

Sanjay Kumar Singh
Rajarshi School of Management and Technology, Varanasi, India

Vishakha Singh
Independent Researcher

Righa Tandon
Department of Computer Science and Engineering, Jaypee University of Information Technology, Solan, Himachal Pradesh, India

Shrikant Tiwari
Department of Computer Science and Engineering, Shri Shankaracharya Technical Campus, Bhilai, Chhattisgarh, India

Sandeep S. Udmale
Department of Computer Engineering and IT, Veermata Jijabai Technological Institute (VJTI), Mumbai, Maharashtra, India

Shefali Varshney
Department of Computer Science and Engineering, Jaypee University of Information Technology, Solan, Himachal Pradesh, India

Madhushi Verma
Department of Computer Science Engineering, Bennett University, Greater Noida, India

Rohit Verma
Manipal Institute of Technology, Manipal, Karnataka, India

Rimmy Yadav
CT Institute of Management & Information Technology, Jalandhar, Punjab, India

Preface

The Internet of things (IoT) is a new revolution of the Internet era that is rapidly discovering the research path in numerous academics and industrial domains, particularly in the healthcare industry. A remarkable proliferation of wearable devices, as well as smartphones, has assisted the IoT-enabled technology to evolve the healthcare industry for offering numerous and potentially innovative services to today's digital world. The smart healthcare industry improves the medical treatment process, efficiently manages the multiple resources, and eases the insurance claiming process, etc. The technology-enabled system identifies the current needs and presents future directions to medical organizations, manufacturers, service providers, and application developers. As a result, this is the most promising and demanding paradigm for providing the various medical services to a vast world population. Besides, it allows the medical facility providers to reach remote locations, correspond to experts, and understand the medicinal positions in a diversified environment. As an effect, it reduces the costs as well as saves time of patients and allows the doctors as well as supporting staff to attain more users in time.

The smart healthcare industry has a big market and provides ample opportunities for the business. The medical equipment manufacturers, Internet service providers, insurance companies, and software developers are the main stakeholder of the intelligent healthcare industry. The design, development, and deployment of a well-organized intelligent healthcare system require reliable hardware and software platforms with IoT-enabled technologies like communication networks. This healthcare system gets evolved over time for providing the new advantageous services to end-users, and thus, stakeholders must provide support for the scalability of the system. Nowadays, the healthcare industry has extended its services to agriculture, pet, and wild animals for tracking, monitoring, identifying diseases, etc. It presents the new challenges like designing a new device/sensor for providing the IoT-enabled solution to a specific animal.

Objective

The IoT-based healthcare industry provides a broad range of opportunities to identify and examine the new architectures and methods, theories and algorithms, intelligent analysis approaches, and construction of the practical application. Thus, this book aims to present a common platform to academicians as well as practitioners to observe the recent advances in the health industry through IoT, which consists of IoT architectures, platforms, data analysis methods, and health industry applications. The book introduces state-of-the-art-based approaches based on IoT and intelligent methods, challenges, opportunities, and future research directions. The book presents a potential thought and comprehensive review of emerging methodology for health IoT data analytics that provide valuable information for creating new knowledge for the future to develop novel prediction methods for the health industry.

Organization of book

This book is organized into four sections. Each section helps to understand the importance of IoT in the health industry and various research as well as business opportunities. The summary of each section is as follows.

Section I reveals the history of IoT, the importance of IoT in healthcare with challenges for the implementation of IoT-based system and business opportunities. The personal and clinical smart architectures are described with a variety of sensors. The importance of a variety of sensors in designing the human healthcare system is explored along with different data types. The overview of data analysis methods is presented with challenges. Also, computational techniques are introduced as an arsenal tool to meet the continuously rising and demanding needs of the healthcare industry.

Section II presents the role of various services offered by the IoT-based healthcare system to patients, physicians, hospitals, and insurance companies. Also, there have been various decision-making tools, and methods are available for supporting healthcare decision making. However, multicriteria decision analysis techniques are exploited to provide effective services for a healthcare decision process. As a consequence, the analytic method requires analysis and mining of useful information from substantial medical data by handling the semantic interoperability issues in an electronic health record system. Besides, the extension of IoT services is discussed for offering the security and privacy to electronic health records as well as a smart healthcare system and assists the insurance companies to avoid fraud claims.

A potential roadmap for developing the healthcare applications by enabling seamless integration of IoT with existing healthcare infrastructure is focused in Section III. Smart healthcare systems designed for remote monitoring and personalized healthcare are highlighted with a challenge to the medical frontier. Thus, challenging applications such as critical heart monitoring and pain monitoring are discussed in detail. Further, an assistive solution for visually impaired is introduced to reduce the fraud cases and attacks by stray animals.

Finally, the potential of transforming the animal health paradigm through the rapid expansion of IoT in the animal sector is covered in the last section by introducing the Internet of animal health things. Smart animal farming and management methods are focused on providing the novel perspectives and opportunities for exploitation of the technology in the animal sector. As a result, the IoT provides a promising solution to livestock management, control, and prevention of infectious diseases like COVID-19, and developing the digital pathology.

Health IoT data analytics

Internet of things in the healthcare industry

Sandeep S. Udmale[a], Anil Kumar Pandey[b], Ravi Shankar Singh[c], and Sanjay Kumar Singh[c]

Department of Computer Engineering and IT, Veermata Jijabai Technological Institute (VJTI), Mumbai, Maharashtra, India[a] Banaras Hindu University (BHU), Varanasi, Uttar Pradesh, India[b] Department of Computer Science and Engineering, Indian Institute of Technology (BHU), Varanasi, Uttar Pradesh, India[c]

1 Introduction

The Internet is connecting a number of physical objects at an exceptional rate to recognize the initiative of the Internet of things (IoT) [1]. A basic example in the healthcare domain of such an object is tracking of medical equipment like nebulizers, wheelchairs, ventilators, etc. IoT has played a crucial role in other fields for enhancing the quality of human life. These applications broadly include smart homes and cities, industrial automation, transportation, etc. [1–6].

The IoT assists the physical objects in observing and hearing the environment and executing various tasks by communicating in coordination [1, 5]. The sharing of valuable information helps to make a combined decision. The IoT framework takes advantage of technologies like sensor networks, pervasive computing, embedded devices, etc. for transforming the physical objects from traditional to smart. Thus, a domain-specific IoT-based application is developed for a particular objective with intelligent objects, advanced technologies, and analytic services [1, 5, 7, 8].

Over the years, IoT has demonstrated significant contributions in the personal lives of human beings as well as various businesses. As a result, it has allowed the improvement in the quality of life and noteworthy progress in the financial system [1, 5, 7]. For example, the smart cities enable the ambulance to travel with minimum hurdles, inform the doctors, and support staff for the monitoring of patients and necessary prepreparation. To recognize this prospective escalation, advanced communication technologies, services, and applications are required to develop simultaneously for providing customer satisfaction and compete with market demands. Besides, the smart devices are required for customer ever-increasing needs to achieve the objective in terms of availability anywhere and anytime [1, 5, 7–10].

Further, the communication between the heterogeneous objects requires a standardized architecture for providing the integrated approaches to many real-world problems. Thus, these architectures are the backbone for the design and development of IoT-based solutions with competitive environments between the firms for the delivery of quality products and services. Also, the Internet, as well as security and privacy architectures, are the critical role players due to the association of heterogeneous objects in the IoT-based solutions [1, 5, 7–11].

Based on the previous discussion, this chapter provides an overview of IoT in the healthcare industry. Also, the analysis of future business opportunities for the different contributors to the healthcare

IoT Based Data Analytics for the Healthcare Industry. https://doi.org/10.1016/B978-0-12-821472-5.00020-X

industry is presented. Further, the quality of service measures is expressed for the design and development of a smart healthcare system.

The remaining chapter is planned into five sections. The origin of IoT is discussed in Section 2, and Section 3 provides information about the importance of IoT in the healthcare industry. Sections 4 and 5, respectively, present the business opportunities and quality of services requirements of IoT. Section 6 concludes the chapter.

2 Origin of Internet of things

The IoT was introduced in 1999 by Ashton et al. [12] with radio-frequency identification (RFID) technology by considering it as uniquely identifiable interoperable connected objects. Brock coined the term "Auto-ID" by establishing the Auto-ID center at the Massachusetts Institute of Technology (MIT) [13]. Any type of identification technologies are represented by it for the different types of applications, and thus, relevant Electronic Product Code (EPC) was initiated in 2003 by the center. As a result, moving objects can be monitored, and therefore, the EPC network provides the big representation of the IoT paradigm to an overall commercial mainstream. As an effect, the new era of technology has started in academic as well as industry [14]. The evolution of IoT is illustrated in Fig. 1 [15].

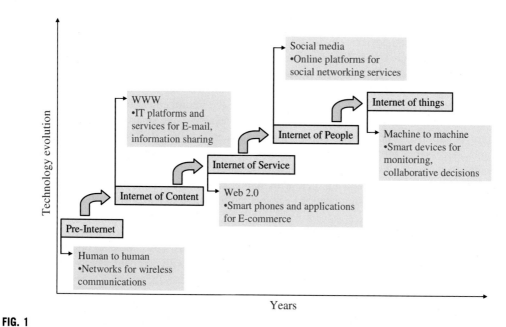

FIG. 1

Evolution of the IoT [15].

The suggestion about IoT was given by the International Telecommunication Union (ITU) for the integration of technologies like sensor networks, pervasive computing, embedded devices, etc. to connect the objects in the world for perceiving, sensing, tagging, controlling things over the Internet [14, 16]. Thus, the IoT framework consists of multiple technologies to perform the task of interaction and communication through a wide variety of networking devices. Also, it is supported with storage as well as processing capabilities of fog/edge and cloud computing. As a consequence, commercial IoT-based systems are designed and developed in the domain like healthcare, transportation, industrial, etc. with motivation to provide the quality of life to a human being [5, 6, 10].

3 Internet of things in healthcare industry

Enjoying the quality of life is the primary goal of a human being. Generally, health is the primary concern to achieve the quality of life and the various health-related issues get initiated mostly by aging. Thus, the rise in the aging population has introduced new challenges in the healthcare industry. The rehabilitation after the medical emergency for older people is a new challenge as well as the process faces the obstacles. The training or therapy of recovery demands a long-time commitment from the patients. The availability and ease of access to rehabilitation resources and services are relatively tricky due to a rise in the population in current society [14, 17, 18].

One promising solution to mitigate the issue, as mentioned earlier, is to implement an IoT-based healthcare industry for providing smart medical services. IoT-based medical solutions assist in providing intelligent services at remote locations and offer favorable treatment to patients. Thus, IoT-based creative medical solutions are vital in maximizing the utilization of available resources and improving the overall healthcare system. The healthcare activities are performed by connecting the resources as a network over the Internet to complete the task like diagnosis, monitoring, etc. [14, 19, 20].

IoT-based intelligent technology has demonstrated extraordinary progress in the healthcare system by developing numerous applications like patient monitoring, smart ambulance, remote surgeries, etc. [21]. Also, it has shown a significant improvement in the animal healthcare system. Thus, this intelligent healthcare system is viewed as a subsystem of a smart city [22]. However, the healthcare industry has a dedicated framework for providing uninterrupted healthcare services from hospitals to end-users. This advancement illustrates the effectiveness and promising future of IoT-based technology in the healthcare industry [14, 19–21].

Generally, an IoT-based healthcare system mainly consists of three components: end-users, server, and things. End-users include the patients, doctors, and hospital supporting staff. They operate dedicated health applications for communication with the help of smartphones, personal computers/laptops, tablets, etc. The server is responsible for providing the various healthcare services like database management, data analysis, security, and privacy, etc. to end-users. Things are the objects that are connected through the networks [14, 19–21].

Further, the smart healthcare industry not only provides the environment to patients but also doctors, hospitals, insurance companies, and others get benefited. The smart healthcare industry supports hospitals for managing various resources like medical equipment; store the consolidated records of patients, and transfer them as per the requirement to different locations. A trustworthy environment is provided to insurances company for improving the process of insurance [14, 19–21].

4 Business opportunity in IoT-based healthcare industry

The IoT provides great business opportunities to the Internet service provider, medical instrument manufacturer, and software industry for the development of health-related application. It is observed that the number of connected machines has grown in the last few years, and also, machine-to-machine (M2M) communication has enhanced. It is anticipated that approximately 200 billion IoT smart objects are deployed globally at the end of 2020. As a result, M2M communication will get enhanced and nearly 45% of Internet traffic compose of M2M traffic by the end of 2022 [1, 23, 24].

Further, IoT-based services have shown substantial business potential and also a significant impact on economic growth. This effect has been seen in healthcare and related industries. Thus, it is expected that the mobile health application through electronic media will attain more than $1.0 trillion global business by the end of 2025. These applications are widely accepted and have displayed a great impact on society [1, 24]. For example, the Government of India has launched the Aarogya Setu applications during the COVID-19 pandemic situation to enable medical awareness, prevention, and monitoring services to the world-largest populated country. It has helped the people to understand the current pandemic condition in their respective areas and assist the government in addressing the mass population with various beneficiary decisions, notification, and collection of data.

This projected business illustrates the effect of IoT-based applications on the global economy in the upcoming years. It is important to note that the above-mentioned statistics are sufficient to point the prospective of IoT-based business in the healthcare industry in the future. As a result, the traditional healthcare business will get transformed into the smart healthcare industry. This transformation will be seen in manufacturing firms, insurance companies, hospitals, and also the start of application development for the animal's health monitoring.

5 Quality of service in IoT-based healthcare industry

The vision of IoT-based healthcare industry can be achieved by providing the quality of services to end-users. Attaining the criterion of quality of services is a crucial challenge of IoT-based healthcare industry. It includes measures like availability, reliability, mobility, performance, management, scalability, interoperability, security, and privacy [1–4]. Application developers, as well as service providers, have to consider these challenging measures for executing the services efficiently and they are discussed as follows:

- **Availability**: The IoT-based system presents anywhere and anytime services to end-users, and thus, availability must be perceived through the software as well as hardware layers. The IoT applications must offer the software availability to the end-user by simultaneously providing the multiple services irrespective of the locations. The continuous existence of devices refers to hardware availability and must also be compatible with different platforms, protocols, and functionalities [1–4].
- **Reliability**: The proper functioning of the IoT-based system with respect to its specification refers to the reliability and aims to deliver the uninterrupted services to the users. This measure is essential in the healthcare system under emergency conditions. Thus, the reliability must be achieved by implementing the IoT system through the software as well as hardware levels [1–4].

- **Mobility**: Most of the healthcare applications are utilized through smartphones by the users and, thus, provide the continuous desired services to end-users and one of the requirements is while moving across the network. As a result, the major challenge of mobility is to offer services while moving from one network to another [1–4].
- **Performance**: The various underlying technologies are employed to construct the IoT-based healthcare system for providing valuable health services to users. Thus, the performance evaluation of individual technologies needs to be assessed with the help of multiple metrics. The performance of the IoT system includes device form factor, network speed, power consumption, processing rate, etc. Besides, user satisfaction with respect to services and cost is an essential parameter of evaluation [1–4].
- **Management**: The large numbers of smart devices are connected through the Internet and supervise this device for the monitoring of the faults, accountability, configuration, security, and performance. The effective management of applications and devices is required for the success and growth of the IoT system [1–4].
- **Scalability**: Scalability is an ability of the IoT system to accept the new devices and update the functionality and services for the customers without affecting the existing platforms. The addition of new devices and operations is a difficult task in diversified IoT platforms. Besides, the design of IoT applications must support the upgradation of applications for the new services [1–4].
- **Interoperability**: Various IoT platforms handle a large number of heterogeneous things, and as a result, providing end-to-end interoperability is an open issue. The device manufacturer and software developer must ensure the interoperability for the delivery of various services irrespective of the hardware platforms available toward the end-users. Thus, it is an important parameter for the designing and development of IoT systems [1–4].
- **Security**: The IoT systems are designed and developed on diversified platforms and utilize heterogeneous communication networks to exchange information between the large numbers of smart objects. Thus, security is a big challenge for providing IoT-based solutions to users. Besides, the lack of standard architectures and protocols generates additional hurdles in implementing security solutions [1–4].
- **Privacy**: The IoT system has collected a large number of user's data. Hence, providing the privacy to access the profile information to the owner as well as authorized users like doctors is an extremely important and difficult task. In addition, privacy needs to be maintained while exchanging the user's sensitive information [1–4].

6 Conclusion

The IoT is an emerging technology that is quickly able to discover the road in our lives to improve our quality of life. IoT technology has transformed the healthcare system by modernizing the hospital management and services for the patients. Besides, the healthcare industry provides ample business opportunities to manufacturers, software firms, insurance companies, and various service providers. Further, the challenges in the design and development IoT system are presented from the quality of service point of view.

References

[1] A. Al-Fuqaha, M. Guizani, M. Mohammadi, M. Aledhari, M. Ayyash, Internet of things: a survey on enabling technologies, protocols, and applications, IEEE Commun. Surv. Tutor. 17 (4) (2015) 2347–2376.

[2] J.A. Stankovic, Research directions for the internet of things, IEEE Internet Things J. 1 (1) (2014) 3–9.

[3] S. Chen, H. Xu, D. Liu, B. Hu, H. Wang, A vision of IoT: applications, challenges, and opportunities with China perspective, IEEE Internet Things J. 1 (4) (2014) 349–359.

[4] Z. Sheng, S. Yang, Y. Yu, A.V. Vasilakos, J.A. Mccann, K.K. Leung, A survey on the IETF protocol suite for the internet of things: standards, challenges, and opportunities, IEEE Wireless Commun. 20 (6) (2013) 91–98.

[5] S. Li, L.D. Xu, S. Zhao, The internet of things: a survey, Inf. Syst. Front. 17 (2) (2015) 243–259.

[6] L. Atzori, A. Iera, G. Morabito, The internet of things: a survey, Comput. Netw. 54 (15) (2010) 2787–2805.

[7] A.H. Ngu, M. Gutierrez, V. Metsis, S. Nepal, Q.Z. Sheng, IoT middleware: a survey on issues and enabling technologies, IEEE Internet Things J. 4 (1) (2017) 1–20.

[8] P.P. Ray, A survey of IoT cloud platforms, Futur. Comput. Inform. J. 1 (1) (2016) 35–46.

[9] O. Salman, I. Elhajj, A. Chehab, A. Kayssi, IoT survey: an SDN and fog computing perspective, Comput. Netw. 143 (2018) 221–246.

[10] A. Whitmore, A. Agarwal, L. Da Xu, The internet of things—a survey of topics and trends, Inf. Syst. Front. 17 (2) (2015) 261–274.

[11] L.D. Xu, W. He, S. Li, Internet of things in industries: a survey, IEEE Trans. Ind. Inform. 10 (4) (2014) 2233–2243.

[12] K. Ashton, That "internet of things" thing, RFID J. 22 (7) (2009) 97–114.

[13] D.L. Brock, The electronic product code (EPC), Auto-ID Center White Paper MIT-AUTOID-WH-002, 2001, pp. 1–21.

[14] Y. Yin, Y. Zeng, X. Chen, Y. Fan, The internet of things in healthcare: an overview, J. Ind. Inf. Integr. 1 (2016) 3–13.

[15] A. Khanna, S. Kaur, Evolution of internet of things (IoT) and its significant impact in the field of precision agriculture, Comput. Electron. Agric. 157 (2019) 218–231.

[16] International Telecommunication Union, The internet of things, International Telecommunication Union (ITU), Internet Reports (2013).

[17] M.A. Feki, F. Kawsar, M. Boussard, L. Trappeniers, The internet of things: the next technological revolution, Computer 46 (2) (2013) 24–25.

[18] X. Li, R. Lu, X. Liang, X. Shen, J. Chen, X. Lin, Smart community: an internet of things application, IEEE Commun. Mag. 49 (11) (2011) 68–75.

[19] J. Qi, P. Yang, G. Min, O. Amft, F. Dong, L. Xu, Advanced internet of things for personalised healthcare systems: a survey, Pervasive Mob. Comput. 41 (2017) 132–149.

[20] S.M.R. Islam, D. Kwak, M.H. Kabir, M. Hossain, K. Kwak, The internet of things for health care: a comprehensive survey, IEEE Access 3 (2015) 678–708.

[21] M.M. Dhanvijay, S.C. Patil, Internet of things: a survey of enabling technologies in healthcare and its applications, Comput. Netw. 153 (2019) 113–131.

[22] M. Dohler, C. Ratti, J. Paraszczak, G. Falconer, Smart cities, IEEE Commun. Mag. 51 (6) (2013) 70–71.

[23] J. Gantz, D. Reinsel, The digital universe in 2020: big data, bigger digital shadows, and biggest growth in the far east, IDC iView: IDC Analyze the Future 2007 (2012) (2012) 1–16.

[24] J. Manyika, M. Chui, J. Bughin, R. Dobbs, P. Bisson, A. Marrs, Disruptive Technologies: Advances That Will Transform Life, Business, and the Global Economy, vol. 180, McKinsey Global Institute, San Francisco, CA, 2013.

IoT healthcare architecture

M. Jayashankara[a], Sandeep S. Udmale[b], Anil Kumar Pandey[c], and Ravi Shankar Singh[d]

*Department of Computer Science and Engineering, P.E.S. College of Engineering, Mandya, Karnataka, India[a]
Department of Computer Engineering and IT, Veermata Jijabai Technological Institute (VJTI), Mumbai,
Maharashtra, India[b] Banaras Hindu University (BHU), Varanasi, Uttar Pradesh, India[c] Department of Computer
Science and Engineering, Indian Institute of Technology (BHU), Varanasi, Uttar Pradesh, India[d]*

1 Introduction

Modern society demands an efficient and reliable healthcare infrastructure, providing quality facilities and simultaneously reducing healthcare costs. In the current scenario, hospital and patient management is performed by healthcare staff. This represents a bottleneck in providing quality healthcare facilities and sometimes becomes a source of errors in healthcare practices and facilities.

The recent advancements in design and development of the Internet of Things (IoT) technologies have allowed the development of various smart systems to facilitate and enhance procedures in many fields, such as agriculture, refinery industries, and so on. The IoT has been introduced in the medical field to improve processes in biomedicine and healthcare. Many applications, such as tracking and identifying people and medical equipment, real-time patient monitoring and assistance, hygiene monitoring, and others have been developed for patients, doctors, and hospital staff, to improve the healthcare environment [1, 2].

The IoT for healthcare consists of a set of physically connected objects such as sensor nodes, medical devices, and people. The IoT system can connect any of these objects/people at any place and time to gather data and exchange information through wired or wireless networks [3]. Patients and healthcare providers today are showing greater interest in healthcare-related sensor devices, and technological developments have led to many more healthcare-related sensor devices now available in the marketplace for smart healthcare monitoring [3].

Further, patients are examined for health-related issues using smart health monitoring and the health data is recorded to provide the information to the physician for further action. However, if the physician is not available to attend the patient in a timely manner or if the situation requires an expert opinion, then the patient's health data can be forwarded to another doctor for further necessary action using the IoT system. The doctor evaluates the patient information and prescribes any medicines or precautionary measures needed to improve the patient's health and maintain quality of life. This process reduces the cost of healthcare and also speeds up the diagnostic procedure.

Ubiquitous healthcare architecture using the IoT is helping to provide more confident and comfortable lives to people in different age groups through monitoring of health-related issues continuously [4].

IoT Based Data Analytics for the Healthcare Industry. https://doi.org/10.1016/B978-0-12-821472-5.00011-9

To provide and maintain effective health monitoring systems, ubiquitous health-related technologies are required. Thus, the use of wired or wireless network communicating devices to transfer information on health-related issues of patients to concerned caretakers is becoming common. Further, for accurate monitoring and reporting of the medical conditions of patients, many sensor devices, including pulse oximeter, heart rate, glucometer, and blood pressure sensors, among many others, are required and widely available in today's healthcare world market [5].

The construction of reliable healthcare IoT systems can be a challenging task. First, development of an effective sensor device to be worn or used is a difficult exercise, because the device must collect the data from a changing environment, as shown in Fig. 1. In addition, managing large amounts of heterogeneous data and storing it for mining the useful information later is always a challenge. Once the data is stored, data protection becomes a critical task. Hence, security is another major issue for IoT healthcare systems.

The organization of this chapter is as follows: Section 2 provides information about the general healthcare IoT system. An overview of the system design of IoT-based healthcare systems is presented in Section 3. Sections 4 and 5 discuss IoT-based healthcare applications and various sensors. Section 6 provides information on the different challenges for the IoT-based healthcare architecture, followed by conclusions in the final section.

2 General healthcare IoT systems

Generally, the healthcare IoT consists of the major components: the communicating devices, the IoT gateway, personal computer systems, cloud computing, and end users, as shown in Fig. 1.

The communicating devices collect the significant information on the patient and transfer it to the IoT gateway. The available and most commonly used communicating devices are sensors, actuators, and endpoints. Sensors are employed to gather the information from nearby points, and the collected

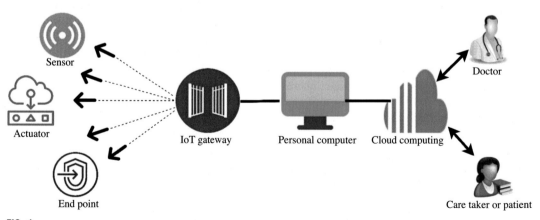

FIG. 1

Communicating devices, Internet of Things gateway, personal computer systems, cloud computing, and end users.

information is forwarded to the next stage, called the IoT gateway. Multiple types of sensors are utilized for various applications and all the sensors are connected to the network for data-processing purposes. It is a small memory used to store information and connected to a network and other devices for communication purpose. There are many different sensors, including temperature, pressure, proximity, accelerometer and gyroscope, infrared (IR), optical, and gas sensors, as shown in Fig. 2.

Many different types of sensors are observed in IoT-based smart healthcare systems to address multiple health issues. These diverse health applications employ sensors such as the pulse oximeter, electrocardiogram, thermometer, fluid level sensor, and sphygmomanometer to measure and record patient data, as shown in Fig. 3.

The actuator is a transducer that is mainly used to translate energy from one form to another form. As a result, an actuator is widely observed in IoT systems for data collection purposes. Basically, the operation of the actuator is opposite to that of the sensor: it takes an electrical input and converts it into physical action. The endpoint is another category of device, which assembles the information to monitor the services and processing of data and assist in the analysis [6].

In the next stage, the IoT gateway is introduced for connecting the communicating devices and the computing environment. The IoT gateway is a software program that efficiently collects data and securely transfers it to the next level for performing the analysis. Some of the fundamental processing and analysis is performed with the help of personal computers, mobiles, and other devices. In some

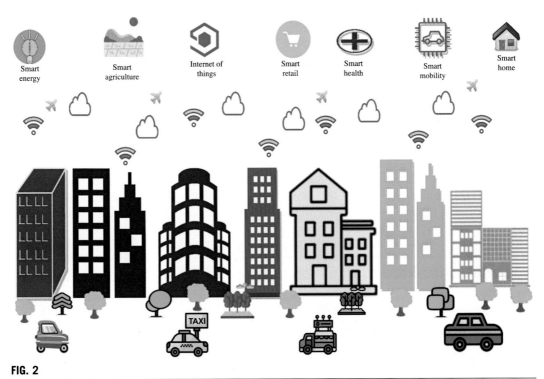

FIG. 2

General types of sensors used in Internet of Things.

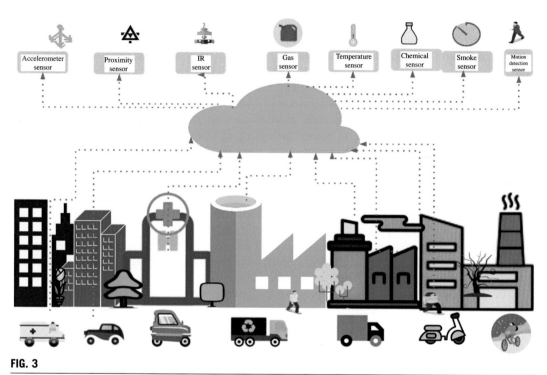

FIG. 3

Types of sensors used in the Internet of Things healthcare architecture.

applications, such as heart-rate monitoring, the information is processed on individual computing devices, such as mobiles, to provide the end results to users. However, in critical applications, like real-time patient monitoring, the information is transferred to a high-end computing environment for processing and storage purposes, with multiple inputs, as shown in Fig. 4. These operations are performed in the cloud computing environment.

Cloud computing is a web-based service for end users, such as patients or patient caretakers and physicians when applied to healthcare IoT. Generally, sensors used in the IoT generate a large amount of data, and hence a cloud computing environment is required to manage this large volume.

3 IoT-based healthcare architecture

During the last few years, researchers have developed various medical sensor devices for providing smart healthcare to end users. The aim of this activity has been to design low-cost sensor devices to provide basic functionality within the healthcare architecture. Multiple healthcare architectures are proposed in the literature, and they are mainly classified as personal IoT in healthcare (IoTH) and clinical IoTH. Personal IoTH does not require any involvement of physicians and includes devices such as smart watches used by individuals for self-monitoring. Clinical IoTH involves the participation of physicians and includes tools such as glucose monitors built particularly for monitoring patients

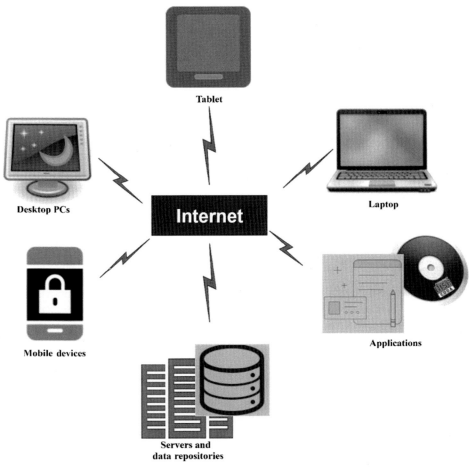

FIG. 4

Interaction of computing devices with the IoT.

under the guidance of a physician. Thus the architecture of IoT-based healthcare consists of a mobile health architecture system and a communication protocol such as 6LoWPAN (IPv6-based Low-Power Personal Area Network). The architecture of mobile health using 6LoWPANs and other related protocols is described in the following paragraphs.

3.1 m-Health architecture

Mobile-health (m-health) is a low-cost mobile health architecture that consists of three important stages. The first stage is the collection of information, in which sensing devices collect the information and parameters required for health tracking. Examples include blood glucose detection systems, temperature monitoring systems, and ECG heart-rate monitoring systems. The second stage is the storage information stage, in which the information related to health is stored. The third stage is the processing stage, in which the stored data is processed using various machine learning techniques [7, 8]. The m-health architecture is shown in Fig. 5.

The sensor collects the information, and in the second stage, the collected information is stored, and finally, the stored information is sent for processing. Communication protocols, such as LoRaWAN (Low Range Wide Area Network), Bluetooth, ZigBee [9], and Constrained Application Protocol (CoAP), are used in this process.

3.2 6LoWPAN architecture

The 6LoWPAN [10] protocol is a standard developed for low-power devices and also for low-cost communication wireless networks. It is an IPv6-based protocol, and allows IPv6 packets, which need higher bandwidth and power to process the data, to be carried on top of low-power networks.

The architecture and overall operation of 6LoWPAN is illustrated in Fig. 6. There are three types of 6LoWPAN networks: Simple LoWPAN, Extended LoWPAN, and Ad-hoc LoWPAN. The Simple LoWPAN and Extended LoWPAN connect to a network, and Ad-hoc LoWPAN connects to a wireless network. Ad-hoc operates without connection to the internet. Simple LoWPAN consists of an edge router, connected to the internet via a backhaul network. The Extended LoWPAN consists of two edge routers, connected to the internet via a backhaul network [10, 11]. The edge router is used to forward the packet from LoWPAN to Network and Network to LoWPAN.

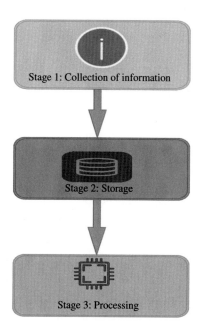

FIG. 5

Structure of m-health.

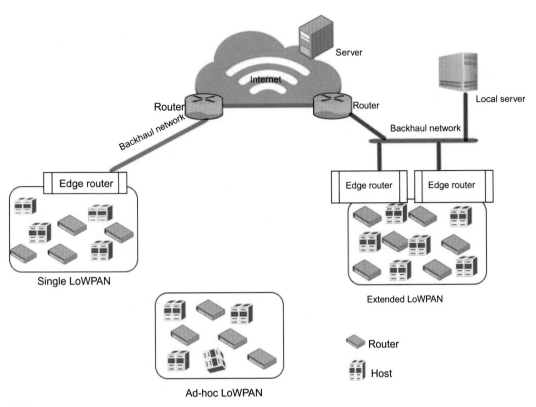

FIG. 6

Communication between Simple, Extended and Ad-hoc LoWPAN.

Adapted from V. Kumar, S. Tiwari, Review article routing in IPv6 over Low-Power Wireless Personal Area Networks (6LoWPAN): a survey, J. Comput. Networks Commun. 2012 (2012), http://doi.org/10.1155/2012/316839.

The protocol stack of LoWPAN consists of the application protocol, used for a particular application: for example, IoT sensors are deployed here. In the next level layer, the User Datagram Protocol (UDP) is used for data transformation. The subsection transport layer also consists of Internet Control Message Protocol (ICMP) for message control, and examples include ICMP echo and ICMP Destination Unreachable, etc. The adaptation layer placed between the IPv6 and Physical layer/MAC layer is as shown in Fig. 7. The protocol stack of IPv6 with 6LoWPAN is also called a 6LoWPAN protocol stack, which is shown in Fig. 7.

The adaptation layer is introduced to provide an efficient usage of 6LoWPAN using IPv6. There is no compatibility exit in IPv6 with 6LoWPAN, because the size of the packet frame in IPv6 is 1280 octets, whereas LoWPAN uses 128 octets as per the IEEE 802.15.4 specification. A 6LoWPAN adaptation layer is suggested between IPv6 or Network layer and IEEE 802.15.4 MAC/IEEE 802.15.4 PHY layer to modify the usage of octets, with 128 octets instead of 1280 octets. The bottom level layer is the Physical layer, which provides the communication for the medium and above the Physical layer is the MAC layer, which also helps to move data between one point to another. Fig. 7 Illustrates that

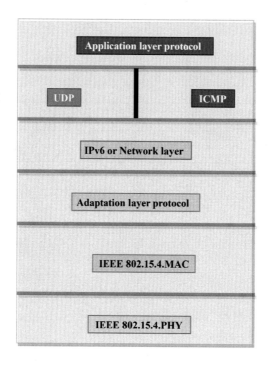

FIG. 7

6LoWPAN architecture using protocol stack.

Adapted from V. Kumar, S. Tiwari, Review article routing in IPv6 over Low-Power Wireless Personal Area Networks (6LoWPAN): a survey, J. Comput. Networks Commun. 2012 (2012), http://doi.org/10.1155/2012/316839.

6LoWPAN as an adaptation layer is placed between the IPv6 and IEEE 802.15.4 MAC/IEEE 802.15.4 PHY layer, and Fig. 8 displays health sensor devices in place of the application layer protocol.

The data is sensed by the sensors embedded with the 6LoWPAN packet, and it will be transferred to a gateway. The next gateway will translate the 6LoWPAN [11] packet to the IPv6 packet and then forward it to the network for further action, and finally it will be sent to the server for processing. The overall process is shown in Fig. 9.

3.3 The architecture of the CoAP

The CoAP is an application layer and web-based protocol designed for constrained devices like sensors. Sensors have a small memory and limited processing power. The CoAP [12] is similar to the HyperText Transport Protocol (HTTP) protocol. CoAP is used as the Representative State Transfer (REST) architecture. CoAP is demonstrated with a layered architecture in Fig. 10.

The transfer of data between two communicating devices is efficient and reliable in CoAP. CoAP is used to move the data, which is collected by a sensor node, to the other networking devices. The CoAP uses two sublayers called the Request/Response Layer and Messaging Layer. CoAP can handle four kinds of messages: Confirmation (CON), Non-Confirmation (NON), Acknowledgement (ACK), and

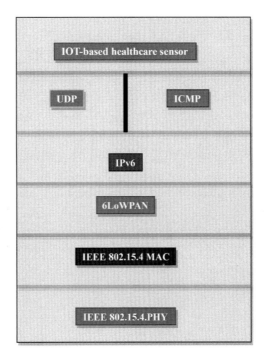

FIG. 8

6LoWPAN protocol stack architecture.

Adapted from V. Kumar, S. Tiwari, Review article routing in IPv6 over Low-Power Wireless Personal Area Networks (6LoWPAN): a survey, J. Comput. Networks Commun. 2012 (2012), http://doi.org/10.1155/2012/316839.

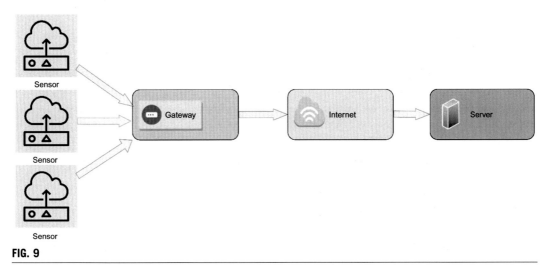

FIG. 9

6LoWPAN-based healthcare system for communication.

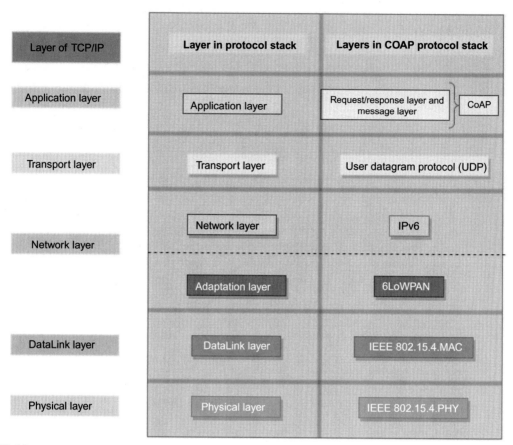

FIG. 10

CoAP protocol stack level for healthcare architecture.

Adapted from J.-W. Kim, J. Barrado, D.-K. Jeon, An energy-efficient transmission scheme for real-time data in wireless sensor networks, Sensors 15(5) (2015) 11628–11652, http://doi.org/10.3390/s150511628.

Reset. The CON and ACA provide the exchange of messages for reliable communication. The layered level architecture and a comparison of the layered protocol stack and CoAP protocol stack are given in Fig. 10.

3.4 IEEE 11073 architecture for healthcare system

The IEEE has introduced the IEEE 11073 standard, which captures health and fitness devices, for example, activity monitors and blood glucose monitors, among others.

This architecture supports Low Power Wireless Personal Area Network in the Bluetooth Low Energy (BLE) network. In addition, it has CoAP for efficient and reliable communication for sensors (constrained) devices [13].

The various components of the IEEE 11073 architecture are healthcare devices, healthcare server, and a web client. The healthcare devices are connected through a 6LoWPAN gateway. When 6LoW-PAN receives the healthcare data, the gateway performs the conversion of the packet between the Bluetooth interface and an Ethernet interface to communicate with the internet [7].

The overall operation is shown in Fig. 11, along with the healthcare server, web client, and 6LoW-PAN devices.

A summary of m-health, 6LoWPAN, CoAP, and IEEE 11073 architecture is given in Table 1.

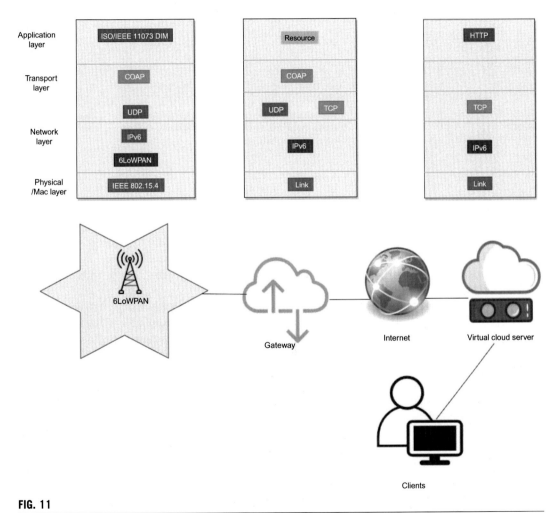

FIG. 11

The architecture of protocol stack with 6LoWPAN.

Adapted from L.M. Dang, M.J. Piran, D. Han, K. Min, H. Moon, A survey on Internet of Things and cloud computing for healthcare, Electronics 8(7) (2019) 768, http://doi.org/10.3390/electronics8070768.

Table 1 Summary of m-health, 6LoWPAN, CoAP, and IEEE 11073 architecture.				
Various communications protocols with architectures	**Architecture**	**Architectural stages**	**Uses**	**Examples**
m-Health	Low-cost mobile health architecture	There are three different architectural stages: data collection, storage, and processing	Efficient and accurate	Blood glucose system and temperature check system
6LoWPAN	Low-power devices architecture	The packet is transferred from LoWPAN to Network and vice versa	Efficiency is high, reachability is high, and reliability is high	Used as a communication protocol
CoAP	Constrained device architecture	As a web-based protocol, it is an application layer architecture	Reliability is high Efficiency is high	Different types of sensors
IEEE 11073 standard	The architecture is validated using IEEE11073	The exchange of information performed; used as a communication standard	Reliability is high Efficiency is good	Sensors and actuator

4 System design for IoT healthcare applications

Nowadays, many different IoT applications can be found in various domains, especially in the healthcare industry. The design of an IoT-based healthcare system [14] consists of different elements: sensors, mobile computing devices, doctor and patient application platforms, along with computers, analytical methods, and connectivity, as shown in Fig. 12.

We can identify a large number of IoT sensors related to the healthcare industry, such as air quality sensors, pulse oximeters, thermometers, etc. to read a patient's related information in the form of data. Mobile computing devices such as smartphones are used to read the information in the form of signals, and basically these systems consist of a software and hardware component. In the end-user stage, the physician accesses the information about their patients on a laptop or tablet. The platforms assist in reading patient information and performing analytics tasks, to take some necessary action such as prediction of a disease, diagnosis, and event-based actions if necessary [15, 16].

Finally, the information is transferred from sensors to microcontroller and microcontroller to sensors. IoT healthcare systems provide improved connectivity for better communication. The complete operation is described in Fig. 12. The various health architectures are discussed in this section.

4.1 Blood glucose control management system

One of the prominent health problems in today's world is that of diabetes, observed in many people worldwide, regardless of age. Prevention is always the best possible solution to this problem, and it is recommended to regularly check the blood glucose level. However, the traditional method of checking

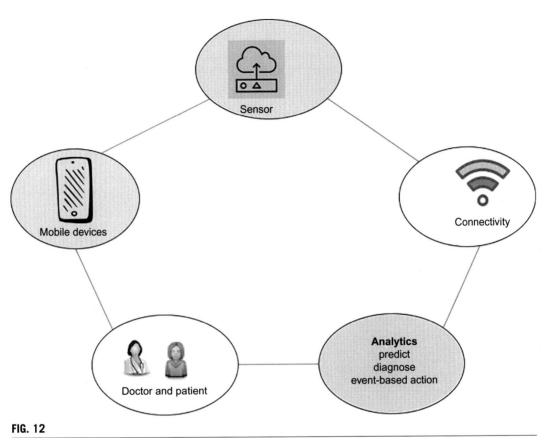

FIG. 12

System design and architecture of IoT healthcare system.

blood glucose is a time-consuming process, whereas the noninvasive IoT-based method can help in monitoring and managing the blood glucose level quicker and easier, at an affordable cost.

Blood glucose measurement can be invasive or noninvasive. In the invasive method, a sample of blood is taken from the patient's finger. This method can cause infection, and it is painful as well as time consuming.

In the noninvasive method, the blood glucose level is checked by using an opto-physiological sensor without damaging the skin or introducing any infection. The sensor uses an accelerometer and photodiode to measure the glucose level of blood.

4.2 Body temperature checking system

Checking and monitoring patient body temperature is very important in healthcare; with this type of system, which observes, transmits, and measures any variation in the body temperature, the result can be referred directly to a doctor, and the doctor can take any necessary action. When the body temperature of patients increases, exceeding the normal range, the doctor is informed by use of the sensor device and Raspberry Pi.

The main component of this system is a temperature sensor. Two commonly used body temperature sensors are the LM35 and TMP36 [17]. The LM35 from Texas Instruments is an integrated circuit temperature device with $\pm 1/4°C$ at room temperature with a range of $-55°C$ to $+150°C$. An illustration of the LM35 is given in Fig. 13.

4.3 Electrocardiogram for monitoring heart rate

The electrocardiogram (ECG) is used to measure the electrical activities of the heart to distinguish heart function. Many IoT-based systems measure heart function. It is important to note that the traditional procedure is a costly one, from both the facility and the patient's points of view. Therefore, low price and cost-effective ECGs to monitor heart rate have now been developed [18].

In this system, ECG sensors sense and collect the data. The collected data is sent to the microcontroller unit, from where it is transferred to the central server via the ZigBee network, as illustrated in Fig. 14. Finally, analysis is performed on the server and the results are stored. Subsequently, it will be forwarded to the next level end user if needed.

Wearable ECGs are available on the market to detect heart rate. It is easy to detect an irregular heart rate, which is known as arrhythmia. The sensor captures the data, and the captured data is transmitted to the mobile device using Bluetooth technology. With the help of the Pan Tompkins algorithm, the mobile device is able to detect the arrhythmia. It uses the discrete wavelet transform (DWT) algorithm for feature extraction, and also the support vector machine (SVM) is used for the classification process [19–21].

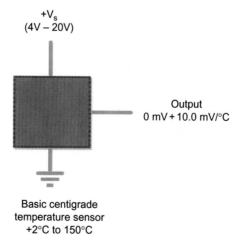

FIG. 13

Block diagram of LM35.

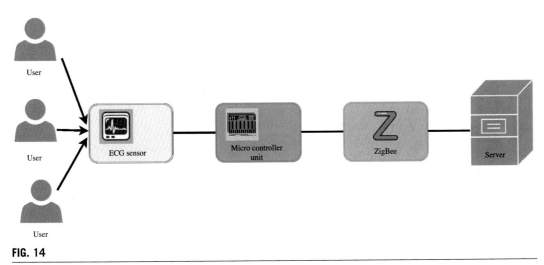

FIG. 14

ECG heart monitor system overview.

4.4 **Heart-rate checking system**

Another important device used to check the regularity of the heart is the heart-rate checking system [22]. This system is very much needed in hospitals in critical situations. Many IoT devices are found in this heart-rate checking system, including Fitbit, Garmin, etc. These devices help to check as well as to monitor heart rate over a limited time at low cost.

A summary of the overall design of IoT healthcare architecture is shown in Table 2.

Table 2 Summary of the overall design of IoT healthcare.

System design of various IoT healthcare systems	Uses	Scalability	Reliability	Efficiency
Blood glucose control management system	To monitor blood glucose levels constantly	Low	Medium	Medium
Body temperature checking system	To measure body temperature	High	Medium	Low
Electrocardiogram for monitoring heart function	To monitor the functioning of the heart	High	High	High
Heart-rate checking system	To measure the heart rate during a limited time	High	Medium	Low

5 Sensors used in IoT healthcare architecture

IoT-based healthcare already plays a vital role in health management. Today people are more conscious of their health. Generally, in the traditional system, any health-related issues like blood pressure problems, heart-related issues, or blood glucose problems have to be resolved in a medical facility; the patient has to spend significant money on health and it is a time-consuming process. To mitigate this situation, IoT based on healthcare architectures is essential [23].

Further, in IoT systems based on healthcare architecture, sensors always play a crucial role. Different types of sensors can be used in this role, and some of them are listed in the following sections [24, 25].

5.1 Accelerometer

Electromechanical devices are capable of measuring both static and dynamic acceleration forces, which are changes in linear velocity, or velocity divided by time. The static acceleration forces are useful for continues gravity forces and dynamic acceleration forces are useful for sensing vibration. The main healthcare uses of accelerometers are for measuring the blood glucose level and to record movements, especially in monitoring the elderly or patients with dementia [26]. The MMA7260QT accelerometers can detect body movements such as in fall detection and motion sensing.

5.2 Gyroscope

A gyroscope sensor is used in most of the IoT-based healthcare architectures. It measures an object's angular velocity as well as the tilt orientation of the object. It is more powerful than an accelerometer, because the accelerometer measures only linear action whereas the gyroscope [26] measures tilt and angular velocity.

5.3 Magnetometer

A magnetometer provides information on magnetic fields. Similar to the gyroscope, it is useful for finding the relative orientation of an object. Thus it is useful in human fall detection, which is its main application in healthcare [14, 23, 26].

5.4 Electrocardiogram

Electrocardiograms are used to analyze heart conditions and can easily detect heart arrhythmias by detecting the electrical activity of the heart. An arrhythmia is nonregular activity of the heart rhythm and can include heart rates both too slow and too fast. Arrhythmias mean that the heart is not pumping enough blood to the body, and they may create other severe problems in the heart, brain, and other parts of the body [27]. Currently, low-priced and small models of electrocardiograms are available. The AD8232 Heart Rate Monitor from Analog Devices is one example of an electrocardiogram sensor.

5.5 **Temperature sensor**

The main healthcare application of a temperature sensor is to measure the temperature of the human body [28]. The temperature sensor is also used in many other fields, such as food processing and medical manufacturing. Thus different types of temperature sensors, such as thermistors, infrared sensors, and thermometers, have been developed. Some of the temperature sensor modules available on the market are the Texas Instruments LM35 and TMP75, and the Analog Devices TMP36. Some different temperature sensors are shown in Fig. 15.

5.6 **Heart-rate sensor**

Monitoring the heart rate is useful in many settings; by knowing the heart rate, various physical exercises can be adapted to different people. The heart-rate sensor measures the heart rate in terms of beats per minute with the help of an optical LED light sensor and LED light sensor. The light glows through the person's skin and the sensor monitors how much is reflected. The reflection of light varies depending upon how the blood pulse varies in the body. The variation of the light reflection is recorded as a heartbeat.

5.7 **Blood pressure sensor**

The infrared (IR) sensor is a typical sensor used in healthcare system design architecture. IR can be used to sense some properties of the surroundings by emitting infrared radiation. These IR sensors can be used to check the blood flow and blood pressure.

Finally, for proper functioning of sensors, hardware support is required. Data from the sensors described so far in this chapter (accelerometers, gyroscopes, electrocardiograms, temperature sensors, heart-rate sensors, and blood pressure sensors) are sent to data processors to be communicated to higher-level networks and all of these sensors are placed on hardware. The hardware support can be Arduino, Raspberry Pi with a microcontroller, and others.

The ECG sensed data problem can be resolved by using an Arduino hardware platform. Raspberry Pi can be used to obtain the data from a heart-rate sensor, body temperature sensor, and body

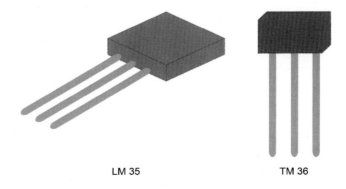

LM 35   TM 36

FIG. 15

Types of temperature sensor.

movement sensor. The microcontroller is used as a hardware platform for IoT based on healthcare architecture. A summary of types of sensors used in IoT healthcare architecture is shown in Table 3.

6 Challenges in IoT healthcare architecture

IoT healthcare architecture now plays an important role in global development, and in this regard, many researchers have developed as well as proposed IoT healthcare architecture development frameworks. However, the development of a viable IoT healthcare architecture faces major challenges:

1. *Security issues*: A communications protocol to transfer data between the various levels of architecture is required and Internet Protocol (IP) is widely used. The implementation of secure IP-based protocol is always challenging.
2. *Privacy issues*: Low-power constrained devices are used in the basic architectural level stage and data collected using them needs to be maintained carefully. As a result, privacy is another major issue in the development of an IoT healthcare architecture.
3. *Data management issues*: The various devices interact in the smart system and, as a result, the exchange of information between different architectural stages occurs. Hence the amount of data generated is huge. The processing of generated data is another important challenge.
4. *Storage management issues*: In an IoT healthcare architecture, there are different levels of stages and each stage has a significant contribution. A large amount of data is transferred through different stages and hence the storage of these data in the cloud is a major issue. Once it is stored, retrieval of data for each level of architectural stage is needed. Thus retrieval of data is another challenge.
5. *Network management issues*: A number of devices are connected to the IoT Gateway protocol and hence the performance of network usage is important. The execution time of data through different architectural stages is another major challenging issue. Different devices used in the IoT healthcare

Table 3 Summary of types of sensors used in IoT healthcare architecture.

Types of sensors	Functionality	Applications
Accelerometer	Measure linear acceleration	Used in medical field for bionic limbs and also for other artificial body parts
Gyroscope	Measure angular velocity or tilt orientation of the object	Used in patient monitoring systems such as movement and position monitoring and also in diagnostic and treatment equipment positioning applications
Magnetometer	Measure magnetic induction	Medicine, aircraft, and spacecraft
Electrocardiogram	To trace and diagnose heart disease	Used for finding the heart rate in the human body; also applicable for health check-ups in remote places
Temperature sensor	To detect and measure temperature or heat	Food processing, medical devices, and alcohol breathalyzer
Heart-rate sensor	Measure heart rate	Patient monitoring, robotics biofeedback control
Blood pressure sensor	To measure blood pressure	Monitor patients remotely, where hospitals are not available

Table 4 Summary of challenges in IoT healthcare architecture.

Sl no.	Types of challenges	Architectural issues
1.	Security	IP-based protocol needed to provide security for different stages of IoT healthcare architecture
2.	Privacy	Data maintenance needed
3.	Data management and monitoring	Processing of data needed. Monitoring of data discussed
4.	Storage management	A large amount of data is generated, so storage management issues are challenging
5.	Network management	Minimum latency and execution time are important architectural issues

architecture are distributed in nature, whereas the nature of the cloud is centralized. So, data transmission between these two, as well as selecting the architectural elements and accessing capability, is another important challenge.

6. *Data monitoring issues*: As the resources are limited and low power-constrained devices are used in the system design, the collection, processing, and storage of these data represent a challenging task. Continuously monitored data is another related issue in an IoT healthcare architecture.

A summary of the challenges is given in Table 4.

7 Conclusions and future work

In this work, sensor-based IoT healthcare architectures are discussed. These architectures are used in smart healthcare systems, allowing various health-related issues to be identified and necessary action to be taken quickly. Architectures with various communication protocols are discussed and compared. System design for IoT-based healthcare architecture is outlined and various types of sensors are described. In addition, various applications with multiple sensors are summarized and compared. Security through different stages of an IoT healthcare architecture can be treated as future work. In addition, achieving low latency through different levels of architectural stages with various compression techniques is an open issue.

References

[1] B. Dorsemaine, J.P. Gaulier, J.P. Wary, N. Kheir, P. Urien, Internet of Things: a definition and taxonomy, in: Proceedings-NGMAST 2015: The 9th International Conference on Next Generation Mobile Applications, Services and Technologies, 2016, pp. 72–77, https://doi.org/10.1109/NGMAST.2015.71.

[2] F. Firouzi, A.M. Rahmani, K. Mankodiya, M. Badaroglu, G.V. Merrett, P. Wong, B. Farahani, Internet-of-Things and big data for smarter healthcare: from device to architecture, applications and analytics, Futur. Gener. Comput. Syst. 78 (2018) 583–586, https://doi.org/10.1016/j.future.2017.09.016.

[3] Y.J. Fan, Y.H. Yin, L. Da Xu, Y. Zeng, F. Wu, IoT-based smart rehabilitation system, IEEE Trans. Ind. Inform. 10 (2) (2014) 1568–1577, https://doi.org/10.1109/TII.2014.2302583.

[4] A. Zanella, S. Member, N. Bui, A. Castellani, L. Vangelista, M. Zorzi, Internet of things for smart cities, IEEE Internet Things J. 1 (1) (2014), https://doi.org/10.1109/JIOT.2014.2306328.

[5] L. Catarinucci, et al., An IoT-aware architecture for smart healthcare systems, IEEE Internet Things J. 2 (6) (2015) 515–526, https://doi.org/10.1109/JIOT.2015.2417684.

[6] Y. Zhang, L. Sun, H. Song, X. Cao, Ubiquitous WSN for healthcare: recent advances and future prospects, IEEE Internet Things J. 1 (4) (2014) 311–318, https://doi.org/10.1109/JIOT.2014.2329462.

[7] M.S. Jassas, A.A. Qasem, Q.H. Mahmoud, A smart system connecting e-health sensors and the cloud, in: Canadian Conference on Electrical and Computer Engineering, 2015, pp. 712–716, https://doi.org/10.1109/CCECE.2015.7129362. vol. 2015-June.

[8] T.N. Gia, N.K. Thanigaivelan, A.M. Rahmani, T. Westerlund, P. Liljeberg, H. Tenhunen, Customizing 6LoWPAN networks towards Internet-of-Things based ubiquitous healthcare systems, in: NORCHIP 2014-32nd NORCHIP Conference: The Nordic Microelectronics Event, 2015, https://doi.org/10.1109/NORCHIP.2014.7004716.

[9] L.M. Dang, M.J. Piran, D. Han, K. Min, H. Moon, A survey on Internet of Things and cloud computing for healthcare, Electronics 8 (7) (2019) 768, https://doi.org/10.3390/electronics8070768.

[10] A. Özdemir, An analysis on sensor locations of the human body for wearable fall detection devices: principles and practice, Sensors 16 (8) (2016) 1161, https://doi.org/10.3390/s16081161.

[11] Q. Li, J.A. Stankovic, M.A. Hanson, A.T. Barth, J. Lach, G. Zhou, Accurate, fast fall detection using gyroscopes and accelerometer-derived posture information, in: 2009 Sixth International Workshop on Wearable and Implantable Body Sensor Networks, Berkeley, CA, 2009, pp. 138–143, https://doi.org/10.1109/BSN.2009.46.

[12] A network management architecture for 6LoWPAN network, (Online). Available from: https://www.researchgate.net/publication/239765239_A_network_management_architecture_for_6LoWPAN_network. (Accessed 7 March 2020).

[13] S.N. Han, Q.H. Cao, B. Alinia, N. Crespi, Design implementation and evaluation of 6LoWPAN for home and building automation in the Internet of Things, in: Proceedings of IEEE/ACS International Conference on Computer Systems and Applications AICCSA, 2016-July, 2016, https://doi.org/10.1109/AICCSA.2015.7507264.

[14] D. Azariadi, V. Tsoutsouras, S. Xydis, D. Soudris, ECG signal analysis and arrhythmia detection on IoT wearable medical devices, in: 2016 5th International Conference on Modern Circuits and Systems Technologies, MOCAST 2016, 2016, https://doi.org/10.1109/MOCAST.2016.7495143.

[15] G. Montenegro, N. Kushalnagar, J. Hui, D. Culler, Transmission of IPv6 packets over IEEE 802.15. 4 networks, in: Internet Proposed Standard RFC 4944, 2007, p. 130.

[16] W. Colitti, K. Steenhaut, N. De Caro, B. Buta, V. Dobrota, Evaluation of constrained application protocol for wireless sensor networks, in: IEEE Workshop on Local and Metropolitan Area Networks, 2011, https://doi.org/10.1109/LANMAN.2011.6076934.

[17] H.W. Kang, C.M. Kim, S.J. Koh, ISO/IEEE 11073-based healthcare services over IoT platform using 6LoWPAN and BLE: architecture and experimentation, in: Proceedings-2016 International Conference on Networking and Network Applications, NaNA 2016, 2016, pp. 313–318, https://doi.org/10.1109/NaNA.2016.26.

[18] S. Hadiyoso, K. Usman, A. Rizal, Arrhythmia detection based on ECG signal using Android mobile for athlete and patient, in: 2015 3rd International Conference on Information and Communication Technology (ICoICT), Nusa Dua, 2015, pp. 166–171, https://doi.org/10.1109/ICoICT.2015.7231416.

[19] S.B. Baker, W. Xiang, I. Atkinson, Internet of Things for smart healthcare: technologies, challenges, and opportunities, IEEE Access 5 (2017) 26521–26544, https://doi.org/10.1109/ACCESS.2017.2775180.

[20] A. Harris, H. True, Z. Hu, J. Cho, N. Fell, M. Sartipi, Fall recognition using wearable technologies and machine learning algorithms, in: IEEE International Conference on Big Data, 2016.

[21] H. Mansor, M.H.A. Shukor, S.S. Meskam, N.Q.A.M. Rusli, N.S. Zamery, Body temperature measurement for remote health monitoring system, in: 2013 IEEE International Conference on Smart Instrumentation, Measurement and Applications, ICSIMA 2013, 2013, https://doi.org/10.1109/ICSIMA.2013.6717956.

[22] Internet of Things: Remote Patient Monitoring Using Web Services and Cloud Computing | Semantic Scholar, (Online). Available from: https://www.semanticscholar.org/paper/Internet-of-Things%3A-Remote-Patient-Monitoring-Using-Mohammed-Lung/3fb5c8c63628a00cc74c26687855edd95b6b98a7. (Accessed 12 March 2020).

[23] N. Jagtap, J. Wadgaonkar, K. Bhole, Smart wrist watch, in: 2016 IEEE Students' Conference on Electrical, Electronics and Computer Science, SCEECS 2016, 2016, https://doi.org/10.1109/SCEECS.2016.7509273.

[24] V. Bhelkar, D.K. Shedge, Different types of wearable sensors and health monitoring systems: a survey, in: Proceedings of the 2016 2nd International Conference on Applied and Theoretical Computing and Communication Technology, iCATccT 2016, 2017, pp. 43–48, https://doi.org/10.1109/ICATCCT.2016.7911963.

[25] S.M.R. Islam, D. Kwak, M.H. Kabir, M. Hossain, K.S. Kwak, The internet of things for health care: a comprehensive survey, IEEE Access 3 (2015) 678–708, https://doi.org/10.1109/ACCESS.2015.2437951.

[26] IEEE Xplore Full-Text PDF, (Online). Available from: https://ieeexplore.ieee.org/stamp/stamp.jsp?arnumber=7113786. (Accessed 12 March 2020).

[27] D. Ugrenovic, G. Gardasevic, CoAP protocol for Web-based monitoring in IoT healthcare applications, in: 2015 23rd Telecommunications Forum TELFOR 2015, 2016, pp. 79–82, https://doi.org/10.1109/TELFOR.2015.7377418.

[28] M. Ryan Fajar Nurdin, S. Hadiyoso, A. Rizal, A low-cost Internet of Things (IoT) system for multi-patient ECG's monitoring, in: ICCEREC 2016-International Conference on Contro, Electronics Renewable Energy and Communications 2016 Conference Proceedings, 2017, pp. 7–11, https://doi.org/10.1109/ICCEREC.2016.7814958.

Characteristics of IoT health data

Ritesh Sharma[a], Sandeep S. Udmale[b], Anil Kumar Pandey[c], and Ravi Shankar Singh[a]

Department of Computer Science and Engineering, Indian Institute of Technology (BHU), Varanasi, Uttar Pradesh, India[a] Department of Computer Engineering and IT, Veermata Jijabai Technological Institute (VJTI), Mumbai, Maharashtra, India[b] Banaras Hindu University (BHU), Varanasi, Uttar Pradesh, India[c]

1 Introduction

The Internet of Things (IoT) is a concept of connecting physical objects having electronics, software, and sensors over the Internet to make them communicate and exchange data with each other without much human intervention. For example, IoT-based accident detection and rescue system detect the accident with the help of accelerometer sensor, and the Global System for Mobile (GSM) module sends messages about the location to the respective rescue team that reduces the time required by the ambulance to reach the hospital. Therefore it helps in reducing the loss of life due to the accidents. The IoT provides appropriate solutions for various applications such as health, manufacturing, electricity, agriculture, and security.

In the last few years, IoT has attracted full attention from researchers due to its capability of proving the various solutions and benefits to different fields. As a result, there are now numerous applications, services, and prototypes that are designed and developed based on IoT. Fig. 1 illustrates that health is projected as the most attractive application of IoT. A lot of benefits are achieved by integrating IoT to the healthcare sector. For example, Internet of Things in Healthcare (IoTH) decreases the time taken to diagnose a health condition, reduces the chance of readmission for the same health problem, and also lowers the death rate in the rural area, which is far from the reach of the doctor. It also enables patients to monitor their progress and provide continuous feedback to the doctors, which enhances the satisfaction of both patients and doctors.

Health is defined as a state of being free from illness or injury. Health is a significant part of human life. A healthy body keeps our mind healthy and helps us to make important decisions about our business, education, professional, etc. Thus IoTH has been categorized into two subcategories: personal and clinical [1]. Personal IoTH does not require any involvement of physicians and includes devices such as smartwatches that are used by individuals for self-monitoring. Clinical IoTH involves the participation of physicians and includes tools such as glucose monitors that are particularly build for monitoring patients under the guidance of a physician.

Health data are any data related to the medical conditions of an individual or population. A lot (volume) of different types (variety) of health data are captured in real time (velocity) by the network of sensors while maintaining the correctness (veracity) of data. These four V's, namely, volume, variety, velocity, and veracity, are known as the characteristics of IoT health data.

IoT Based Data Analytics for the Healthcare Industry. https://doi.org/10.1016/B978-0-12-821472-5.00013-2

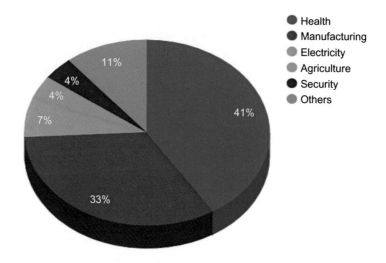

- Health
- Manufacturing
- Electricity
- Agriculture
- Security
- Others

FIG. 1

Projected market share of IoT applications by 2024.

This chapter mainly focuses on the characteristics of IoT health data and discusses them in details. Thus Section 2 presents a brief introduction to the characteristics of IoT health data. In Section 3 a variety of health data, which are one of the four characteristics of IoT health data, are presented in detail. The conclusion is provided in Section 4.

2 Characteristics of IOT health data

In 2001 Gartner analyst Doug Laney introduces three V's (volume, velocity, and variety) to define the characteristics of data. But later, as people started collecting more and more data, some of the data were not correctly architected. Therefore IBM introduced the fourth V (veracity) [2]. Hence the characteristics of IoT health data are volume, variety, velocity, and veracity, and they are demonstrated in Fig. 2.

2.1 Volume

The volume deals with the amount of data. With the incorporation of IoT into healthcare over time, health-related data are continuously accumulating, which results in an unbelievable quantity of data. Besides the already existing healthcare data include clinical trial data, personal medical records, and genomic sequences. Newer forms of big health data, such as imaging and biometric sensor reading data, are also fueling this exponential growth. Medical imaging is an area where the growing volume of data is noticeable. As per the report by IBM, every day, the world creates 2.5 quintillion bytes of data. Out of this large volume, 30% of the data are medical images. Also the simultaneous advances in cloud computing and virtualization approaches are facilitating the development of platforms for more effective storage and processing of large volumes of data.

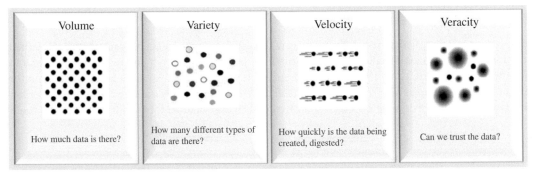

FIG. 2

Characteristics of IoT health data.

2.2 **Velocity**

Velocity in the context of data refers to two related concepts: One is the speed at which new data are created, and the other is the speed at which the data are analyzed. Due to the integration of IoT to healthcare, there is a noteworthy mismatch between the rate at which data are evolving and the speed at which it is being analyzed. It is not possible to analyze every single data stream lightning fast; moreover, it is not required to do so. Therefore, by defining which data sources are essential to analyze in real time, which data sources are necessary to examine in days, and which data sources can be delayed for weeks, the available resources can be utilized efficiently. For example, data related to readmission reports or patient collection rates can be delayed for weeks without any negative impact. Meanwhile, critical data like bedside heart monitors and operating room monitors for anesthesia have to be analyzed in real time as delay in them can lead to severe issues.

2.3 **Variety**

Variety comes with increasing volume and velocity. This third "V" describes the massive diversity of data that is produced by healthcare organizations every day. Health data are classified as structured or unstructured. Structured data can be easily queried, stored, recalled, and analyzed by a machine. Traditional data analytics work very well with structured data, which are present in a relational database with a well-formed schema. However, this does not apply to unstructured data (signal data, video data, image data, text data, etc.). As a result, analysis becomes challenging and requires the decent methodology to examine the unstructured data for mining the useful information.

2.4 **Veracity**

Veracity deals with quality and truthfulness of the health data. Healthcare systems are providing various decisions related to human life. Most of the health data is in an unstructured format and thus is highly variable. Moreover, data are also imprecisely collected; thus it also has biases, noise, and abnormality problems. Therefore veracity is of great concern in the case of health data for developing the reliable healthcare system.

3 Type of health data

Health data are classified as structured or unstructured [3]. Structured health data obey well-defined structure and are standardized, whereas unstructured health data do not follow any fixed structure or pattern. As structured data are standardized, therefore it is easily transferable between health information systems, whereas this is not easily applicable in the case with unstructured data. Fig. 3 displays that most of the health data are available or collected in an unstructured form. Thus unstructured health

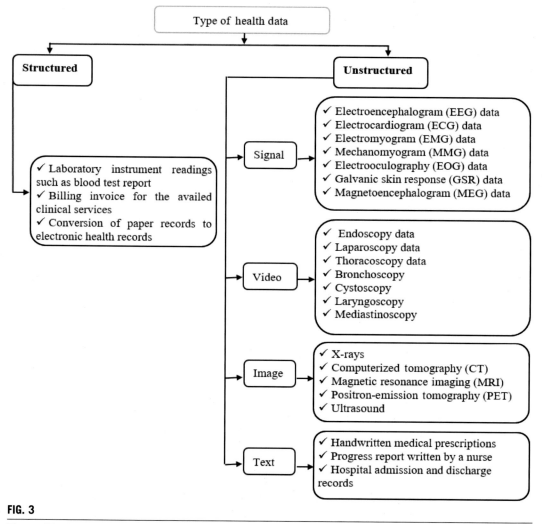

FIG. 3

Types of health data.

data are further categorized into a signal, video, image, and text data based on their importance in different health applications.

3.1 Signal data

In terms of health, signal data are nothing but biosignal data that are present in living beings and can be continually measured. The well-known biosignals present in living beings are electroencephalogram (EEG), magnetoencephalogram (MEG), electrocardiogram (ECG), electromyogram (EMG), mechanomyogram (MMG), electrooculography (EOG), and galvanic skin response (GSR) [4, 5]. They are introduced and utilized by the researchers and practitioners for developing the smart health systems due to the advancement in signal processing.

3.1.1 Electroencephalogram

The brain cells communicate with each other via tiny electrical signals called brainwaves. The EEG measures these signals using small metal electrodes. These electrodes are connected using wires to a computer that records the brain's electrical activity through wavy lines.

EEG results

Hunan brain has five brainwaves, namely, gamma, beta, alpha, theta, and delta. Each one has a distinct electrical pattern (as shown in Fig. 4). The nature of the illness is identified by analyzing results from these five brainwaves. Both high and low values of these brainwaves denote improper functioning of the brain (as shown in Table 1), and therefore the optimum value of these brainwaves is required [6].

3.1.2 Magnetoencephalogram

It records magnetic fields that are produced by the electrical currents of the brain. Like EEG, MEG is also useful in diagnosing various brain disorders, but MEG is superior to EEG in most of the cases as MEG signals do not get distorted by the skull and intervening soft tissues. There are two issues in recording the magnetic fields in MEG; one is that the magnetic fields are very minute, so recording them is not an easy task, and the other is to shield out the magnetic fields of MEG from earth's stronger magnetic fields. To cope with the first issue, a superconducting quantum interference detector, which is a highly sensitive magnetic field meter, is used. To deal with the second issue, the MEG is enclosed inside a magnetically shielded hall, which attenuates the external magnetic noise [7].

MEG results

MEG is currently used for preoperative brain mapping and for detecting the location of the seizure focus, which is required for the surgical treatment of epilepsy. Results from the MEG test are combined up with the magnetic resonance imaging (MRI). The MEG and MRI together form a map known as the magnetic source image (MSI) map that displays areas of normal and abnormal activity in the brain.

3.1.3 Electrocardiogram

Cells present in the upper right chamber of the heart generate an electrical impulse that in turn triggers our heart. An ECG records the strength and timing of these signals. There are two ways to interpret ECG: One is ECG pattern recognition, and the other is to understand the exact electrical vectors recorded by an ECG. Most of the time the combination of both is used to get a more precise

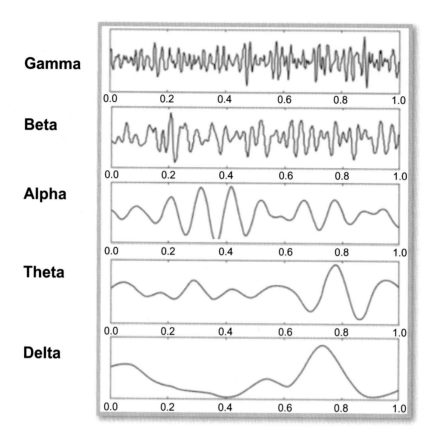

FIG. 4

EEG signal.

Adapted from P.A. Abhang, B.W. Gawali, S.C. Mehrotra, Technological basics of EEG recording and operation of apparatus, in: Introduction to EEG- and Speech-Based Emotion Recognition, Elsevier, 2016, pp. 19–50.

Table 1 Summary of high and low values of brainwaves.

Brainwave	Frequency range (Hz)	High levels	Low levels
Delta	0–4	• Brain injuries • Learning problems • Inability to think	• Inability to rejuvenate the body • Inability to revitalize the brain • Poor sleep
Theta	4–8	• Depressive states • Impulsive activity	• Higher stress levels • Poor emotional awareness
Alpha	8–12	• Too much daydreaming • Inability to focus	• Anxiety symptoms • Higher stress levels
Beta	12–40	• Inability to feel relaxed • High adrenaline levels	• Depression • Lack of attention
Gamma	40–100	• Anxiety • Stress	• Depression • Learning issues

Table 2 Major types of ECG.		
Resting ECG	**Ambulatory ECG**	**Cardiac stress ECG**
This type of ECG usually takes 5–10 min. The patient has to lie down for this type of ECG, and no movement is allowed during the test, as electrical impulses generated by other muscles may interfere with those produced by the heart	The patient has to wear a portable recording device for at least 24 h. This type of ECG is used for people whose symptoms are intermittent (stop-start) and may not show up on a resting ECG and for people recovering from heart attack to ensure that their heart is functioning correctly	This type of ECG takes about 15–30 min. It is used to record ECG while people ride on an exercise bike or walk on a treadmill

understanding [8]. There are several different ways to perform ECG out of which three significant types are resting ECG, ambulatory ECG, and cardiac stress ECG, and the details are shown in Table 2.

ECG results

Generally, for a healthy person, the heart rate must be between 50 and 100 beats a minute. If the heartbeat is faster or slower or irregular, then it is a sign of improper functioning of the heart. ECG recording shows proof of a previous heart attack. Also, it detects inadequate oxygen and blood supply to the heart. Moreover, ECG provides clues about the enlargement of the walls of the heart.

3.1.4 Electromyogram

EMG is a diagnostic test used to identify disorders in the muscles. EMG records the electrical activity of muscles when they are at rest or when they are contracted. During an EMG process a pin having electrode is inserted into the muscle. The patient is requested to make movements that will cause the muscle to contract. A similar diagnostic test is presented to identify the disorder in nerves, which is known as nerve conduction velocity (NCV). It examines how well sensory nerves are working by measuring the speed of electrical impulses through the nerves. NCV is performed with the help of surface patch electrodes that are placed at various locations over the nerve. One electrode is employed to stimulate the nerve by applying electrical impulse, and other electrodes record the resulting electrical activity. The time taken by impulses to travel between electrodes and the distance between electrodes are recorded, which are used in calculating NCV.

EMG results

Generally, abnormal results in EMG are observed in two ways. First the muscle shows the electrical activity in a relaxed position. Second the muscle displays unusual electrical activity during contraction. Abnormal EMG results indicate a problem with the nerves that control the muscle or damage of muscle.

3.1.5 Mechanomyogram

MMG records the acoustic waves generated by contracting muscle fibers. The waves propagate through human fat to the skin surface and able to detect vibration or sound using a microphone or an accelerometer. The MMG provides an alternative to the EMG for monitoring muscle activity.

MMG is superior to EMG as it has a higher signal-to-noise ratio than the EMG. Thus it can be used to monitor muscle activity from deeper muscles without using invasive measurement techniques [9].

MMG results

Like EMG results, abnormal MMG results may indicate a problem with the nerves that control the muscle or damage of muscle.

3.1.6 Electrooculography

EOG detects eye movement by placing electrodes near the eye. The human eye has a positive charge at the front (cornea) and a negative charge at the back (retina); therefore it acts as an electrical dipole. The difference in electrical charge is measured by placing two electrodes in the electrical field of the eye. As the eye moves, the EOG detects changes in the electrical charge between the cornea and retina (as shown in Fig. 5). While using the EOG signal, the frequency of the signal is of less interest. For interpretation of the signal, shape and amplitude of the signal are analyzed [10].

3.1.7 Galvanic skin response

There is an increase in eccrine sweat gland activity when saddening, threatening, joyful, etc. actions are observed. GSR measures the change in electrical activity, which takes place due to the shift in sweat gland activity (as shown in Fig. 6). An increase in sweat gland activity can take place due to both positive ("joyful") and negative ("scary") events. Therefore the GSR signal does not represent the type of emotions [11]. The hands have a high density of sensitive sweat glands; thus GSR data are collected from the finger, wrist, or palm. A well-known application of GSR is a lie detection test.

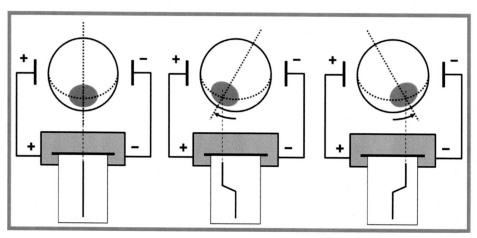

FIG. 5

Changes according to the eyeball movement for EOG measurement.

Adapted from J.-J. Yang, G. Gang, T. Kim, Development of EOG-based human computer interface (HCI) system using piecewise linear approximation (PLA) and support vector regression (SVR), Electronics 7(3) (2018), 38, doi:10.3390/electronics7030038.

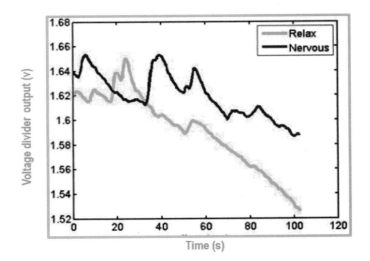

FIG. 6

GSR signal.

Adapted from M.V. Villarejo, B.G. Zapirain, A.M. Zorrilla, A stress sensor based on Galvanic Skin Response (GSR) controlled by ZigBee, Sensors 12(5) (2012) 6075–6101, doi:10.3390/s120506075.

From the earlier discussion, it is clear that EEG and MEG both are used to identify disorders that involve the brain, but MEG is more superior. ECG is used to identify disturbances in the heart. EMG and MMG are used to identify complications that affect the muscles, but MMG is more preferred. EOG detects the eye movement. GSR records electrical activity in the skin, which is due to variation of moisture level in the body as a result of sweating. This significant information assists in identifying the various abnormality of the human body. In addition, the various signals are collected for addressing particular health issue, and thus same signal may or may not provide the satisfactory results for other health problem.

3.2 Video data

Mostly, video data of the internal parts of a body are captured using the surgical and nonsurgical procedures for diagnostic and treatment purposes. A large amount of medical video data is available from various types of devices like endoscope, laparoscope, thoracoscope, bronchoscope, cystoscope, laryngoscope, and mediastinoscope for concentrating on particular health concern.

3.2.1 Endoscopy

It is a nonsurgical procedure used to view the inner lining of the upper digestive system. Endoscopy is performed using a device known as an endoscope (as shown in Fig. 7). It is a flexible tube having a video camera and light attached to it. The tube is inserted in through the mouth, and video of the digestive system is displayed on the television screen.

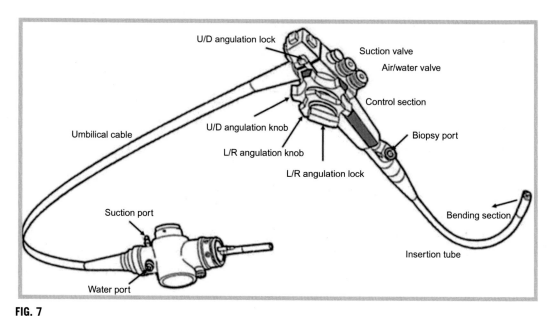

FIG. 7

Endoscope.

Adapted from A. Khanicheh, A.K. Shergill, Endoscope design for the future, Tech. Gastrointest. Endosc. 21(3) (2019) 167–173, W.B. Saunders, doi:10.1016/j.tgie.2019.05.003.

The need for endoscopy

- To determine the cause of nausea, vomiting, etc.
- For collecting the tissue samples to test for bleeding, cancers of the digestive system, etc.
- To pass special tools through the endoscope in the digestive system. For example, an ultrasound may be attached to the endoscope to create images of the hard-to-reach organs such as the pancreas and wall of the stomach.

3.2.2 Laparoscopy

It is a keyhole surgical procedure that is employed to look inside the abdomen and pelvis. Laparoscopy is performed using a device known as a laparoscope (as shown in Fig. 8). It is a flexible tube having a video camera and light attached to it. The tube is injected in through small opening made near the belly button, and video of the abdomen is displayed on the television screen.

The need for laparoscopy

Laparoscopy is performed when abdominal problems cannot be diagnosed with imaging techniques such as ultrasound, CT scan, and MRI scan. A doctor recommends laparoscopy to examine pancreas, gallbladder, appendix, small intestine, liver, stomach, etc. By observing these areas with a laparoscope, a doctor can detect the following:

- the degree to which cancer has progressed,
- fluid in the abdominal cavity,

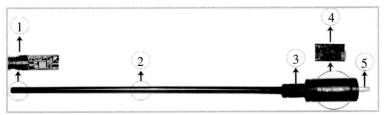

1. Video camera
2. Stainless steel tube
3. Handle of poly acetal (Delrin)
4. Analogue to digital converter
5. USB male connector

FIG. 8

Laparoscope.

Adapted from S. Jawale, G. Jesudian, P. Agarwal, Rigid video laparoscope: a low-cost alternative to traditional diagnostic laparoscopy and laparoscopic surgery, Mini-invasive Surg. 2019 (2019), doi:10.20517/2574-1225.2019.12.

- tumor in the stomach,
- liver disease.

3.2.3 Thoracoscopy

It is a knothole surgical procedure that is utilized to observe the chest cavity. Thoracoscopy is performed using a device known as a thoracoscope. It is a flexible tube having a video camera and light attached to it. The tube is injected in through small opening made near the lower end of the shoulder blade between the ribs, and video of the chest cavity is displayed on the television screen.

The need for thoracoscopy

- to check for the suspicious area in a chest,
- to treat small lung cancers,
- to obtain fluid samples from the lungs to diagnose infections.

3.2.4 Bronchoscopy

It is a nonsurgical procedure that is operated to view the human lungs. Bronchoscopy is performed using a device known as a bronchoscope. It is a flexible tube having a video camera and light attached to it. The tube is inserted in through nose or mouth, and the video of the lungs is displayed on the television screen.

The need for bronchoscopy

- for the diagnosis of a lung problem,
- for the removal of any obstruction in the lungs,
- for the biopsy of tissue from the lung.

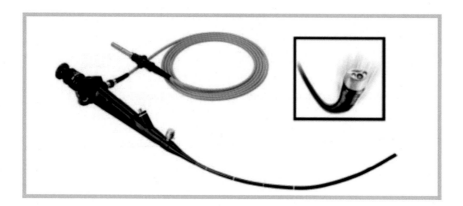

FIG. 9

Flexible urethrocystoscope.

Adapted from D. Georgescu, E. Alexandrescu, R. Mulţescu, B. Geavlete, Cystoscopy and urinary bladder anatomy, in: Endoscopic Diagnosis and Treatment in Urinary Bladder Pathology, Elsevier, 2016, pp. 1–24.

3.2.5 Cystoscopy

It is a nonsurgical procedure that is used to look inside the bladder and urethra. Cystoscopy is performed using a device known as a cystoscope (as shown in Fig. 9). It is a flexible tube having a video camera and light attached to it. The tube is inserted in through urethra, and a video of bladder is displayed on the television screen.

The need for cystoscopy

- to diagnose bladder diseases,
- to diagnose an enlarged prostate,
- to determine the cause of repeated urinary tract infections.

3.2.6 Laryngoscopy

It is a nonsurgical procedure that is employed to look at the larynx (voice box). Laryngoscopy is performed using a device known as a laryngoscope (as shown in Fig. 10). It is a flexible tube having a video camera and light attached to it. The tube is inserted in through nose or mouth, and a video of the voice box is displayed on the television screen.

The need for laryngoscopy

- to diagnose a constant cough, throat pain, or continuous bad breath;
- to discover a possible blockage of the throat;
- to remove foreign objects swallowed by mistake;
- to diagnose voice problems, such as weak and hoarse voice.

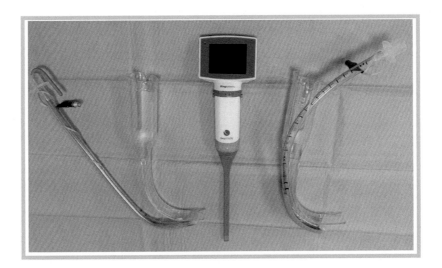

FIG. 10

Interchangeable channeled (right) and nonchanneled (left) blades of the King Vision video laryngoscope.

Adapted from J. Votruba, et al., Video laryngoscopic intubation using the King Vision™ laryngoscope in a simulated cervical spine trauma: a comparison between non-channeled and channeled disposable blades, Diagnostics 10(3) (2020) 139, doi:10.3390/diagnostics10030139.

3.2.7 Mediastinoscopy

It is a keyhole surgical procedure that is used to look inside the chest between the lungs. Mediastinoscopy is performed using a device known as a mediastinoscope, which is a flexible tube having a video camera and light attached to it. The tube is inserted in through small opening made just above the sternum, and video is displayed on the television screen.

The need for mediastinoscopy

- to check if lung cancer has spread to lymph nodes,
- to check for certain infections like tuberculosis and sarcoidosis.

From the earlier discussion, it is clear that endoscopy, bronchoscopy, cystoscopy, and laryngoscopy are nonsurgical procedures, whereas laparoscopy, thoracoscopy, and mediastinoscopy are keyhole surgical procedure.

3.3 Image data

Medical imaging is the technique used to capture images of the internal parts of a body for diagnostic and treatment purposes. A large amount of medical image data is available from various types of devices like X-rays, computed tomography (CT) scan, magnetic resonance imaging (MRI), positron emission tomography (PET), and ultrasound. A contrast agent (known as a dye) is given to the patient before taking a medical image of the interior of the body in X-ray, CT, MRI, and PET to improve the quality of the images. Depending on the body part to be examined, the dye is inhaled, swallowed, or

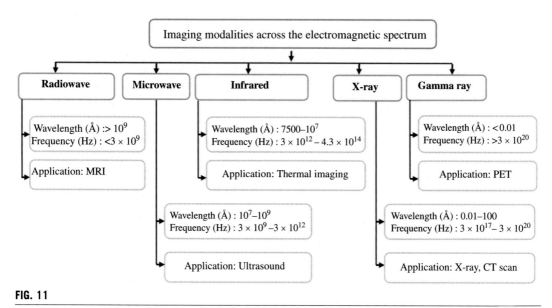

FIG. 11

Association between the electromagnetic spectrum and the imaging modalities.

injected into a vein. Based on the electromagnetic spectrum, the diagnostic imaging modalities utilize portions of the spectrum as illustrated in Fig. 11 [12].

3.3.1 X-rays

X-rays are the most commonly used medical imaging technique. X-rays are a form of electromagnetic radiation like the light. X-rays can progress smoothly through soft tissue such as muscles, but they cannot progress smoothly through hard tissue such as bones. As a result of this, bones appear white, and other substances like fat and air appear gray and black, respectively.

The need for X-rays

- to check fractures and infections in bones,
- to measure bone density
- to check for cavities in teeth,
- to check for lung cancer,
- to examine breast tissue for cancer.

3.3.2 CT scan

It is a sequence of X-ray images of organs taken from nonidentical angles. These images are analyzed by computer and then assembled into a three-dimensional model of organs.

The need for CT scan

- to detect the area of masses and tumors,
- to estimate the extent of internal injuries and internal bleeding,

- to check how well treatment is working,
- to study the blood vessels and other internal structures.

3.3.3 Magnetic resonance imaging

It is also applicable to all the parts of the body on which CT is suitable. The only difference between these two is that MRI uses radio waves, whereas CT uses X-rays, and MRI is superior to CT in regard to the detail of the image.

The need for MRI

- to detect anomalies of the brain;
- to detect cysts, tumors, and other irregularities in various parts of the body;
- to detect injuries or abnormalities of the joints;
- to identify particular types of heart problems.

3.3.4 Positron emission tomography

It allows the doctor to check for metabolic changes occurring at the cellular level in an organ or tissue. These variations are difficult to identify with CT scan and MRI. The CT scan and MRI only notice the changes in the later stage when disease harms the structure of our organs. PET uses special dye containing radioactive tracers. This dye is absorbed by organs that have high chemical activity. As a result the organs are shown up as bright spots on PET.

The need for PET

- to detect the particular surgical site before the surgical operation of the brain;
- to assess the brain after trauma to detect bleeding, blood clot, and proper oxygen flow;
- to detect the spread of cancer to other parts of the body from the original cancer site;
- to judge the effectiveness of cancer treatment.

3.3.5 Ultrasound

It uses high-frequency sound waves to look at organs inside the body. Most people associate ultrasound scans with pregnancy. However, the test has many other uses.

The need for ultrasound

- to evaluate the fetus and related structures in pregnant women;
- to evaluate the urinary bladder, vascular structures, and lymph nodes;
- to evaluate the peripheral vascular structures;
- to evaluate the blood vessels and the gastrointestinal tract.

Ultrasound and MRI are safest among all imaging techniques as they do not expose the human body to radiations. The dye is given to the patient before doing an MRI scan, which may cause allergy in some of the patients; hence among ultrasound and MRI, ultrasound is safer. In PET, body remains radioactive for around 6 h; thus PET is most dangerous.

4 Conclusion

The idea of IoT is improving the standard of life by allowing for the automation of everything around us. Among the variety of applications enabled by the IoT, healthcare is the most attractive and vital. In this chapter the authors have concentrated mainly on the characteristics of IoT health data (four V's, i.e., volume, variety, velocity, and veracity). Among these four V's, veracity is the most crucial one as the veracity of data is difficult to verify. Moreover, due to the integration of IoT with healthcare, a lot (volume) of different types (variety) of health data are produced in high velocity, which hinders the ability to cleanse data before analyzing it. This noncleanliness of data before analyzing it magnifies the issue of veracity further. Therefore veracity still remains the biggest challenge to deal with, and integration of IoT with healthcare will be fruitful when the appropriate solution is provided for the same.

References

[1] H. Habibzadeh, K. Dinesh, O. Rajabi Shishvan, A. Boggio-Dandry, G. Sharma, T. Soyata, A survey of healthcare internet of things (HIoT): a clinical perspective, IEEE Internet Things J. 7 (1) (2020) 53–71.

[2] Understanding the Many V's of Healthcare Big Data Analytics, (Online). Available from: https://healthitanalytics.com/news/understanding-the-many-vs-of-healthcare-big-data-analytics. Accessed 15 February 2020.

[3] H.J. Kong, Managing unstructured big data in healthcare system, Healthc. Inform. Res. 25 (1) (2019) 1–2.

[4] A. Ray, S.M. Bowyer, Clinical applications of magnetoencephalography in epilepsy, Ann. Indian Acad. Neurol. 13 (1) (2010) 14–22.

[5] M.R.M. Irfan, N. Sudharsan, S. Santhanakrishnan, B. Geethanjali, A comparative study of EMG and MMG signals for practical applications, in: 2011 Int. Conf. Signal, Image Process. Appl, vol. 21, 2011, pp. 106–110.

[6] Your 5 Brainwaves: Delta, Theta, Alpha, Beta and Gamma | Lucid., (Online). Available from: https://lucid.me/blog/5-brainwaves-delta-theta-alpha-beta-gamma/, 2020. Accessed 16 February 2020.

[7] M. Hämäläinen, R. Hari, R.J. Ilmoniemi, J. Knuutila, O.V. Lounasmaa, Magnetoencephalography theory instrumentation and applications to noninvasive studies of the working human brain, Rev. Mod. Phys. 65 (2) (1993) 413–497.

[8] M. Lemay, Data Processing Techniques for the Characterization of Atrial Fibrillation, EPFL (2008) https://doi.org/10.5075/epfl-thesis-3982.

[9] Z.F. Yang, D.K. Kumar, S.P. Arjunan, Mechanomyogram for identifying muscle activity and fatigue, in: Proceedings of the 31st Annual International Conference of the IEEE Engineering in Medicine and Biology Society: Engineering the Future of Biomedicine, EMBC 2009, 2009, pp. 408–411.

[10] E. Moon, H. Park, J. Yura, D. Kim, Novel design of artificial eye using EOG (electrooculography), in: Proceedings – 2017 1st IEEE International Conference on Robotic Computing, IRC 2017, 2017, pp. 404–407.

[11] J. Romano Bergstrom, S. Duda, D. Hawkins, M. McGill, Physiological response measurements, in: Eye Tracking in User Experience Design, Elsevier Inc., 2014, pp. 81–108

[12] Electromagnetic Spectrum, http://hyperphysics.phy-astr.gsu.edu/hbase/ems3.html fbclid=IwAR0pDNt3BENvXEu9mLUku34hN2IRxtysA4AmaLf_fM3BXouqzExKZ2Ca14U, 2020. Accessed 25 April 2020.

Health data analytics using Internet of things

4

Vishakha Singh[a], Sandeep S. Udmale[b], Anil Kumar Pandey[c], and Sanjay Kumar Singh[d]

Independent Researcher[a] Department of Computer Engineering and IT, Veermata Jijabai Technological Institute (VJTI), Mumbai, Maharashtra, India[b] Banaras Hindu University (BHU), Varanasi, Uttar Pradesh, India[c] Department of Computer Science and Engineering, Indian Institute of Technology (BHU), Varanasi, Uttar Pradesh, India[d]

1 Introduction

We live in a world where data are more powerful than the deadliest weapons and more precious than the rarest resources. But in its raw form, it is practically worthless. It becomes useful only after being processed. In a typical Internet of things (IoT) architecture, there is a sensor layer, a network layer, and a decision layer. The sensor layer collects data from its environment, which is then transmitted by the network layer to the processing layer. Here, data analysis is performed, which helps the decision layer in deriving conclusions. Thus, data analytics is the backbone of IoT.

The healthcare sector has always been an intriguing area for researchers working in the field of IoT. Also, the global shortage of medical professionals has caused such researches to grow. Let us discuss some important aspects of health IoT data analytics (IoTDA).

1.1 Types of health IoTDA

In the healthcare sector, IoTDA is done in two ways:

1. **Time series analytics**: This helps in finding out patterns in the data collected by continuously monitoring the patients.
2. **Prescriptive analytics**: This is done for various tasks, such as giving a diagnosis and suggesting future remedial treatments.

1.2 Platform for health IoTDA

An important aspect of IoTDA is to determine an appropriate platform for data management. There are two such platforms used for this purpose, namely, Cloud and Fog [1]. Their appropriateness is determined by the motive of data analysis. For example, a fog-based platform is preferred in the case where low latency and low response time are desirable, that is, for handling emergencies [1–3]. We can use a cloud-based platform in other cases.

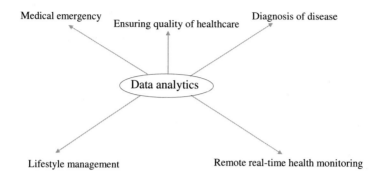

FIG. 1

Utility of health IoTDA.

1.3 Utility of health IoTDA

A robust and quick IoTDA may lead to the optimization of various healthcare processes, shown in Fig. 1, and enumerated as follows [2]:

1. **Diagnosis of disease**: The physiological data such as blood pressure, body temperature, etc., collected by the wearable IoT devices along with various test reports (urine tests, blood tests, magnetic resonance imaging [MRI] scans, etc.), can be used for expediting the process of disease diagnosis.
2. **Lifestyle management and preventive health care**: Routine analysis of data related to the behavioral aspects of a person, such as sleep patterns, food habits, amount of physical activity per day, etc., helps in the lifestyle management of chronic patients. Such an analysis may also alert a seemingly healthy individual regarding the diseases that he/she might suffer from, due to his/her current lifestyle.
3. **Medical emergency**: An emergency (e.g., an elderly falls and faints in his/her house) can be efficiently handled if an anomaly in the physiological data is swiftly detected. This is particularly useful for the elderly and patients suffering from terminal diseases.
4. **Remote real-time health monitoring**: In the healthcare sector, one of the major applications of IoTDA is real-time health monitoring. IoT sensors collect and send physiological data of the patient(s) [1]. Then, this data get analyzed in real time. Whenever an anomaly in the normal pattern is detected, a detailed report is forwarded to his/her healthcare professional for further analysis. This is particularly useful for monitoring individuals with chronic diseases, like hypertension.
5. **Ensuring quality of health care**: The use of recommender systems can help a person in finding the best physician to treat his/her ailment. This is achieved by forming a list of healthcare providers according to the ratings given by their patients.

This chapter deals with the techniques that are being used for IoTDA in the healthcare sector. These methods must be very accurate because this affects the health of an individual. As of now, we have many intelligent systems that perform better than actual medical practitioners in the case of disease diagnosis [4, 5], etc. But even then, several hurdles must be crossed to expand the market for such IoT-based services. These points have been discussed in detail in the forthcoming sections.

The rest of the chapter is organized as follows. Some of the prominent researches in this field are briefly discussed in Section 2. Various methods used in health IoTDA are enumerated in Section 3. The challenges in this field are mentioned in Section 4. Lastly, the conclusion and future scope of research in this area are discussed in Section 5.

2 Literature review

In this section, we have discussed various researches that have been done in the field of health IoTDA. The researchers in this field have extensively used deep learning algorithms [6, 7] for the diagnosis of various diseases including cardiovascular ailments. In [5, 8], the authors propose the use of convolutional neural networks (CNNs) for the diagnosis of myocardial infarction (MI) and congestive heart failure, respectively. Similarly, Shashikumar et al. [9] have developed a deep learning model to detect atrial fibrillation. Juarez-Orozco et al. [10] used a machine learning algorithm, called LogitBoost, for the detection of myocardial ischemia. In [11], Chen et al. have proposed an algorithm that uses CNN for predicting the risk of a person being susceptible to the chronic diseases prevalent in his/her community.

In the case of real-time health monitoring, Tuli et al. [3] have used ensemble learning for the detection of cardiovascular diseases using the data collected by a wearable device. Detection of a medical emergency is also an important aspect of IoTDA in the health sector. Yacchirema et al. [12] proposed an IoT system based on the decision tree learning algorithm, for sending a swift alert to the caregivers whenever an elderly falls in his/her indoor environment. Similarly, in [13], Kau and Chen suggest the use of a cascade classifier for detecting such incidents. Another important application of health IoTDA is to suggest lifestyle changes to a person for enhancing his/her quality of living. For example, in [14], Gachet Páez et al. have proposed a recommender system for suggesting lifestyle changes to a person for avoiding any probable chronic diseases.

All the aforementioned researches use machine learning in one or the other form. These methods have been described thoroughly in the following section.

3 Methods used for health IoTDA

Before the emergence of techniques such as machine learning, various health-related data and reports were analyzed manually by the medical professionals. But nowadays, the use of technology has led to the automation of such mundane tasks. In this section, we have discussed how machine learning is being used for health IoTDA.

Machine learning refers to the creation of intelligent applications, which are trained to perform specific tasks by themselves. For health IoTDA, researchers have used supervised as well as unsupervised machine learning algorithms/classifiers. However, in this section, we have focused more on the supervised algorithms since research in the case of unsupervised classifiers is still in the budding stage. The performance of these algorithms is usually measured in terms of some metrics [15], such as accuracy, F1 score, precision, recall, etc. Let us have a look at some of these algorithms.

3.1 Unsupervised learning techniques

These types of classifiers are trained on unlabeled data (in which data items do not belong to any group or class explicitly). Thus, they can be useful in cases where we need to find hidden patterns and trends in the dataset. The clustering approaches, such as the k-means algorithm, can be used for various purposes in health IoTDA. For example, in [16], the authors have proposed an algorithm based on k-means clustering to group patients based on the course of progression of chronic kidney disease (CKD). This can help in the identification of subtypes of the said disease. Pereira and Silveira [17] proposed an approach to detect anomaly in electrocardiogram (ECG) signals using three clustering algorithms, namely, hierarchical, spectral, and k-means clustering.

Due to the limited applicability of unsupervised learning algorithms in health IoTDA, the research in this area remains modest. Although with the emergence of new requirements in this field, it is expected to pick up the pace.

3.2 Supervised learning techniques

1. **Logistic regression**: This is one of the simplest among the traditional machine learning classifiers. Logistic regression uses a sigmoid function to return a set of probabilities, which represent the likelihood of a data point belonging to a set of classes. Then, based on a threshold or some other criteria, the data point is finally classified. Obasi and Shafiq [18] have used several classifiers for predicting MI. They found that the accuracy of 59.7% was achieved by the logistic regression classifier.

2. **Support vector machine (SVM)**: This algorithm finds a hyperplane that classifies the data points among various classes. For this purpose, there could be many such eligible hyperplanes. But SVM chooses a plane that has the maximum margin (distance between the data points of different classes). It is one of the most powerful traditional machine learning algorithms. In [19, 20], the authors have found that the accuracy of detection of MI using SVM is approximately 85%.

3. **Decision tree classifier**: This is one of the most popular methods used for classification problems. In a decision tree, all the internal nodes represent features of the input data, and the leaf nodes represent the possible predictions. An illustration of this method, for the detection of MI, is shown in Fig. 2. Troponin I is an enzyme that is present in negligible amounts in the bloodstream of healthy individuals. However, its concentration increases in case of some cardiovascular pathology. According to the University of Washington's Department of Laboratory Medicine, the following are the ranges of Troponin I levels present in our bloodstream [21].
   - **Below 0.04 ng/mL**: Normal range.
   - **Between 0.04 and 0.39 ng/mL**: Probable heart disease(s) like heart valve defects, coronary heart disease [22], etc.
   - **Above 0.40 ng/mL**: Probable heart attack.

4. **Deep learning:** Deep learning refers to the use of neural networks [23] comprising of a large number of hidden layers. These layers perform a series of computations on the physiological data (blood reports, MRI scans, etc.) collected by IoT sensors. Subsequently, it can predict the ailment

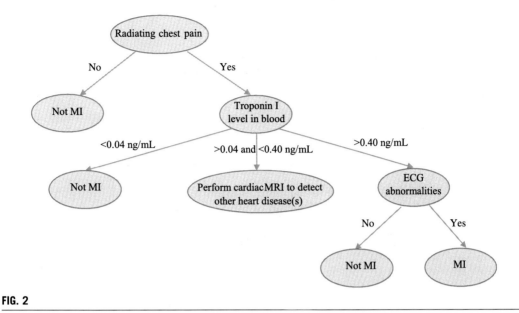

FIG. 2

A decision tree for diagnosing myocardial infarction.

from which a person might be suffering. Some popular deep learning architectures are given as follows [24]:

- **CNNs:** This is a very popular and efficient algorithm used for classification problems that involve images. It can be used for diagnosing diseases such as MI from ECG readings [3, 5, 8]. In [5], the authors found that applying CNN on ECG readings for the detection of MI can give up to 93.53%–95.22% accuracy.
- **Recurrent neural networks (RNNs):** It deals with sequence-based input (text and speech data). For example, it can be used for interpreting blood reports or patient's text/speech for diagnosing a disease [25].

    An illustration of the working of a CNN-based model in disease diagnoses is shown in Fig. 3.

    Table 1 shows the accuracy of various classifiers concerning the detection of MI. An enormous advantage of using various deep neural network approaches is that there is a lesser dependency on the hand-engineered input features, unlike the other approaches. Due to IoT, we now have access to a humongous amount of data, and when the data are abundant, deep neural networks are known to perform better than other classifiers [26]. In Fig. 4, we have shown an approximate comparison of performance of various machine learning approaches according to the size of the training dataset [27].

5. **Recommender systems:** These systems use various methods like collaborative filtering, content-based filtering, etc. and are trained mainly using deep neural networks [28]. An interesting application of recommender systems is to suggest lifestyle changes to a person [14]. It can also be

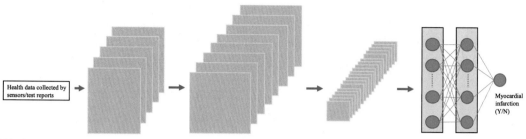

FIG. 3

Data analytics using deep learning.

Table 1 Accuracy of detection of MI.	
Classifier	**Accuracy (%)**
Logistic regression	60
Naive Bayesian	62
SVM	85
CNN	93

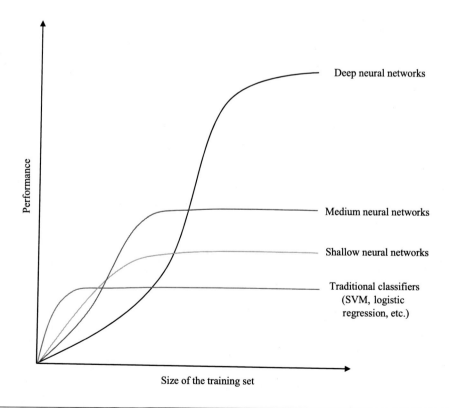

FIG. 4

Comparison of different approaches.

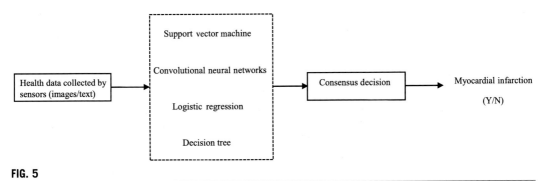

FIG. 5

Ensemble learning.

used for matchmaking of doctors and patients [29]. Another application of recommender systems is clinical quality control [30]. For example, Hoens et al. [31] proposed a system where patients can rate doctors regarding a particular consultation/treatment. This system then displays doctors' names along with their ratings. The biggest issue of such an application is the generation of false ratings of a person using fake reviews.

6. **Ensemble learning:** This technique uses a set of classifiers (or ensemble), which gives any decision based on consensus. Let us consider an example where a particular ensemble tries to diagnose a disease based on some input (say, blood report). It is said to have high confidence in the classification if the majority of classifiers predict the same disease. Otherwise, if half of them predict one disease and the rest of them agree on another, this means that the ensemble has low confidence in its decision. Note that high confidence in a decision does not imply that it is correct. Tuli et al. [3] used ensemble deep learning algorithms in their proposal for health IoTDA. An illustration of the ensemble method is shown in Fig. 5.

4 Bottlenecks of the health IoTDA

The process of IoTDA seems to be perfectly smooth. But there are a few difficulties that arise time and again during the actual implementation. Some of them are stated as follows.

4.1 Performance of IoTDA methods

Health is the most significant aspect of our lives. So, the algorithms used to perform health IoTDA must be robust. If these algorithms misclassify and give the wrong diagnosis, it could devastate a person monetarily (in form of catastrophic health expenditure), physiologically (if treatment leads to other health-related side effects), and psychologically (treatment may create emotional turmoil within a person and affect his mental health). Also, if an IoTDA algorithm fails to detect a life-threatening situation, for example, when an individual suffers from a heart attack, it could be fatal.

Two important performance metrics that are used to evaluate a machine learning algorithm are accuracy and F1 score. The IoTDA algorithms employed must have high accuracy and high F1 score [15].

4.2 Latency and response time

Some tasks in the healthcare sector, like real-time health monitoring of the chronic patients, medical emergency detection systems, etc., require low response time and latency. Thus, IoTDA must be fast. This problem is very prevalent in cloud platforms. Therefore, the researchers have now started using various edge computing techniques like Fog [3], which significantly decreases the latency and response time by bringing data processing closer to the end-user.

4.3 Vulnerability to cyberattacks

In 2017, Newsweek [32] reported that nearly half of the US firms (that use IoT) were targeted by cyber-criminals. The healthcare sector is an easy target because the security layers are generally weak and penetrable. The confidential nature of health data could attract hackers who may use it to blackmail the healthcare institutions as well as their patients. There is also a possibility that such an attack may compromise the integrity of a medical device. This reduces the trust of people toward this technology. Various countries have passed strict laws to prevent companies that possess the personal information of the citizens, from being inconsiderate toward data security.

4.4 Data management infrastructure

As of now, the infrastructure for storing IoT data is insufficient. In other words, it is incapable of handling huge volumes of IoT data generated every day. This may lead to crucial data getting overlooked in the data analysis stage. Such a scenario may lead to some serious ramifications, and the health as well as well-being of an individual may get affected.

4.5 Data localization

The restrictive laws enacted by various countries have led to data localization. In other words, the data of citizens are stored within the borders. So the data analysis is dependent on the infrastructure present within that country. This acts as a challenge to the globalization of IoT. Some countries like Russia, Nigeria, etc. have enacted such laws [33].

---

5 Conclusion

In this chapter, we have discussed various machine learning algorithms for efficient IoTDA in the healthcare sector. We saw that supervised learning algorithms like deep learning, decision trees, etc. have become quite famous among the research fraternity. We concluded that deep learning is the best approach for IoTDA, as of now. Then we elucidated some of the challenges that exist in the area of health IoTDA. We could try to tackle these very issues, starting with trying to increase the scalability of the existing data management infrastructure to accommodate the huge volumes of data. It should also be made resilient to cyberattacks. Another important aspect that needs attention is that a person must have the right to get his data deleted from any database whenever he/she desires.

As a future topic of research in this area, we may focus on various unsupervised learning approaches like clustering algorithms. This can be particularly helpful in finding trends and patterns in the infection rate, recovery rate, etc., in case of pandemics such as the COVID-19 crisis [34].

References

[1] A.A. Mutlag, M.K.A. Ghani, N. Arunkumar, M.A. Mohamed, O. Mohd, Enabling technologies for fog computing in healthcare IoT systems, Futur. Gener. Comput. Syst. 90 (2019) 62–78.

[2] T. Saheb, L. Izadi, Paradigm of IoT big data analytics in healthcare industry: a review of scientific literature and mapping of research trends, Telematics Inf. 41 (2019) 70–85.

[3] S. Tuli, N. Basumatary, S.S. Gill, M. Kahani, R.C. Arya, G.S. Wander, R. Buyya, Healthfog: an ensemble deep learning based smart healthcare system for automatic diagnosis of heart diseases in integrated IoT and Fog computing environments, Futur. Gener. Comput. Syst. 104 (2020) 187–200.

[4] L.E. Juarez-Orozco, O. Martinez-Manzanera, F.M. Van Der Zant, R.J.J. Knol, J. Knuuti, 241 Deep learning in quantitative PET myocardial perfusion imaging to predict adverse cardiovascular events, Eur. Heart J. Cardiovasc. Imaging 20 (suppl. 3) (2019) jez145-005.

[5] U.R. Acharya, H. Fujita, S.L. Oh, Y. Hagiwara, J.H. Tan, M. Adam, Application of deep convolutional neural network for automated detection of myocardial infarction using ECG signals, Inf. Sci. 415 (2017) 190–198.

[6] J. Ker, L. Wang, J. Rao, T. Lim, Deep learning applications in medical image analysis, IEEE Access 6 (2017) 9375–9389.

[7] S. Chilamkurthy, R. Ghosh, S. Tanamala, M. Biviji, N.G. Campeau, V.K. Venugopal, V. Mahajan, P. Rao, P. Warier, Deep learning algorithms for detection of critical findings in head CT scans: a retrospective study, Lancet 392 (10162) (2018) 2388–2396.

[8] U.R. Acharya, H. Fujita, S.L. Oh, Y. Hagiwara, J.H. Tan, M. Adam, R. San Tan, Deep convolutional neural network for the automated diagnosis of congestive heart failure using ECG signals, Appl. Intell. 49 (1) (2019) 16–27.

[9] S.P. Shashikumar, A.J. Shah, Q. Li, G.D. Clifford, S. Nemati, A deep learning approach to monitoring and detecting atrial fibrillation using wearable technology, in: 2017 IEEE EMBS International Conference on Biomedical & Health Informatics (BHI), IEEE, 2017, pp. 141–144.

[10] L.E. Juarez-Orozco, R.J.J. Knol, C.A. Sanchez-Catasus, O. Martinez-Manzanera, F.M. van der Zant, J. Knuuti, Machine learning in the integration of simple variables for identifying patients with myocardial ischemia, J. Nucl. Cardiol. 27 (1) (2018) 1–9.

[11] M. Chen, Y. Hao, K. Hwang, L. Wang, L. Wang, Disease prediction by machine learning over big data from healthcare communities, IEEE Access 5 (2017) 8869–8879.

[12] D. Yacchirema, J.S. de Puga, C. Palau, M. Esteve, Fall detection system for elderly people using IoT and big data, Procedia Comput. Sci. 130 (2018) 603–610.

[13] L.-J. Kau, C.-S. Chen, A smart phone-based pocket fall accident detection, positioning, and rescue system, IEEE J. Biomed. Health Inf. 19 (1) (2014) 44–56.

[14] D. Gachet Páez, M. de Buenaga Rodríguez, E. Puertas Sanz, M.T. Villalba, R. Muñoz Gil, Healthy and well-being activities' promotion using a Big Data approach, Health Inf. J. 24 (2) (2018) 125–135.

[15] K.P. Shung, Accuracy, Precision, Recall or F1?, 2018, Available from: https://towardsdatascience.com/accuracy-precision-recall-or-f1-331fb37c5cb9.

[16] D.T.A. Luong, V. Chandola, A k-means approach to clustering disease progressions, in: 2017 IEEE International Conference on Healthcare Informatics (ICHI), IEEE, 2017, pp. 268–274.

[17] J. Pereira, M. Silveira, Learning representations from healthcare time series data for unsupervised anomaly detection, in: 2019 IEEE International Conference on Big Data and Smart Computing (BigComp), IEEE, 2019, pp. 1–7.

[18] T. Obasi, M.O. Shafiq, Towards comparing and using machine learning techniques for detecting and predicting heart attack and diseases, in: 2019 IEEE International Conference on Big Data (Big Data), IEEE, 2019, pp. 2393–2402.

[19] N.A. Bhaskar, Performance analysis of support vector machine and neural networks in detection of myocardial infarction, Procedia Comput. Sci. 46 (4) (2015) 20–30.

[20] S. Ghumbre, C. Patil, A. Ghatol, Heart disease diagnosis using support vector machine, in: International Conference on Computer Science and Information Technology (ICCSIT) Pattaya, 2011.

[21] Z. Villines, What is the Normal Range for Troponin Levels?, 2019, Available from: https://www.medicalnewstoday.com/articles/325415.php#normal-troponin-range.

[22] 2018. Magnetic Resonance Imaging (MRI)—Cardiac (Heart), 2018, Available from: https://www.radiologyinfo.org/en/info.cfm?pg=cardiacmr.

[23] J. Brownlee, A Tour of Machine Learning Algorithms, 2019, Available from: https://machinelearningmastery.com/a-tour-of-machine-learning-algorithms/.

[24] J. Brownlee, When to Use MLP, CNN, and RNN Neural Networks 2018, 2018, Available from: https://machinelearningmastery.com/when-to-use-mlp-cnn-and-rnn-neural-networks/.

[25] Q. Sun, M.V. Jankovic, L. Bally, S.G. Mougiakakou, Predicting blood glucose with an LSTM and Bi-LSTM based deep neural network, in: 2018 14th Symposium on Neural Networks and Applications (NEUREL), IEEE, 2018, pp. 1–5.

[26] D. Li, X. Li, J. Zhao, X. Bai, Automatic staging model of heart failure based on deep learning, Biomed. Signal Process. Control 52 (2019) 77–83.

[27] T. Shimpi, Difference Between ML and Deep Learning With Respect to Splitting of the Dataset Into Train/Cross-Validation/Test Sets, 2019, Available from: https://medium.com/@tanmayshimpi/difference-between-ml-and-deep-learning-with-respect-to-splitting-of-the-dataset-into-375d433ee2c8.

[28] P. Sánchez, A. Bellogín, Building user profiles based on sequences for content and collaborative filtering, Inf. Process. Manag. 56 (1) (2019) 192–211.

[29] Q. Han, M. Ji, I.M.d.R de Troya, M. Gaur, L. Zejnilovic, A hybrid recommender system for patient-doctor matchmaking in primary care, in: 2018 IEEE 5th International Conference on Data Science and Advanced Analytics (DSAA), IEEE, 2018, pp. 481–490.

[30] M. Singh, V. Bhatia, R. Bhatia, Big data analytics: solution to healthcare, in: 2017 International Conference on Intelligent Communication and Computational Techniques (ICCT), IEEE, 2017, pp. 239–241.

[31] T.R. Hoens, M. Blanton, A. Steele, N.V. Chawla, Reliable medical recommendation systems with patient privacy, ACM Trans. Intell. Syst. Technol. (TIST) 4 (4) (2013) 67.

[32] A. Dellingar, Internet of Things Safety: Nearly Half of U.S. Firms Using IOT Hit by Security Breaches, 2017, Available from: https://www.newsweek.com/iot-security-internet-things-safety-breaches-businesses-how-protect-621230.

[33] M. Newton, J. Summers, Russian Data Localization Laws: Enriching Security and the Economy, 2018, Available from: https://jsis.washington.edu/news/russian-data-localization-enriching-security-economy/.

[34] T. Zoumpekas, COVID-19 Cluster Analysis, 2020, Available from: https://towardsdatascience.com/covid-19-cluster-analysis-405ebbd10049.

Computational intelligence in Internet of things for future healthcare applications

5

Vandana Bharti, Bhaskar Biswas, and Kaushal Kumar Shukla

Department of Computer Science and Engineering, Indian Institute of Technology (BHU), Varanasi, Uttar Pradesh, India

1 Introduction

The Internet of things (IoT) has opened the way for a new boom of research in the healthcare sector. Emerging IoT advancements in healthcare urgently need robust networking systems that are immune to threats and disruptions, like high-mobility, disasters, system failures, cyberattacks, etc. This also calls for safer IoT environments, enhancement of human interaction in the field of IoT, and the development of smart apps in the broad context of IoT, since there is a need for safe, efficient, and effective communication in a networking system for performance sensitive and critical surgeries such as remote surgery. New opportunities and challenges lie in allowing computational intelligence (CI) in the broad context of IoT to establish a fertile ground for research and innovation. This calls for new computational approaches that incorporate the latest advances in communications, data analytics and artificial intelligence to solve theoretical and practical problems in IoT systems. Fig. 1 illustrates the integrated approach to healthcare IoT.

The rapid growth in the semiconductor industry has resulted in an explosion of usage of reasonably priced, processor-based sensors. These systems when integrated with powerful communication technology such as Zigbee, radio frequency identification (RFID), 3G, 4G and Inseton, give rise to a new paradigm for IoT. Basically, IoT is a group of interconnected objects such as sensors, wireless sensor networks (WSNs), medical devices, robots, plants, animals, vehicles, smart watches, humans, and many others. Hence, IoT can be defined as an interconnection of distinctively addressable physical things with different degrees of sensing, processing, actuation, and sharing capabilities that interoperate and communicate through wired or wireless networks as their joint platform [1]. Nowadays, peoples are showing a keener interest in healthcare-related sensor devices, and today the development of technology has provided many more healthcare-related sensor devices in the market for smart healthcare monitoring [2]. Thus IoT research communities are motivated to design sensors for smart farming, animal health monitoring, and human health monitoring. Recently, authors in [3] have introduced a well-conceptualized IoT architecture, shown in Fig. 2, for agro-industrial and environmental applications. Subsequently they are focused on enhancing the network lifetime by minimizing the energy consumption of sensors in the IoT network. It seems that the fourth revolution in healthcare systems is taking place along with the fourth revolution in industry, and the technical foundation of modern healthcare systems is an integration of computers, Internet, and computer networks embedded with

IoT Based Data Analytics for the Healthcare Industry. https://doi.org/10.1016/B978-0-12-821472-5.00018-1

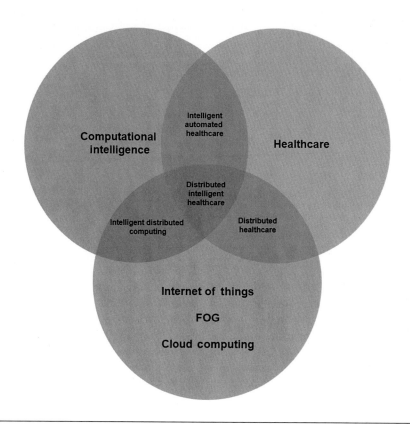

FIG. 1

Integrated approach to intelligent healthcare IoT.

CI that is called a medical cyber physical system [4]. Thus, the Internet applications are closely integrated into the daily lives of human beings and cover a wide domain from email to search engine, e-marketing to e-health, and e-learning to e-booking, etc. The debut of these applications has made our lives far more convenient than in the past. At this point, the IoT is not yet well defined; however, it has led to significant improvements in providing healthcare services. In addition, the full capacity of IoT in healthcare is yet to be explored and the challenges like interoperability, scalability, usability, privacy, security, and quality of services still need to be addressed by researchers. Since heterogeneous devices are interconnected in IoT and give rise to voluminous healthcare data, data storage, resource management, service creation, and resource and power management are other challenging aspects that cannot be efficiently handled by IoT alone. Therefore, researchers are directed toward the intelligent future internet that leads to the Future Internet of Things (FIoT).

Healthcare devices and sensors are suffer from limitations of power and constrained resources along with the requirement to offload computation. To overcome these limitations, the cloud, which is rich in extensive processing and provides sufficient resources, has integrated with IoT and given rise to a new terminology, that is, Cloud of Things (CoT), which seems to be a possible solution. Moreover, the cloud

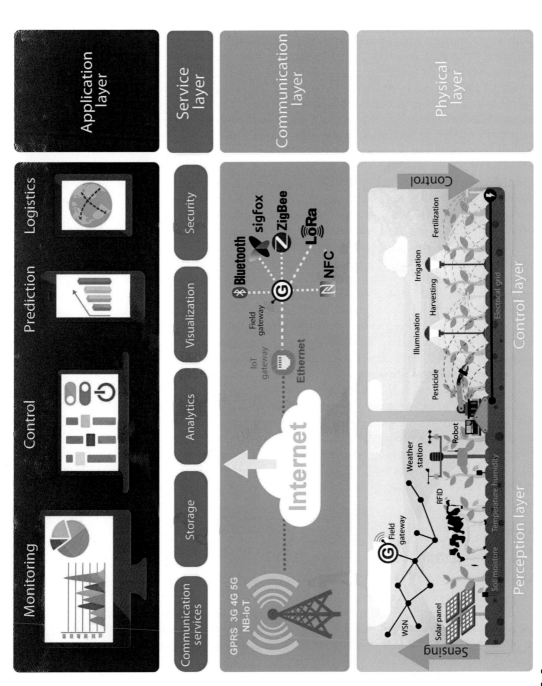

FIG. 2

IoT architecture for applications in agro-industry [3].

is accessible through public internet because it is distant from the IoT nodes and restricts its applications to real-time latency sensitive services, such as telemedicine, and also makes sensitive personal healthcare information from a patient vulnerable to adversary attack. Fog and mobile edge computing help to get rid of these issues in CoT by integrating cloud resources and edge devices. Fog plays an important role in smart healthcare contributor modules, such as mobile healthcare, ubiquitous healthcare, smart farming, etc. by providing fast response, data privacy, and data filtering. It acts as a middleware in a variety of modern and intelligent healthcare. Fog resources are placed between the cloud and healthcare IoT layers. Fog also monitors sensors and edge devices beneath it on their energy consumption [5, 6]. CI is an indispensable branch of computer science that comprises nature-inspired computation, fuzzy logic, and neural networks, which aims to enable computers to "think on their own" [7]. Nowadays CI is quite popular in the healthcare sector due to its ability to solve complex problems efficiently. The major goal of any healthcare IoT system is to provide efficient and secure consultation without any physical presence. Additionally, to process the data in a decentralized environment and to automate the overall prediction and analysis, CI is also integrated with these systems. Generally, the overall smart healthcare system (human healthcare, agriculture healthcare, and animal healthcare) consists of a complex optimization problem, scheduling problem, and optimal feature engineering with a predictive model and security algorithm incorporated within. Some of the real-time data scheduling in the distributed framework can be formulated as a dynamic optimization problem. It is widely accepted that the nature-inspired optimization of CI has been successfully applied to solve recent popular real-life application problems [8]. IoT marketing trends have evolved steadily over the past decades. More precisely, the IoT market is expected to reach $309 billion per annum in 2020 [9]. In another technical report, it is pointed out that by 2025 IoT's potential economic impact will reach $2.7 trillion to $6.2 trillion per annum [10]. Even though the market value projections for IoT and related applications vary from researcher to researcher depending on the specific domain, it is quite evident that IoT will have a major impact on industries, economies, and, most importantly, the healthcare sector in the near future.

Therefore, a fast and convenient environment can be expected by utilizing CI with IoT in our daily lives. Especially in the healthcare sector, proper data acquisition, data analysis, and optimal use of computation resources for accurate decision play a vital role. Recent developments in IoT, distributed computing, CI, multicriteria decision making and the healthcare sector motivate us to describe the concept of CI in health IoT. Subsequently, we also cover the different realms of healthcare, including human healthcare, agriculture healthcare, and animal healthcare. Finally, this chapter presents conclusions and future directions.

2 IoT for healthcare

Currently, IoT is considered to be a crucial aspect of smart healthcare. It is expected to transform the healthcare sector to the same degree it has transformed the way industries work and has influenced the lives of countless people around the world. IoT allows the development of a centralized network of interconnected devices that can generate and share data within a single framework. Data obtained from the Internet of Medical Things (IoMT) apps make it easier to provide patient services effectively and creatively. Additional information collected from IoT monitoring devices analyzed with the assistance of CI healthcare applications will allow medical professionals to develop more focused and personalized treatment services for their patients and improve the well-being of the world's population. In terms

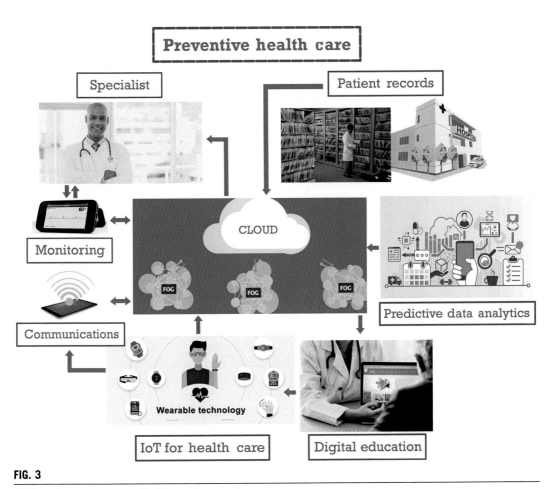

FIG. 3

IoT-based smart preventive healthcare framework.

From J. Bravo, I. González, Call for papers: towards preventive health care through digital technologies, J. Biomed. Inform. 95 (2019) 103217, https://doi.org/10.1016/j.jbi.2019.103217.

of enhancing medical facilities, this means that a regular hospital can be turned into a smart hospital. Fig. 3 demonstrates the key aspects of the IoT-based smart preventive healthcare framework. In addition to human healthcare, IoT for agriculture healthcare is one of the fastest growing fields in IoT due to the continuous increase in the world's population and the increasing demand for food supply. Agriculture has to turn to new technology to meet these increasing needs. New IoT-based smart farming applications will enable the agriculture industry to reduce waste and increase productivity, from optimizing fertilizer use to enhancing the efficiency of farm vehicle paths. The sensing of soil humidity and nutrients, the monitoring of water consumption, and customized fertilizer for plant growth are some simple IoT applications in agriculture.

2.1 IoT for human healthcare

Healthcare services are often challenging because many diseases can arise unexpectedly. IoT has been widely used to interconnect available medical resources and provide reliable, effective, and smart healthcare services to chronic disease patients. Today's internet-connected apps are designed to improve efficiencies, reduce the cost of treatment, and achieve improved healthcare outcomes. As computing capacity and wireless capabilities increase, companies are capitalizing on the potential of Internet of Medical Things technologies (IMoT). These applications play a central role in tracking and preventing diseases such as COVID-19 for government agencies, patients, and clinicians and they are poised to evolve the future of care. The IoT has provided a variety of medical possibilities, as ordinary medical tools can gather useful additional data while they are linked to the Internet, provide unique information into symptoms and trends, enable remote care, and simply provide patients with better preventive treatment and more control over their lives. Fig. 3 demonstrates the key aspects of the emerging preventive healthcare framework. Several IoT instances of healthcare, showing the monitoring of various diseases, are proposed. Some of these are described in the following sections.

2.1.1 COVID-19 monitoring

The ongoing COVID-19 outbreak has prompted IoT healthcare providers to rapidly find solutions to meet the rising demand for high-quality virus protection devices. The rapid spread of COVID-19 has taken over the entire health ecosystem including pharmaceutical companies, drug makers, COVID-19 vaccine developers, health insurers, and hospitals. Applications such as telemedicine include remote patient monitoring, and interactive medicine is expected to gain traction during this time, along with inpatient monitoring. Further, digital contact tracing came to public attention during the COVID-19 pandemic, which is a form of contact tracing that depends on tracking systems, most often based on mobile devices, to establish the connection between the infected patient and the user. Such accomplishments have demonstrated the efficacy and exciting future of IoT in healthcare systems. Despite the obvious successes, there is also ambiguity, and there is still a technical challenge in the question of how to set up smart IoT-based healthcare systems quickly and systematically.

Artificial intelligence (AI) along with the IoT has successfully contributed to the battle against COVID-19. Since there is no specific treatment for coronaviruses, global monitoring of COVID-19-infected humans is desperately required. The IoT serves as a platform for public-health organizations to access data for the monitoring of the COVID-19 pandemic, such as the "Worldometer." It gives a real-time report on the total number of people reported to have COVID-19 across the world. These smart disease monitoring systems may provide for continuous reporting and surveillance, end-to-end communication, tracking, and alerts. IoT and telemedicine will help provide not only affordable healthcare, but also help in collecting of data on the monitoring of drugs and vaccines that are currently being tested worldwide.

A lot of work has recently been published on COVID-19. Recently, the deep convolutional neural network-based COVID-Net was proposed for the detection of COVID-19 cases from chest X-ray (CXR) images [11]. The authors explored how COVID-Net makes predictions in an effort to gain a deeper insight into the crucial factors associated with COVID cases, which can help clinicians improve screening, as well as enhance confidence and consistency while using COVID-Net for rapid computer-aided screening. Ghoshal et al. [12] introduced a Bayesian Convolutional Neural Network for estimating the uncertainty of diagnosis in COVID-19 prediction, using patient X-ray images with COVID-19, acquired from an online COVID-19 dataset [13], and non-COVID-19 images, acquired from Kaggle's Chest X-Ray Images (Pneumonia). The experiment revealed that Bayesian inference enhanced the detection accuracy of the standard VGG16 model from 85.7% to 92.9%. The authors also generated saliency maps to demonstrate the locations of the deep network, improve the understanding of deep learning outcomes, and facilitate a more informed decision-making process. Recently, the authors in [14] introduced a schematic of an app for COVID-19 contact tracing, as shown in Fig. 4. In this app, contacts between Person A and all persons using the application are traced by Bluetooth with low-energy connections with other app users. Person A requests the SARS-CoV-2 test with the application, which causes immediate notification of those in close contact with each other of the positive test result of that person. The application recommends isolation for Person A and quarantine of the individual's contacts.

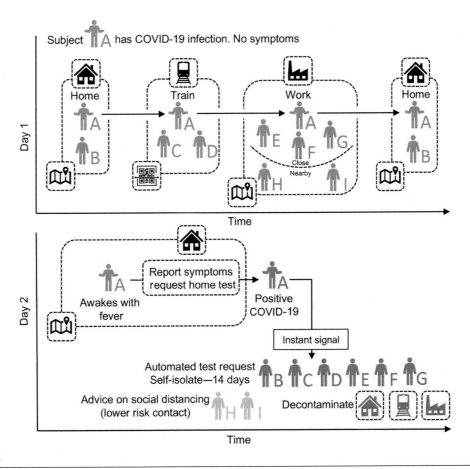

FIG. 4

A schematic of app-based COVID-19 contact tracing [14].

2.1.2 Cancer monitoring

As computers and tools become smarter when interacting with each other, AI systems, such as IBM's Watson, as well as robotic surgeons, can support doctors from diagnosis to treatment for cancer. In general, the earlier a doctor can recognize symptoms, the faster they can reach a diagnosis and start treatment. A number of early cancer signs are unclear and unrecognizable, so it is understandable that cancers can go undiagnosed in the first instance. The argument lies therein: AI and IoT will boost care for cancer, but they will function together. A patient monitoring system work flow on IoT is shown in Fig. 5.

Recently, the researchers have also been investigating the integrated framework of IoT, fog computing, CI, and cancer diagnosis. An IoT-based fog computing model for cancer detection and monitoring is proposed in [15] in which they used a mobile application interface to capture the symptoms of the patient and further applied a neutrosophic multicriteria decision-making approach for examining and forecasting of disease based on the reported symptoms. Similarly, in another work, the author also proposed an IoT-based healthcare framework for cancer care services along with the treatment options [16]. Further, authors also worked to monitor cancer patients in a secure home environment for which a multisensory IoT framework was proposed. Using an intelligent IoT sensor, the patient's physiological as well as mental status data were collected and shared with physicians for visualization and better understanding of current patient status for real-time decision-making. To ensure the secure transmission of private data, a blockchain and off-chain-based framework was adopted [17].

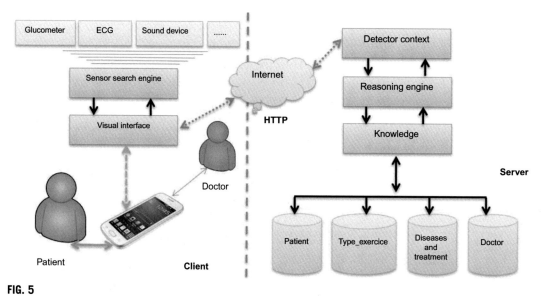

FIG. 5

Patient monitoring system based on IoT [18].

2.1.3 Depression monitoring

Depression and anxiety are two common psychological conditions that arise as a consequence of excessive tension and distress experienced every day by individuals. It is difficult to avoid the down periods of life and their consequences, but people respond to them in many different ways. The potential source of depression and anxiety may be a variation of psychological, biological, and social causes. Depression is one of the prevalent causes of major depressive disorder (MDD), which can lead to thoughts of death or suicide if not treated in a timely manner. Many studies have come up with a number of speculations linked to the occurrence, treatment, and control of depression and anxiety in individuals.

Recent research suggests that MDD can be monitored by a smartwatch device that patients use every day to monitor their moods and emotions. Wearable technology has a significant capacity for doing more than tracking steps; in this scenario, it may be used to measure the symptoms of depression in real time. Like other IoT wellness devices, a depression app might provide more insight into the condition for patients and healthcare providers. A fear induction task for 20 s was demonstrated in [19], using a wearable sensor on young children using machine learning, which resulted in a high fraction of accuracy that points to this diagnosing approach for children with internalizing disorders. In [20], the authors attempted to design a prototype of a wearable device to assist individuals with MDD by using speech recognition to determine their positive and negative phases by an emotional user interface. Further, a deep regression network known as DepressNet was presented in [21] to assimilate depression representation with a visual explanation, which provides a clinical prediction of the depression severity from facial images.

2.2 IoT for agriculture healthcare

Generally, crops and animal healthcare are covered under the single-term "agriculture healthcare" due to their interdependency. However, some researchers consider these to be different domains. Recently, the industrial IoT has disrupted many industries, and the agriculture industry is no exception. In addition, this CI has been successfully applied and has gained popularity in almost every field, from complex optimization problems to data analysis. Thus two popular and recent techniques are combined together to provide an intelligent solution in the agriculture domain and have become an emerging research area that has attracted different communities, such as web developers, data analysts, agronomists, and scientists. Many sectors throughout the world are interdependent on agriculture and due to the unprecedented rise in human population, there is an immense pressure on the agricultural sector to increase the production of agricultural commodities. It is also true that the agriculture sector faces many different challenges, such as crop disease infestations, locust attacks, limited storage services, lack of irrigation and drainage facilities, weed management, crop quality, etc., resulting in low exploitable yield. Therefore, it is necessary to employ the recent advanced technology in agriculture, including sensor networks, IoT, CI, and advanced imaging techniques like hyperspectral imaging systems, for use in smart farming, smart crop management, smart irrigation, smart greenhouses, and intelligent livestock disease surveillance, etc. Different applications of IoT in agriculture are shown in Fig. 6.

Further, it is found that advanced imaging techniques have been analyzed widely for facilitating sustainable agriculture by mapping and monitoring the agricultural situation, retrieval of biophysical parameters, and development of a decision support system. For crops like oilseed rape, the technology

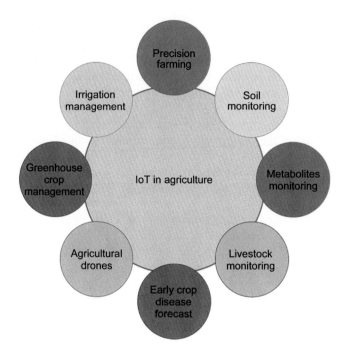

FIG. 6

Applications of IoT in agriculture.

has been found to be useful for disease forecasting, detection of fungal pathogens, monitoring damages induced by infestation, predicting seed yield, weeds, and macronutrient analysis for monitoring fertilizer application [22]. Therefore, the agriculture sector is introducing IoT and integrating recent advanced technology to stabilize the demand and supply of agriculture products, which is also necessary for national growth. Fig. 7 gives an overview of smart agriculture in a wide perspective.

2.2.1 Precision agriculture

Precision agriculture (also called precision farming, satellite farming, or site-specific crop management) is now one of the popular research areas among computer science, IoT, and agricultures. They are motivated to integrate CI with IoT and agricultures to ensure that the crop and soil receive exactly the macronutrients that are essential for healthy crops and high productivity. For this, agriculture can be connected to IoT using sensor networks. These sensors are of wide applicability: measuring soil moisture, humidity, monitoring greenhouse parameters, etc. With the aid of the Internet, mobile communications, and sensor networks, agronomists and data analysts provide timely cultivating guidance including pest control, disease, and catastrophic climate alerts to farmers throughout the country, regardless of geographical constraints. This approach helps farmers, data analysts, and experts to better understand the crop growth model for improved farming practices. Precision farming is also one of the emerging applications of IoT and CI in agriculture that aims to ensure productivity, sustainability, and environmental conservation. Further, it can provide quick and intelligent decisions for more efficient

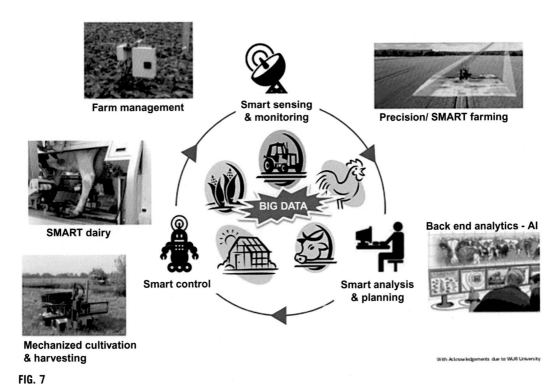

Farm management

Smart sensing & monitoring

Precision/ SMART farming

SMART dairy

BIG DATA

Back end analytics - AI

Smart control

Smart analysis & planning

Mechanized cultivation & harvesting

With Acknowledgements due to WUR University

FIG. 7

Overview of smart agriculture using IoT [23].

and effective farming by analyzing sensor generated data. Several techniques, such as vehicle tracking, irrigation management, field observation, inventory monitoring, and livestock monitoring, come under precision agriculture. Through these techniques, water and nutrient level can also be monitored in real time.

2.2.2 Crop health monitoring

Farm productivity is highly influenced by the different environmental conditions that need to be understood for forecasting crop performance. Currently, crop recommendations are mostly based on manual field-based agriculture studies that analyze crop performance under diverse conditions such as soil quality and environmental conditions, and the quality of manually collected data is low. Recently IoT has provided a common platform to agronomists and researchers to maintain real-time data and give an immediate solution to farmers by analyzing generated data from a variety of sensors deployed in the field. IoT also provides easy access to data generated from wireless sensor network nodes. In addition, this cloud and fog-based framework helps to manage big data of agriculture information in terms of soil properties, fertilizer distribution, and image cultivation. However, GPS and GIS are used for PA, which is costly, therefore low-cost technological solutions such as wireless communication cameras, sensor networks, etc. are used to monitor crop conditions and growth for a longer time interval without any failures; the data is sent to the cloud where different CI-based techniques are applied. By following this approach, remote decisions related to the potential of new crops can be made, which helps in field crop healthcare management [24]. Different sensors are deployed in the field to monitor the overall crop health, including the Intellia INT G01 (Soil Moisture Sensor), INT-PH1 (soil PH sensor), MQ4 (natural gas sensing), and MQ7 (carbon monoxide sensor). Other wireless nodes used in the agriculture domain are listed in Table 1.

These sensors are generally deployed in the agriculture field, as shown in Fig. 8, and collect data that are further transmitted through communication channels to the cloud-based server. On the server, these data are analyzed using an intelligent framework that acts as a decision support system. Also, based on the measured values of the parameters monitoring system, devices are automatically switched ON and OFF, such as water pump for irrigation system, fan for air circulation, etc. In recent work, researchers have developed a cloud-enabled CLAY-MIST measurement index used for real-time crop data monitoring to assess the comfort levels of a crop and send notifications to the farmers as per the livestock safety index classes. This allows farmers to take necessary action for healthy crops [26]. Similarly, CropSight [27] is another innovation in modern agriculture to closely monitor the dynamic crop performance and agricultural condition to predict crop production. It also helps to estimate the appropriate time and area for chemical applications.

Climate is also one of the crucial aspects that affect crop health to enhance the quality and quantity of crop production, it is necessary to have the proper knowledge of climate conditions. Hence, IoT solutions facilitate this process by placing sensors inside and outside agricultural land and collecting environmental data to give real-time weather conditions. The entire IoT ecosystem consists of sensors that can perceive environmental conditions in real-time, such as humidity, rainfall, and temperature. These collected data help farmers to choose the right crops that can be grown and sustained under

Table 1 Wireless sensors used in agriculture.

Sensor	Measured parameter
107-L (temperature sensor)	Plant temperature
LT-2M (leaf temperature sensor)	Plant temperature
LW100 (leaf wetness/rain sensor)	Plant wetness, plant moisture, plant temperature
YSI 6025 (chlorophyll sensor)	Photosynthesis
TPS-2 (portable photosynthesis sensor)	Carbon dioxide, photosynthesis, plant moisture
EC250, Hydra Probe II (soil sensor)	Soil temperature, soil moisture, salinity level, conductivity
CI-340 (hand-held photosynthesis)	Photosynthesis, plant wetness, plant temperature, plant moisture, carbon dioxide, and hydrogen level in plants
XFAM-115KPASR (pressure sensor)	Air temperature, air pressure, and air humidity
CM100 (compact weather sensor)	Air temperature, air pressure, air humidity, and wind speed

FIG. 8

The agro-weather sensors deployed in an agriculture field [25].

particular climatic conditions. There are numerous sensors available to measure all these parameters and they can be configured to meet smart farming needs. These sensors monitor the conditions of the crop and the weather surrounding it and send an alert message for unfavorable or disturbing weather conditions. This also averts the need for the physical presence of farmers in such unfavorable and disturbing climate conditions, and therefore this approach enhances productivity.

2.2.3 Early disease diagnostic system and pest control

Insects such as locusts damage crops and thus adversely affect their productivity. Besides these pests that proliferate rapidly and destroy crops, the macronutrients of the soil are also a major hurdle in farming and crop quality. Earlier farmers used traditional methods to control all these things, which were not efficient. Therefore, systems are now being developed that provide the functionality of pest warning in the form of alarms; also, by using deep learning and fuzzy-based methods, they can even detect the type of disease and categorize insects that damage the crops [28]. Recent works in this area show that advanced imaging techniques such as hyperspectral images and multispectral images can also be utilized for macronutrient analysis as well as disease prediction in crops [22, 29] and they have opened a challenging research area for researchers of agriculture, distributed computing, and CI. In a similar way, a cyber-physical system-based potato monitoring and management system was designed that made decisions based on the collected spectral signature analysis [30]. Fertilization pesticide control is another aspect of enhancing crop health. IoT solutions have applied conservation practices to improve nutrient usage, efficiency, crop quality, overall yield, and economic return while reducing off-site transport of nutrients.

2.2.4 Smart greenhouse

IoT has made it possible for weather stations to change greenhouse climate conditions automatically according to a particular set of instructions. Applications of IoT in greenhouses have eliminated human intervention, making the whole process cost-effective and increasing accuracy at the same time. Nowadays, solar-powered IoT sensors are used for inexpensive greenhouses and monitor its state precisely in real time in terms of different environmental parameters, water consumption, and light levels and provide information in the form of SMS alerts. An IoT smart greenhouse is shown in Fig. 9 in which different devices are used to monitor the crop and the collected data are analyzed over the cloud using CI. Similarly, water is of great importance for healthy crops, so smart irrigation management based on IoT and intelligent system plays a major role. This is one of the research domains in which intelligent techniques can be utilized.

2.2.5 Agricultural drones

The advent of agricultural drones is yet another revolutionary trend. Aerial and ground drones are used to determine crop health and perform crop growth monitoring, planting, and field analysis. With proper strategy and real-time data-based planning, drone technology has given a new edge to the modern agriculture industry. These drones incorporate different thermal or multispectral sensors that identify the areas that need changes in irrigation. As the crops begin to grow, the sensors indicate their health and measure their vegetation index. It is quite evident that smart drones can reduce the burden on the environment as well as farmers. Rather than manual spraying of fertilizers, automated drone techniques help in uniform sprays of fertilizer; thus there has been a significant decrease in chemical flow to the groundwater. The use of robotics is massive in agriculture, as robots can be used for seeding and planting, fertilization and irrigation, and weeding of crops, as well as spraying, harvesting, and shepherding. A GPS-based remote-controlled robot for a similar task was introduced by the authors in [32]. In another work, the authors also discussed the use of designed robots that follow the track of a white line, which is assumed by the robot to be a working place, while other surfaces are assumed to be brown or black. They also pointed out that robots can perform duties like dropping of seeds, plowing, and water supply [33]. Since drones are widely used for monitoring the farm due to their thermal and hyperspectral cameras and are capable of providing continuous real-time data from the fields, they can help farmers in irrigation by showing specific areas with less quantity of water. Therefore, it prevents scarcity of water or water flooding in the farming land and maintains the health of the crops by providing an adequate amount of water.

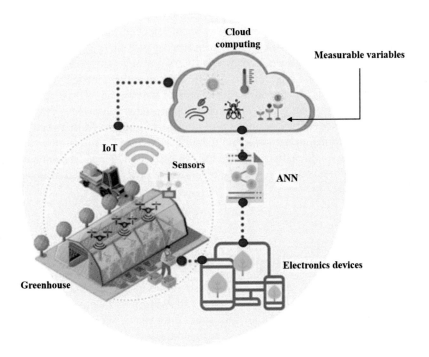

FIG. 9

IoT-based smart greenhouse [31].

2.2.6 Animal health monitoring and tracking

Due to the constantly increasing demands for safe farming products, it is necessary to use nondestructive techniques for food quality assessment, which are related to animal health. Recent advancement in biosensors as well as wearable sensors are considered as a part of the future of animal healthcare and precision livestock farming. Some of the technologies successfully used for accurate health status monitoring and disease diagnosis can be used in animal healthcare with little modification. Precision livestock farming techniques can be integrated with advanced technologies such as sweat and salivary sensing, microfluidics, serodiagnosis, sound analyzers, advanced imaging techniques, and CI and cloud computing to provide real-time health monitoring of animals. A wide range of sensors and wearable technologies can be implanted inside the bodies of animals or patched under their skin to detect their sweat constituents, stress, body temperature, behavior, and the presence of viruses and pathogens. These biosensor devices are also used for animal health treatment, iontophoretic drug delivery, detecting the need for cooling, and heating and antibiotic detection [34]. However, these sensors are useful but still need to meet the emerging challenges in animal healthcare, such as disease surveillance, diagnostics and control, and animal traceability. These challenges are being met by integrating these sensors and forming a network of sensors that is capable of providing animal health information. The integration of novel diagnostic systems with advanced imaging techniques, smart phones, biosensors, and IoTs give new directions to the livestock and agricultural industry. Among the various techniques, microfluidics technology is one of the efficient methods to detect on-farm disease like influenza virus, mostly spread in birds by their saliva, using fluorescence resonance energy transfer (FRET). Similarly, precision livestock farming through various devices allows real-time monitoring of animals such that immediate action can be taken in favor of animal health welfare and production yields. Movement and behavior monitoring of animals can help in analyzing animal activity and well-being. For future animal health welfare and food quality analysis, such as meat, dairy products, and many others, advanced imaging techniques like hyperspectral imaging is a prominent and open area that needs to be explored [35]. Moreover, CI, such as neural networks and fuzzy sets can also be utilized to automate the diagnosis system. Recent work, based on fuzzy classifiers, was presented for canine mammary tumor image classification [36]. Therefore, automated disease diagnostic models for animal healthcare are also an emerging application area.

3 Current issues and challenges in healthcare IoT

IoT-enabled applications are widely accepted by society and researchers in different domains. These provide healthcare services with a new edge, allowing it to take a step forward in modern automated applications. However, there are still some drawbacks, such as data privacy, interoperability, scalability, data storage, etc., that need to be addressed properly. IoT-based healthcare applications need reliable networking solutions that are robust to disturbances and disruptions, including high mobility, high density, disasters, infrastructure failures, cyberattacks, etc. Performance sensitive and mission critical applications such as remote surgery and self-driving, need a framework for networking that is capable of providing more secure, reliable, and efficient communications in different network environments. There are two major obstacles to the introduction of robust IoT. The first challenge arises from the spatial diversity of the entities involved in communications, due to the limitation of the propagation media, and the second is due to the varying temporal characteristics of the environment. In addition, a variety of sensors are used to collect health data and are mostly uploaded to a cloud server for further analysis; therefore, due to different transmission modes, noise is introduced in data and these data are also vulnerable to cyberattacks. Remaining open issues are:

- Data generated from numerous sensors give rise to enormous amounts of data, and different health devices as well as the cloud need to provide an efficient data management of big data.
- Researchers need to consider developing low-cost IoT-enabled drug solutions; however, no studies have considered this subject to date.
- An efficient security scheme needs to be developed for secure, cloud-based health care.
- Task scheduling and optimal resource scheduling are major concerns over cloud-assisted services.
- Generally, health data analysis is performed on the server so that there is a need for efficient technology that reduces the cost of computing and communication in IoT-based services.
- Energy conservation and harvesting represent another big challenge, as wearable devices and sensors are highly resource-constrained and therefore their energy needs to be conserved in a real-time monitoring system.

- Service composition problem is also an open challenging issue, as numerous heterogeneous sensor nodes are connected in an IoT-based application and need to access multiple Internet services to complete the task, so that an efficient intelligent technology is needed.

4 Need of CI in future healthcare IoT

IoT, along with cloud computing, is bringing a revolution to the healthcare industry that includes human as well as agriculture healthcare. It facilitates remote service access from different geographical domains. Patients can take suggestions from doctors and access their reports, emergency services can be scheduled, etc. In the case of agriculture, manual collection of temperature, soil humidity, crop disease diagnosis, to precision livestock farming are automated, minimizing the human intervention. But in addition to all these IoT-based applications, there are major challenges to secure big private data management, task scheduling, and automated and intelligent data analysis. In addition, heterogeneous data in the form of signals, images, and different formats are loaded over the cloud, making it quite difficult to analyze big data in a distributed real-time environment and to make decisions. These different challenges must be resolved by integrating intelligent techniques, because any wrong decision in the health services can be life-threatening. These applications all involved various communications between different parties and operated on data collected from different sensors, smartphones, wearable devices, etc., all of which gives rise to data uncertainty. Therefore, in order to overcome these issues, the CI, which includes evolutionary algorithms, fuzzy logic, neural networks, and their variants such as deep convolutional networks, needs to be integrated with the IoT in a well-conceptualized manner to meet the future needs. Fig. 10 shows the evolution of digital agriculture from past to future.

IoT-based healthcare faces many problems, ranging from complex optimization such as task scheduling, and optimal resource management, optimal data routing problems to multicriteria decision-making, robust data modeling, automated feature learning, and disease diagnostic models. Future healthcare also needs to integrate security and privacy measures into these applications to address the issue of cyberattacks. Efficient and secure automated analysis is not feasible without intelligent technology. Recent trends in research show the efficacy of CI in advanced imaging healthcare. Evolutionary algorithms have great ability to solve a complex optimization problems. One of these optimization problems is the management of hospital resources as well as the crucial decision-making processes, for which the authors presented PROMETHEE II for multicriteria decision making in the hospital emergency department [38]. Since evolutionary algorithms have the characteristics of robustness, flexibility, and scalability, they can help to overcome some of the key architectural challenges of IoT-based systems, such as scalability, interoperability, robustness, and flexibility [39]. Researchers have discussed extensive IoT applications using CI, which cover various complex optimizations such as routing, scheduling, and healthcare applications [40]. Even deep learning as part of CI has also been successfully applied to cancer diagnosis in animals as well as humans, by designing an efficient feature learning method [41]. In another work, researchers have proposed a new enhanced nature-inspired algorithm and also applied it to medical images encryption [42]. Similarly, in a hybrid framework for the classification of breast cancer, a chain-like agent genetic algorithm has been used as an application of IoT [43]. Fuzzy logic is another component of CI, bringing robustness to models for data uncertainty handling. In a recent work, a fuzzy-based

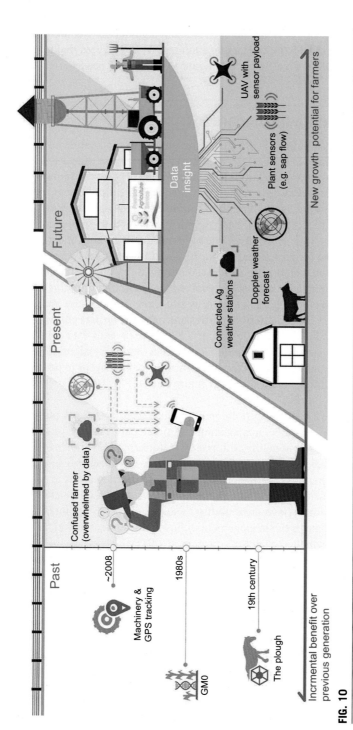

FIG. 10

Evolution of digital agriculture [37].

model was applied successfully on tumor images for cancer prediction [36]. Besides these, other challenging problems of data deployment and partitioning of IoT applications across distributed platforms are being addressed by proposing the PATH2iot framework [44].

The recent success of CI in the healthcare sector shows that there is a tremendous opportunity to integrate CI with healthcare data, IoT, and cutting-edge computing for the development of energy-efficient intelligent healthcare IoT. There are a variety of complex optimization issues, as discussed earlier in the IoT-based application, which can be explored for health data collected using advanced imaging techniques. Security is another open area of IoT-based applications research that needs to be explored individually or in conjunction with CI.

5 Conclusion

The popularity and exponential growth of the Internet have impacted almost every field of work. The computational capabilities of wireless sensor networks, computers, and digital devices have created a revolution in the healthcare sector. Therefore, this chapter presents the concepts of IoT in healthcare. In order to provide a complete conceptual insight, we covered healthcare in two subareas, namely human health and agriculture. Agricultural healthcare covers both animal and crop health care. The recent outbreak of COVID-19 is also covered in this chapter, which is a challenging issue open to research to develop various IoT solutions for tracking, monitoring, and much more.

Moreover, the chapter details the challenges and opportunities for future IoT health with CI. It covers intelligent future health IoT and outlines the future directions from various aspects such as complex optimization problems in scheduling, routing, and multicriteria disease diagnostics, advanced imaging techniques, and wireless sensor networks in agriculture, to agricultural drones, human disease diagnosis to animal and crop disease, and from data privacy to private prediction. Consequently, the security and privacy of confidential medical data, the development of energy-efficient devices, and the development of digital apps for preventive healthcare as well as the robust and efficient data distribution technology over the cloud are a few of the directions that need to be explored. There is also a need to integrate all available sensors and create an efficient online monitoring system so that animal and crop health can be monitored in real time without delays, for the well-being of both humans and animals.

References

[1] M.H. Miraz, M. Ali, P.S. Excell, R. Picking, A review on Internet of things (IoT), Internet of everything (IoE) and Internet of nano things (IoNT), in: 2015 Internet Technologies and Applications (ITA), IEEE, 2015, pp. 219–224.

[2] Y.J. Fan, Y.H. Yin, L. Da Xu, Y. Zeng, F. Wu, IoT-based smart rehabilitation system, IEEE Trans. Ind. Inform. 10 (2) (2014) 1568–1577.

[3] J.M. Talavera, L.E. Tobón, J.A. Gómez, M.A. Culman, J.M. Aranda, D.T. Parra, L.A. Quiroz, A. Hoyos, L.E. Garreta, Review of IoT applications in agro-industrial and environmental fields, Comput. Electron. Agric. 142 (2017) 283–297.

[4] H. Qiu, M. Qiu, G. Memmi, M. Liu, Secure health data sharing for medical cyber-physical systems for the healthcare 4.0, IEEE J. Biomed. Health Inform. (2020), https://doi.org/10.1109/JBHI.2020.2973467.

[5] M. Abdel-Basset, G. Manogaran, A. Gamal, V. Chang, A novel intelligent medical decision support model based on soft computing and IoT, IEEE Internet Things J. 7 (5) (2020) 4160–4170, https://doi.org/10.1109/JIOT.2019.2931647.

[6] M. Aazam, S. Zeadally, K.A. Harras, Health Fog for smart healthcare, IEEE Consum. Electron. Mag. 9 (2) (2020) 96–102.

[7] D. Besozzi, L. Manzoni, M.S. Nobile, S. Spolaor, M. Castelli, L. Vanneschi, P. Cazzaniga, S. Ruberto, L. Rundo, A. Tangherloni, Computational intelligence for life sciences, Fund. Inform. 171 (1–4) (2020) 57–80.

[8] V. Bharti, B. Biswas, K.K. Shukla, Recent trends in nature inspired computation with applications to deep learning, in: 2020 10th International Conference on Cloud Computing, Data Science & Engineering (Confluence), IEEE, 2020, pp. 294–299.

[9] J.-D. Lovelock, Gartner Says Global IT Spending to Reach $3.9 Trillion in 2020, (2020), Available at: https://www.gartner.com/en/newsroom/press-releases/2020-01-15-gartner-says-global-it-spending-to-reach-3point9-trillion-in-2020.

[10] J. Manyika, M. Chui, J. Bughin, R. Dobbs, P. Bisson, A. Marrs, Disruptive Technologies: Advances That Will Transform Life, Business, and the Global Economy, 180 McKinsey Global Institute, San Francisco, CA, 2013 vol.

[11] L. Wang, A. Wong, COVID-Net: a tailored deep convolutional neural network design for detection of COVID-19 cases from chest X-ray images, arXiv paper. arXiv:2003.09871 (2020).

[12] B. Ghoshal, A. Tucker, Estimating uncertainty and interpretability in deep learning for coronavirus (COVID-19) detection, arXiv paper. arXiv:2003.10769 (2020).

[13] J.P. Cohen, P. Morrison, L. Dao, COVID-19 image data collection, arXiv paper. arXiv:2003.11597 (2020).

[14] L. Ferretti, C. Wymant, M. Kendall, L. Zhao, A. Nurtay, L. Abeler-Dörner, M. Parker, D. Bonsall, C. Fraser, Quantifying SARS-CoV-2 transmission suggests epidemic control with digital contact tracing, Science 368 (6491) (2020), https://doi.org/10.1126/science.abb6936.

[15] M. Abdel-Basset, M. Mohamed, A novel and powerful framework based on neutrosophic sets to aid patients with cancer, Futur. Gener. Comput. Syst. 98 (2019) 144–153.

[16] A. Onasanya, M. Elshakankiri, Smart integrated IoT healthcare system for cancer care, Wireless Netw. (2019) 1–16. https://doi.org/10.1007/s11276-018-01932-1.

[17] M.A. Rahman, M. Rashid, S. Barnes, M.S. Hossain, E. Hassanain, M. Guizani, An IoT and blockchain-based multi-sensory in-home quality of life framework for cancer patients, in: 2019 15th International Wireless Communications & Mobile Computing Conference (IWCMC), IEEE, 2019, pp. 2116–2121.

[18] J. Gómez, B. Oviedo, E. Zhuma, Patient monitoring system based on Internet of things, Procedia Comput. Sci. 83 (2016) 90–97, https://doi.org/10.1016/j.procs.2016.04.103.

[19] R.S. McGinnis, E.W. McGinnis, J. Hruschak, N.L. Lopez-Duran, K. Fitzgerald, K.L. Rosenblum, M. Muzik, Wearable sensors and machine learning diagnose anxiety and depression in young children, in: 2018 IEEE EMBS International Conference on Biomedical & Health Informatics (BHI), IEEE, 2018, pp. 410–413.

[20] D. Siegmund, L. Chiesa, O. Hörr, F. Gabler, A. Braun, A. Kuijper, Talis—a design study for a wearable device to assist people with depression, in: 2017 IEEE 41st Annual Computer Software and Applications Conference (COMPSAC), 2 IEEE, 2017, pp. 543–548 vol.

[21] X. Zhou, K. Jin, Y. Shang, G. Guo, Visually interpretable representation learning for depression recognition from facial images, IEEE Trans. Affect. Comput. 11 (3) (2020) 542–552, https://doi.org/10.1109/TAFFC.2018.2828819.

[22] A. Kumar, V. Bharti, V.K.U. Kumar, P.D. Meena, Hyperspectral imaging: a potential tool for monitoring crop infestation, crop yield and macronutrient analysis, with special emphasis to Oilseed Brassica, J. Oilseed Brassica 7 (2) (2016) 113–125.

[23] S. Wolfert, L. Ge, C. Verdouw, M.-J. Bogaardt, Big data in smart farming—a review, Agric. Syst. 153 (2017) 69–80.

[24] C. Antón-Haro, T. Lestable, Y. Lin, N. Nikaein, T. Watteyne, J. Alonso-Zarate, Machine-to-machine: an emerging communication paradigm, Trans. Emerg. Telecommun. Technol. 24 (4) (2013) 353–354.

[25] A. Khattab, S.E.D. Habib, H. Ismail, S. Zayan, Y. Fahmy, M.M. Khairy, An IoT-based cognitive monitoring system for early plant disease forecast, Comput. Electron. Agric. 166 (2019) 105028.

[26] M.S. Mekala, P. Viswanathan, CLAY-MIST: IoT-cloud enabled CMM index for smart agriculture monitoring system, Measurement 134 (2019) 236–244.

[27] D. Reynolds, J. Ball, A. Bauer, R. Davey, S. Griffiths, J. Zhou, CropSight: a scalable and open-source information management system for distributed plant phenotyping and IoT-based crop management, Gigascience 8 (3) (2019) giz009.

[28] L. Sobreiro, S. Branco, J. Cabral, L. Moura, Intelligent insect monitoring system (I 2 MS): using internet of things technologies and cloud based services for early detection of pests of field crops, in: IECON 2019-45th Annual Conference of the IEEE Industrial Electronics Society, 1 IEEE, 2019, pp. 3080–3084 vol.

[29] A. Kumar, V. Bharti, V. Kumar, P.D. Meena, G. Suresh, Hyperspectral imaging applications in rapeseed and mustard farming, J. Oilseeds Res. 34 (1) (2017) 1–8.

[30] C.-R. Rad, O. Hancu, I.-A. Takacs, G. Olteanu, Smart monitoring of potato crop: a cyber-physical system architecture model in the field of precision agriculture, Agric. Agric. Sci. Procedia 6 (2015) 73–79.

[31] A. Escamilla-García, G.M. Soto-Zarazúa, M. Toledano-Ayala, E. Rivas-Araiza, A. Gastélum-Barrios, Applications of artificial neural networks in greenhouse technology and overview for smart agriculture development, Appl. Sci. 10 (11) (2020) 3835. https://doi.org/10.3390/app10113835.

[32] N. Gondchawar, R.S. Kawitkar, IoT based smart agriculture, Int. J. Adv. Res. Comput. Commun. Eng. 5 (6) (2016) 838–842.

[33] S.S. Katariya, S.S. Gundal, M.T. Kanawade, K. Mazhar, Automation in agriculture, Int. J. Recent Sci. Res. 6 (6) (2015) 4453–4456.

[34] S. Neethirajan, Recent advances in wearable sensors for animal health management, Sens. Bio-Sens. Res. 12 (2017) 15–29.

[35] A. Kumar, S. Saxena, S. Shrivastava, V. Bharti, U. Kumar, K. Dhama, Hyperspectral imaging (HSI): applications in animal and dairy sector, J. Exp. Biol. Agric. Sci. 4 (4) (2016) 448–461.

[36] A. Kumar, S.K. Singh, S. Saxena, A.K. Singh, S. Shrivastava, K. Lakshmanan, N. Kumar, R. K. Singh, CoMHisP: a novel feature extractor for histopathological image classification based on fuzzy SVM with within-class relative density, IEEE Trans. Fuzzy Syst. (2020) https://doi.org/10.1109/TFUZZ. 2020.2995968.

[37] Accenture, Digital agriculture: improving profitability. (2017)Available at:https://www.accenture.com/_ acnmedia/Accenture/Conversion-Assets/DotCom/Documents/Global/PDF/Digital_3/Accenture-Digital-Agriculture-Point-of-View.pdf.

[38] T.M. Amaral, A.P.C. Costa, Improving decision-making and management of hospital resources: an application of the PROMETHEE II method in an emergency department, Oper. Res. Health Care 3 (1) (2014) 1–6.

[39] O. Sallent, J. Perez-Romero, R. Ferrus, R. Agusti, On radio access network slicing from a radio resource management perspective, IEEE Wireless Commun. 24 (5) (2017) 166–174.

[40] O. Zedadra, A. Guerrieri, N. Jouandeau, G. Spezzano, H. Seridi, G. Fortino, Swarm intelligence-based algorithms within IoT-based systems: a review, J. Parallel Distrib. Comput. 122 (2018) 173–187.

[41] A. Kumar, S.K. Singh, S. Saxena, K. Lakshmanan, A.K. Sangaiah, H. Chauhan, S. Shrivastava, R. K. Singh, Deep feature learning for histopathological image classification of canine mammary tumors and human breast cancer, Inf. Sci. 508 (2020) 405–421.

[42] V. Bharti, B. Biswas, K.K. Shukla, A novel multiobjective GDWCN-PSO algorithm and its application to medical data security, ACM Trans. Internet Technol., 10.1145/3397679.

[43] S.U. Khan, N. Islam, Z. Jan, I.U. Din, A. Khan, Y. Faheem, An e-Health care services framework for the detection and classification of breast cancer in breast cytology images as an IoMT application, Futur. Gener. Comput. Syst. 98 (2019) 286–296.

[44] D.N. Jha, P. Michalák, Z. Wen, R. Ranjan, P. Watson, Multiobjective deployment of data analysis operations in heterogeneous IoT infrastructure, IEEE Trans. Ind. Inform. 16 (11) (2020) 7014–7024, https://doi.org/10.1109/TII.2019.2961676.

IoT services in health industry

IoT services in healthcare industry with fog/edge and cloud computing

6

Dinesh Kumar[a], Ashish Kumar Maurya[a], and Gaurav Baranwal[b]

Department of Computer Science & Engineering, Motilal Nehru National Institute of Technology, Allahabad, Uttar Pradesh, India[a] Department of Computer Science, Institute of Science, Banaras Hindu University, Varanasi, Uttar Pradesh, India[b]

1 Introduction

A tremendous growth has been observed in Internet-of-Things applications in recent times due to the evolvement of various technologies such as compact and smart sensors, fog computing, edge computing, high-speed networks, software-defined networking (SDN), and wireless sensor networks (WSNs) [1]. IoT has attracted various research communities, industry professionals, and academicians due to its multiple benefits to society. The healthcare industry is one of the many domains in which IoT has been extensively used [2]. The ubiquitous characteristics of IoT help in the remote and uninterrupted monitoring of a person in a healthcare system. Automated patient monitoring, activity tracking, measuring heart rate, calculating calorie burn/intake, etc. are some of the tasks performed by IoT devices attached to sensors in healthcare systems [3]. The generated data from these IoT devices are processed and analyzed either at the fog/edge devices or at the cloud data centers. This whole process forms a distributed and coherent healthcare system using cloud computing, fog or edge computing, and IoT devices. IoT devices generate a massive amount of data. These data need to be processed at some computing facility for making intelligent decisions. This requirement is fulfilled by cloud computing, which has an abundant amount of computing, storage, and network resources [4, 5]. There are other technology drivers in this context, which make these huge data useful like machine learning and big data [6]. These technologies help in optimizing the treatment process, personalizing caring, and identifying patients at risk for chronic diseases [7].

Sometimes, there is a need for ultralow latency in real-time processing of critical events such as heart attack and organ failure. This is done by deploying some computing and storage facilities between medical devices/sensors and cloud data centers to reduce the overall latency, and in this way, fog and edge computing comes into the picture. These technologies are used for real-time monitoring and actuating type services. Along with it, there are also some other benefits of fog and edge computing, such as scalability, usability, reliability, and performance. Hence the major concept of merging IoT with these technologies is to shift from customary ways of caring for the patients, visiting the hospitals, etc. to the smarter ways [8].

Edge and fog computing is beneficial to serve localized healthcare applications because placing the IoT devices near to the user or in the proximity of the user decreases the network latency and response

IoT Based Data Analytics for the Healthcare Industry. https://doi.org/10.1016/B978-0-12-821472-5.00017-X

time. The integration of IoT devices with the edge, fog, and cloud computing can be helping hand to the healthcare industry to reduce the maintenance cost and network processing cost of smart IoT devices also. Cloud computing provides various computing services over the Internet. Integrating it with IoT in the healthcare solves many problems like data management, data storage, privacy, and security. The healthcare industry can also provide a common platform to access shared worldwide health data and can offer on-demand services by integrating the cloud concept with the IoT.

In recent times, there has been a tremendous growth in IoT technology. Among IoT and its various applications, the healthcare domain has emerged as one of the most applicable domains of IoT technology. Diverse research communities, industry professionals, and academicians have started to work in IoT-based healthcare system. Numerous research studies, prototypes, test beds, surveys, review studies, and implementation articles were proposed in the last 5–10 years. This fact can be validated by analyzing the total number of publications in "IoT" and "IoT-based healthcare." Fig. 1 shows the total number of publications published from 2014 to 2019 for the terms "IoT" and "IoT for healthcare."

Implementation of IoT is very challenging with strict and well-defined requirements of a healthcare system. To fulfill a particular type of condition, the use of recent technologies such as fog, edge, and cloud computing is unavoidable and inevitable. In this chapter the role of edge, fog, and cloud computing in an IoT-based healthcare system is presented along with detailed technical aspects of each of the technologies for the realization of a complete and efficient IoT-based healthcare system. Further,

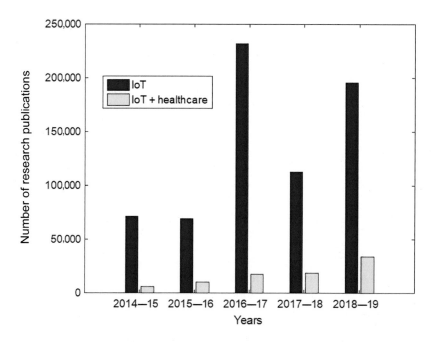

FIG. 1

Popularity of IoT healthcare domain.

for each of these technologies, whether it is fog/edge or cloud computing, the following things have been discussed:

- importance of technology in healthcare services,
- services provided by the technologies in an IoT-based healthcare system,
- benefits of the technology, and
- issues and challenges in the integration of the technology with IoT.

Then, system architecture with three different layers of IoT-based healthcare system with fog, edge, and cloud computing is discussed with the implementation details. This study would help all the stakeholders, researchers, academicians, and policymakers in integrating the IoT technology into healthcare industries in practice.

The rest of the chapter is as follows. In Section 2 an overview of IoT-based healthcare system is presented with a discussion on its need, different IoT services, and its benefits. After that, role of cloud computing is presented in Section 3 with a detailed review of services provided by cloud computing and its benefits. Similar to Section 3, Section 4 discusses the role of fog/edge computing in an IoT-based healthcare system. In Section 5 the integrated IoT-Fog-Cloud system is discussed. Section 6 discusses the research issues and future directions for the digital health system. Section 7 concludes the study.

2 Overview of IoT-based healthcare system

In this section an overview of IoT-based healthcare system is given with a detail description of its need, various components, IoT services in the healthcare industry, and the benefits of using IoT.

2.1 Why digital and smart healthcare system

As discussed in the introduction section, there should be a healthcare system, which supports real-time health monitoring, wheelchair management, wearable device access, ambient assisted living (AAL), indirect emergency healthcare, etc. IoT is the technology through which a digital and advanced health monitoring system can be implemented. The following points focus on the need for a digital and smart healthcare system:

- A large number of peoples have cardiovascular diseases [9]. These diseases have severe consequences such as nerve injury, kidney trauma, or even death. A health monitoring system can be used to lessen the effect of this disease.
- Falling is also the leading cause of disability and other serious injuries. Therefore there is a need for a fall detection system that can prevent these severe consequences due to fall detection.
- According to WHO the life expectancy of persons is increasing per year and will reach 90 years in 2030 [10]. In that domain, technology plays a significant role while providing quality of life to that elderly population. One parameter for that is better health services. The health services like fall detection, predict health attack, and vital sign monitoring are such example of services that can be provided to the persons with IoT [2].
- In the conventional healthcare system, there is noncontinuous health monitoring and nonubiquitous access to the health data generated from the monitoring process. For example, electrocardiography

(ECG) monitoring system in various hospitals does not support mobility, and their measurements are for a short period of time (a couple of minutes). Further, medical doctors and specialists do not analyze these data. To overcome all the aforementioned drawbacks, there is a need for the enhanced, digital, and smart healthcare system, which provides continuous health monitoring and other health services for improving the quality of the healthcare system. This system will help the medical doctors to access the collected data remotely, and real-time analysis can be done using these data. In addition, the system can report the emergency or abnormality (e.g., too high data rate, low blood pressure, and a fall) to doctors or caregivers for a quick response [11].

- As the world's population is increasing at a very fast pace, the demand for healthcare workers, for example, doctors, specialists, nursing staff, and laboratory staff, is also increasing. This is a big challenge for affordable and accessible health for all people. IoT-based digital healthcare systems can tackle this situation by incorporating various technologies. Another issue is increasing medical costs day by day.

2.2 IoT services in digital healthcare system

IoT provides various kinds of services in the healthcare system. These services make the healthcare system more advanced, digital, and smart. There are many services in an advanced and digital healthcare system by IoT technologies. The services are as follows.

2.2.1 Anomaly detection

IoT system helps the medical practitioners to detect the anomalies of a patient using his captured heath data. One of the main advantages of this system is that it evolves and learns with time using machine learning [12]. For example, [13] proposed an IoT-based anomaly detection system based on hierarchical temporal memory (HTM). HTM is a biologically inspired machine learning unsupervised intelligence technology [12]. Using this system, temporal anomalies such as heart attack or stroke can be identified. In the proposed system, biosignals are collected using sensors and sent to the fog nodes for preprocessing and extracting the features from the data. These processed data are then sent to the cloud using a secure channel. The encoder at the cloud server transforms the received data into sparse distributed representation (SDR). The HTM machine learns from the temporal patterns of SDR continuously. HTM also predicts accuracy and predicted anomaly and warning [13].

2.2.2 mHealth service

A person can access their health information on a smartphone application or a cloud dashboard on a computer system. The person can connect with the caretakers and doctors for real-time monitoring [14].

2.2.3 Ambient assisted living

As discussed in the previous section, the aging population is increasing day by day. IoT-based healthcare system can implement AAL by enabling indoor positioning systems and real-time monitoring of that portion of the population [15].

2.2.4 Semantic medical access

In IoT a huge amount of medical data is collected using sensors. This information is analyzed and processed using the semantics and ontologies that would help to make a better and efficient healthcare system [16]. The semantics would help to retrieve valuable and important information from the cloud. These data can be used for providing medical services to needy persons.

2.2.5 Patient-centered care

Patient-centered care (PCC), the term coined by Picker/Commonwealth in 1988, is one of the emerging healthcare models in today's healthcare industry. Institute of Medicine defined the PCC as "Healthcare that establishes a partnership among practitioners, patients, and their families (when appropriate) to ensure that decisions respect patients wants, needs, and preferences and that patients have the education and support they need to make decisions and participate in their own care" [17]. IoT has the capability to provide this service.

2.2.6 Remote health monitoring

Remote health monitoring is one of the fundamental services of a smart and digital healthcare system. The main motivation behind this service is to provide healthcare services outside the hospital or any other healthcare facility. This will reduce the load on crowded hospitals and healthcare workers. Remote health monitoring would be more beneficial for providing healthcare to patients residing in very remote areas.

2.2.7 Context awareness

Context awareness is one of the most crucial requirements of a digital and smart healthcare system. The word "context" defines the understanding and construction of the implied situation of a thing or an entity [18, 19]. IoT devices in the healthcare system record the health data and some other auxiliary information (e.g., contextual information) through monitoring. It is the information regarding the environment and condition of the system. It enables us to understand the health data more accurately for a better diagnosis. For example, seasonal change can be one of the parameters for finding a patient's disease pattern. Context awareness enables this kind of analysis by collecting contextual information. Multiple imputation (MI) method is one way to incorporate context awareness in health data [18].

2.3 Benefits of IoT-based healthcare system

IoT-based healthcare system provides a holistic solution for all health needs of a person like exercise, disease monitoring, safety, or beauty reasons. There are various benefits of IoT-based digital healthcare system. Some of these are as follows.

2.3.1 For doctors

The IoT-based healthcare system saves the time of a doctor by remotely monitoring the health status of a patient. Therefore there is no need to visit the patient after a regular interval. They receive the real-time data of a patient. This helps them to save extra efforts for manually measuring all health parameters.

2.3.2 For patients

There are various benefits of IoT-based healthcare system for patients. IoT provides personalize healthcare by doing data analytics and big data processing of health data generated by sensors. It also provides lifetime monitoring (past, present, and future) of their health. IoT-based systems are easy to use and easily adopted. It tremendously decreases the overall cost spent on their health. This is because a patient only needs to visit a doctor when the health status is below the recommendation. Secondly, there is no need to pay extra money for using each technology because of technology fusion in IoT-based system. Their caretakers of the patient anywhere without any geographical barriers can also access the real-time health data.

In a traditional healthcare system, the biometric parameters are measured manually, and a lot of time is wasted in that process. Moreover, there is a need for a caretaker for continuous manual supervision. IoT-based remote monitoring would remove all these issues [20]. Smart gloves, smartwatches, wheelchair management, and wireless Nano Retina glasses are some examples of IoT devices, which are specifically designed for persons with special needs. Even schools can adopt IoT technology for better management of services provided to children with special needs.

2.3.3 For hospitals and clinics

Generally, hospitals use advanced technology to run their operations like admission, treatment, surgery, or monitoring of patients. IoT can play a significant role in the hospital ecosystem. For example, a hospital can use smart ambulances that can perform on-fleet diagnosis and transmit the health information of the patient to the hospital. This would help the doctors or surgeons to know about the condition of the patient and prepare themselves accordingly. The smart ambulance consists of the medical sensors and secure communication link between ambulance and hospital. With the help of IoT, medical devices, doctors, and medical staff, all can access the information at a single time, which ultimately lead to a better diagnosis. Some other sections of hospitals such as the intensive care unit (ICU), primary care unit, or any other specialized unit can also use IoT technology for a better and efficient healthcare system. IoT can also be beneficial in many ways for small clinics. Doctors could see the lab reports of patients before their visit. The real-time checking of insurance coverage can be provided using IoT. The use of some inexpensive IoT devices can also save some extra cost of patients by not visiting the clinic repeatedly. IoT can also help in mobile clinics [21] where affordable and high-quality healthcare is provided by mobile vehicles in those areas where medical facilities are not available.

2.3.4 For health insurance companies

Health insurance companies play an essential part in an integrated healthcare system. There are several requirements for a health insurance company. For example, there should be efficient access to patient records, which describe their health status. Moreover, there should be a seamless integration of these companies with hospitals. The system should allow smooth cooperation among the patient and insurance companies. IoT technology is able to provide all these requirements due to its various benefits over the conventional healthcare system.

3 Role of cloud computing in IoT-based healthcare systems

Over the last few years, cloud computing has drawn much attention due to its reliable on-demand solutions. Cloud computing is computing over the Internet, where resources like application software, storage, processing units, and CPU cycles are virtually shared to fulfill the demands of users on a pay-per-use basis [4]. National Institute of Standard and Technology (NIST) defines cloud computing by five essential characteristics, three service models, and four deploy models [22]. The cloud services are provided in the form of three service models: software as a service (SaaS), platform as a service (PaaS), and infrastructure as a service (IaaS) [22, 23]. SaaS refers to accessing application resources through a thin client. PaaS denotes accessing platform resources like the support of operating systems and a framework for software development, and IaaS means accessing resources like storage units, processing units, and network resources [22]. The four deployment models of clouds are the private, community, public, and hybrid models. Private cloud is owned, managed, and used by a single organization; some specific communities having common interests use community cloud; the public cloud is open to use by the public; and hybrid cloud is the combination of two or more cloud models, that is, private, community, and public [22]. Every cloud model has its advantages and disadvantages. Thus the selection of an appropriate cloud model depends on the specific scenario and the application to which the cloud environment is considered. There are many economic and technical advantages of cloud computing models from decreasing the infrastructure and operating costs of companies to the optimized utilization of software and hardware resources [24].

3.1 Why cloud for IoT-based healthcare system?

Cloud computing and IoT are two different technologies and have been evolved independently. The characteristics of both technologies are complementary in nature and have been shown in Table 1. Due to their complementary characteristics, one technology can take advantage of other technology in which it is lacking. For example, IoT can be benefited from the virtually unlimited capabilities

Table 1 Complementary characteristics of cloud and IoT.

Characteristics	Cloud computing	Internet of Things
Displacement	Ubiquitous (resources usable from everywhere)	Pervasive (things placed everywhere)
Reachability	Centralized	Limited
Components	Virtual resources	Real-world objects
Storage capabilities	Virtually unlimited	Limited or none
Processing capabilities	Virtually unlimited	Limited
Connectivity	Uses the Internet for delivering services	Uses the Internet as a point of convergence
Big data	Means to manage big data	Source of big data

of processing and storage of cloud to overcome its constraints, and cloud can be benefited from IoT by spreading its scope to provide different services in various real-life scenarios [24].

3.2 Cloud services and its various benefits in healthcare system

By integrating cloud computing with IoT, healthcare services can be significantly improved and can lead to numerous opportunities for the medical field. Fig. 2 shows the integration of IoT and cloud computing for healthcare services. Many medical services can be enhanced through integration of cloud and IoT, for example, collection of patient's data through sensors, transferring the sensed data to cloud for processing and storage, analyzing the stored data at cloud, and performing actions (for instance, prescribing medicines to patients and sending notifications to patients and doctors when needed). Some benefits can be observed while integrating cloud computing with IoT for healthcare applications [25–27].

3.2.1 Computation capabilities

The IoT devices used in the healthcare domain have limited processing capabilities; due to this, it is not possible to perform all the processing on sensed data at the device site. The sensed data are usually transferred to the more powerful device for processing and analysis, but this cannot be done without a suitable infrastructure. Cloud provides virtually unlimited computation capabilities as per requirements or demands, which, together with IoT, can benefit the healthcare applications.

3.2.2 Storage capabilities

The things involved in IoT healthcare applications continuously produce a large amount of nonstructured data, but they cannot store this big data due to their limited storage capacity. To store this big data, IoT healthcare devices can take advantage of cloud, which provides virtually unlimited storage capacity at low cost and on-demand [28].

3.2.3 Communication

The devices in IoT healthcare applications are IP enabled, and the communication among these devices requires dedicated hardware that may be very expensive. Cloud can provide a cost-effective solution to manage and connect these IoT devices from anywhere and anytime using various application portals [29]. The cloud solution of high-speed networks can benefit healthcare applications by increasing the availability of remote IoT devices, their controlling and monitoring, communication, and coordination among the IoT devices.

3.2.4 Scope

The functionalities in the IoT devices are added day by day in healthcare applications that increase the possibilities of the generation of new types of information, opportunities, and risks. For example, integrating cloud to IoT can enable the cloud to give birth to a new paradigm called "Things as a Service" dealing with various real-life scenarios.

It is found that cloud is a very cost-effective and most appropriate solution to handle the big data generated by IoT that further creates the opportunities for data integration, aggregation, sharing, and analysis. Therefore there are many motivating factors encouraging the integration of IoT and cloud computing in healthcare applications.

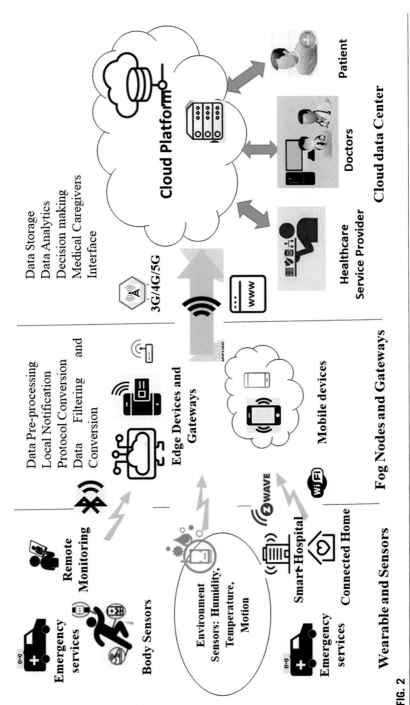

FIG. 2

Integration of IoT, fog/edge, and cloud computing for healthcare services.

3.3 Issues and challenges in the implementation of the technology

In the previous subsection, we have discussed the need for cloud computing in the IoT-based healthcare systems and how the integration of cloud with IoT offers several benefits to healthcare applications. Along with the benefits, several challenges also arise when cloud computing is integrated with the IoT-based healthcare systems. This section is dedicated to the discussion of some of such challenges [25–27].

3.3.1 Security and privacy

The data are very critical in the IoT-based healthcare systems. When these data are stored, processed, and managed in the cloud, some novel challenges such as the adoption of new technologies and trust issues handling in data and service level agreements (SLAs) arise that require some specific attention [30]. Various security techniques must be used to handle data from different types of security attacks and vulnerabilities such as SQL injection, session riding, and session hijacking.

3.3.2 Heterogeneity

The objects used in IoT-based healthcare systems are heterogeneous in nature. It is a big challenge when these objects having different platforms, operating systems, and communication protocols communicate with each other. Some cloud service models provide a few generic solutions to handle the heterogeneity of things in IoT. However, IoT-based healthcare systems require to tackle this challenge in the cloud at different levels that may involve many aspects such as programming interface and middleware.

3.3.3 Standardization and interoperability

Today, most of the things in IoT-based healthcare systems are connected to the cloud via thin client or web browsers that are a generic interface for IoT-based applications. For real-time data processing and efficient communications between things in healthcare domains, there is a demand for a standard interface that can reduce the network delay, overhead of load, and processing of data. As both cloud and IoT implement a nonheterogeneous interface, interoperability is also an issue.

3.3.4 Power and energy efficiency

In IoT-based healthcare applications, there is a lot of data transmission takes place between things to things and things to the cloud. The data transmission between things and the cloud takes place via gateways that are generally smartphones. This process of data transmission drains the battery on the gateways and things. Thus achieving power and energy efficiency for data transmission and processing in such applications is a vital issue.

3.3.5 Big data

The enormous amount of data is generated in the IoT-based healthcare applications. The cloud helps to store these data for a long time and performs big data processing and analysis on it. It is a critical challenge to handle and manage the data produced by healthcare applications as it is highly dependent on the properties of data and its management services [28].

3.3.6 Network communications

The healthcare applications involve continuous data transmission between things and the cloud that consume lots of network bandwidth. As both the cloud and IoT comprise different network technologies, network communication is an important issue. Further, healthcare applications need reliable and fault tolerant continuous transfer of data between things and the cloud. For some applications, relying entirely on remote data centers is also unacceptable because of patient safety in case of network and data center failures.

3.3.7 Legal aspects

Legal aspects are essential as patient data are very critical in healthcare applications. The IoT-based healthcare systems are able to handle different legal issues such as data jurisdiction, intellectual property rights, and contract law. For instance, the cloud service provider has to take care of various national and international laws when handling patient-related data.

4 Role of fog/edge computing in IoT-based healthcare systems

In this section the role of fog or edge computing is explained by discussing its various services and its benefits.

The term "fog computing" was first coined by Cisco in 2012 [31]. Fog computing is similar to edge computing in many ways. However, it has a novel networking architecture and has been proposed in the context of IoT. Fog computing provides the timely response service with local awareness in addition to mobility support and wireless access. Fog computing has an n-tier architecture for the flexible offering. The network devices along the data routing path between an end device and cloud data center can provide the computation and storage services to end devices. A fog device is sometimes called "smart local network equipment." A physical server that is not a local network device but is connected to a local network device is also a fog device. This connection is used to realize the local awareness feature.

According to OpenFog Consortium, fog computing is "A horizontal, system-level architecture that distributes computing, storage, control and networking functions closer to the users along a cloud-to-thing continuum" [32]. Some advantages of fog computing are agility, security, low latency, efficiency, and cognition [32]. A fog device may consist of more than one edge device. A fog node consists of computation, storage, and networking resources for the execution of IoT applications. A fog node can be of two types: type 1 and type 2. Fog nodes of type 1 are the simple devices that are just processing the data. Fog nodes of type 2 are the nodes that aggregate the capacities of edge devices [31]. Examples of fog nodes are gateways, intermediate computer nodes, and network elements such as routers.

In IoT, data are generated from devices. These devices can be sensors, vehicles, roadways, ships, etc. The devices are connected to the network. These data need to be processed with low response time or ultralow latency so that intelligent decision making can be performed [19]. In fog computing the computation, storage, and network services (like cloud services) are shifted from distant cloud data centers to near to the devices/things. This facilitates the serving or execution of real-time applications with very low latency while satisfying other QoS requirements. Fog computing infrastructure may include the edge devices, mobile phones, edge servers, gateways, routers, IoT devices, and virtualized data centers [31]. Therefore the concept of fog computing is very suitable for IoT-based healthcare systems that require real-time processing of health data with low latency.

4.1 Why fog for IoT-based healthcare system

In this subsection, we will justify the importance of fog technology for the healthcare industry:

1. A large amount of data generated in the healthcare diagnosis process should be stored and retrieved in a proper manner. The data are generated continuously (streaming data) that require a lot of network bandwidth for transmission to cloud. This process can congest the backhaul network. Fog computing provides the solution to this problem by storing and processing a large amount of data near to the devices. Therefore a significant portion of data is not transmitted through the network.
2. Fog/edge computing provides a way to provide the IoT services in the healthcare domain with low energy consumption, low latency while maintaining the QoS.
3. The data sensed by various sensors are transmitted to the cloud through the gateway. This is done for the analysis and processing of that data so that intelligent decisions can be made. These data are transmitted through the gateways with a considerable network time between the IoT device and cloud data centers. Gateways are the network devices that just forward the data to the cloud. The earlier network environment is not suitable for time-critical applications where ultralow latency is required. The fact is that the transmission bandwidth of the network is limited for a large amount of data generated by the sensors (e.g., approximately 6 kbps per ECG channel) [33]. This also brings some other problems like higher error rate, high latency, and poor sensor's energy efficiency. Therefore this is unacceptable even for a simple ECG signal propagation where the maximum latency range allowed for an ECG signal is 2–4 s [34].
4. Healthcare applications are one of the most suitable examples of those applications that require real-time processing of big data generated by medical devices and sensors and with ultralow latency.
5. There are a number of studies that proved the effectiveness of fog in a healthcare system [35–38].

4.2 Fog services in the healthcare system

There are various services provided by fog computing in the healthcare domain. These services are as follows.

4.2.1 Data management

In a healthcare system a huge amount of data is generated from the sensor devices and transmitted to the fog layer for data analysis so that intelligent decisions can be made in real time. Generally the data management process contains many subprocesses like data preprocessing, feature extraction, data fusion, and data compression. The data preprocessing removes unwanted data by the filtering process. Then important features are extracted for statistical analysis or artificial intelligence. For example, there is various kind of biosignals in health data such as ECG, electromyography (EMG), and photoplethysmography (PPG). These signals consist of the relevant health information with some unavoidable noises and distortions. These noises can be electrode contact noise, electromagnetic noise, or thermal noise. These noises are removed at fog layer by data filtering unit using various filters, for example, finite impulse response (FIR) filter. Sometimes, data need to be compressed using a lossless or lossy method to minimize the size of massive generated data. In the case of e-health data, the lossless method is preferred to avoid inappropriate disease diagnosis. These compression algorithms do not run on a sensor node due to resource constraints. Therefore fog nodes are ideal choices for this processing of the data.

There is another process in data management that is called data fusion [39]. In this process, data collected from multiple sensors are integrated, and more robust data are retrieved by removing the redundant data. This process helps to decrease the volume of huge data, which ultimately helps for better transmission of data to cloud servers. The data fusion process can be divided into three classes as cooperative, complementary, and competitive [39]. In cooperative data fusion, different sensors from one source provide the data for retrieving new information. For example, both the vital sign heart rate and respiration rate need to determine the medical state of a patient. The complementary data fusion integrated the two or more different types of data from multiple sources. This is done to retrieve more comprehensive information from the collected data. For example, if both the patient's health data and contextual data (data regarding its surroundings) are combined, the health condition of a patient can be determined in a more accurate and better way. The competitive data fusion integrates the data from one source but with two or more sensors to improve the quality and robustness of the data. For example, the heart rate value collected from a respiration signal and ECG signal would be more robust as compared with the case when it is collected from a single sensor.

4.2.2 Local storage

A fog node has two kinds of databases, that is, external database and internal database [40]. The internal database stores the health data in the standard format of health level seven (HL7) [41]. It stores the configuration parameters and data related to fog services and its various algorithms. The external database stores the biosignals and contextual data like heart rate of monitored patients during a period. This storage is limited and has lesser size as compared with the size of the cloud database. Therefore it can store only the limited data at a single time. Generally the most recent data are stored in this database. For a detailed and long history of data, one needs to access cloud storage. Only the external database is synced with the cloud database.

4.2.3 Data analysis

Fog nodes provide the data analysis facility at gateways that decrease the overall latency and size of transmitted data to the cloud. The network bandwidth between the gateways and cloud servers through the backhaul network/Internet is limited. The connection is limited and is not completely reliable. Therefore the processing and analysis of data near to the sensor node improve the reliability and consistency of data. The fog layer stores the data locally and processes it.

4.2.4 Push notification

An IoT-based healthcare system should be reliable and fault tolerant as the results of data analysis and response directly influence the doctor's decision. It should be error free, and the analysis should be processed in real time as the delay may cause serious problems such as a late response from a fall detection system that can result in the death of the person if a doctor is not able to reach at the time in case of serious injury. In this kind of situation, notifications or results from the healthcare system should be available in real time. The latency of the system would depend upon the e-health signal type. For example, the typical maximum latency of an EMG signal is 15.6 ms [42]. An early warning score system

(EWS) can also be utilized to assist health professionals [43]. EWS basically analyzes six basic health parameters including blood pressure, body temperature, pulse rate, level of consciousness, respiratory rate, and oxygen saturation. The measured values of these vital signs are used to calculate the risk score of a patient. The study performed in Ref. [43] shows that EWS can predict any abnormal activity or event regarding the health of a patient well 24 h in advance.

4.2.5 Security

Security is one of the important issues in the healthcare domain. Fog computing could provide security service to some extent in this system. In this process, various encryption and decryption techniques such as AES-128 or CMAC-AES 128 can be applied on both the sensor and fog layer. Different grant or access mechanisms can be implemented using some efficient data structures such as iptable for securing ports [44]. Although there are several complex cryptographic security mechanisms available in the literature, all these cannot be applied at the sensor layer due to limited computation resources at the sensor layer. However, some of these mechanisms can easily run on a fog device. There are several research studies such as [45–47] which have proposed secure end-to-end frameworks for the healthcare system.

4.2.6 Fault detection

The fault detection service provided by fog computing improves the reliability of the healthcare system. It is a mechanism to ensure that sensors and gateways are working properly, thus helping to avoid the interruption of sensor and fog services for a long time. Fog computing is able to provide a fault detection service. When a fog device is not receiving data from a sensor node during some time or a predefined fixed duration (e.g., 10–20 s), it sends some signals to that sensor node. If that sensor node is not replying after receiving certain signals/commands, the fog node informs the system administrator about that sensor. Similar kinds of mechanisms are used to find nonworking or faulty fog devices or gateways. The fog node sends the multicast message to all its neighboring fog nodes. It generates a push notification for the system administrator if it does not receive the acknowledgment from any fog node.

4.3 Benefits of fog computing

Fog computing is suitable for those applications that require low latency, quick response, and real-time execution, for example, healthcare applications. There are various benefits of fog computing in the context of a healthcare system. In this subsection, various benefits of fog computing in the healthcare system are discussed in detail.

4.3.1 Ultralow latency

As discussed earlier an IoT-based healthcare system only integrated with cloud computing cannot fulfill the low latency requirement of various health applications. Fog computing provides a promising solution to this latency problem by providing the computing, storage, and network services near to the IoT device, hence decreasing the overall latency [48].

4.3.2 Energy efficiency

One of the fundamental characteristics of a healthcare system is that sensors or medical devices (wearable or implanted) have minimal battery and computation resources. Therefore these devices in the healthcare industry require more resource efficiency than many other domains. As discussed the concept of fog is implemented in the integrated system by placing fog devices near to the sensor nodes, for example, gateways. During this process the computational load of the sensors or wearable devices is transferred to these fog devices. For example, the ECG extraction algorithm based on wavelet transforms now runs on these smart gateways [49]. This would increase the sensor node lifetime as the node will consume less energy as compared with the previous case [48].

4.3.3 Security and privacy

To secure the data and to maintain privacy, the sensitive data can be stored at a local gateway or any nearby device. In that case, that data would not be transmitted to the cloud through the Internet [50].

4.3.4 Bandwidth optimization

In a fog computing-based healthcare system, the size of the data that need to transmit to the cloud is smaller as compared with the case when there is no fog device. It also reduces the burden on cloud servers and decongests the backhaul network [49, 51, 52].

4.3.5 Enabling context awareness

A fog node in the IoT-based healthcare system has the most recent and updated data regarding the health of patients along with the contextual information. It also stores other contextual and local data such as location, time, and environmental conditions. These data help the system to make a decision based on the context of the data [53].

Some fundamental characteristics of fog networks are interoperability, edge location, low latency, geographical distribution, location awareness, support for online analytics, etc. Therefore, in summary, fog computing is used to enhance the quality of the healthcare system in terms of energy efficiency of nodes, security, proper utilization of network bandwidth, etc. Fog computing also provides real-time analytical results.

4.4 Issues and challenges in the implementation of the technology

There are various issues and challenges while implementing the fog node in a healthcare system. This section discusses these issues and challenges in detail.

There are many use cases of the digital healthcare system, which consist of several devices, different domains with different deployment scenarios. There are certain studies that focus on the standardization of an IoT-based healthcare system, but there is a lot more to do with such kind of research studies. There is a need to explore the interoperability issue in a heterogeneous, IoT service-based healthcare system [40].

In a fog-enabled IoT-based healthcare system, the whole system works for providing smart health. In this system, all the sensors, fog devices, and cloud servers collaborate with each other through computation allocation and data exchange [54]. Here the main issue is to allocate the different and heterogeneous data processing tasks to various entities in the system, which yield maximum performance and minimum resources. In addition, tacking the cooperation between the fog and edge devices is also a significant research problem. There are different tasks with different requirements. Some are compute intensive, whereas some are storage intensive. Allocate of these heterogeneous data processing tasks to edge and fog devices is a significant research problem.

The availability of network infrastructure in rural areas is a huge issue, which can affect the latency requirements of the whole system [55]. One another challenge is the autonomic management of fog devices. Generally, fog devices are interconnected, thus increasing its complexity. This complexity should be solved using minimum maintenance costs and hence improve the usability and efficiency of the system.

5 An integrated IoT-Fog-Cloud system

In this section the integrated IoT-Fog-Cloud is discussed by defining each component in detail. There are three layers in an IoT-based healthcare system. These layers are sensor layer, fog layer, and cloud layer. These layers are discussed in detail.

5.1 Sensor layer

In a healthcare system, at sensor layer, wearable devices or sensor attached to the body of the patient transfer data regarding his health, for example, ECG, EMG, electroencephalography (EEG), glucose level, blood pressure, body temperature, pulse rate, and hemoglobin count. Table 2 gives details about the type, size, and other characteristics of health data generated by various kinds of sensors [56].

Human motions can be detected using accelerometer, gyroscope, and magnetometer. There are two types of sensor nodes as wearables and static. The static sensor nodes are used to collect the contextual information such as room temperature and humidity. Wearable sensor nodes are attached to the body of the person and used to collect the e-health data such as EMG, ECG, blood pressure, and pulse rate. There are three parts of a sensor node, that is, sensor, microcontroller chip, and a wireless communication chip. According to [37] an 8-bit microcontroller chip is more preferred than a 32-bit chip for lightweight computational tasks. In that microcontroller the most suitable operating frequency is 8 MHz because of its low power consumption. While the microcontroller needs a 3-V supply for 8MHz, it required a 5-V supply for operation on 16- and 20-MHz frequency. This extended range of frequency can be supported by adding an external oscillator. This microcontroller can communicate with sensors via 1-MHz energy-efficient SPI wire protocols [37].

For example, a high-quality ECG can be collected using an analog front-end ADS1292 component. There are two channels in it where each channel can collect up to 8000 sample/s. The sample size is 24 bits. The ADS1292 component connects with the microcontroller via SPI, the most energy-efficient wire protocol. The 3-D accelerometer, 3-D gyroscope, and 3-D magnetometer can be used for collecting acceleration and angular velocity.

At the sensor layer, health data are generated and transmitted to a fog or cloud layer. Generally, there are two kinds of data generated by sensors at the sensor layer, e-health data and contextual data. E-health data contain information regarding the health of a person, for example, heart rate, blood pressure, and pulse rate. The contextual data are about the surrounding of a person, for example, room temperature and humidity. Both kinds of data are required for a correct and complete analysis. For example, it is normal of having 100 beats/s when a person is running. But it would be abnormal if a person is sitting in a room and his heart is pumping with 100 beats/s. It is to be noted that measuring

Table 2 Some details about the health data [56].

Application	Signals	Data range	Data rate	Frequency (Hz)	Resolution (bits)
Medical/ health	ECG	0.5–4 mV	6–48 kbps	0–1000	12–16
	Body temperature	32–40 C	2.4–120 bps	0–1	12
	Respiratory rate	2–50 breaths/ min	240 bps	0.1–20	12
	Blood pressure	10–400 mm Hg	1.2 kbps	0–100	12
	Blood flow	1–300 mL/s	480 bps	40	12
	Pathogen detection	0–1	2.4–160 bps	–	12
	Blood pH	6.8–7.8 pH units	48 bps	4	12
	Blood CRP	0–8 mg/L	2.4 bps	–	12
	Glucose concentration	0–20 mM	480–1600 bps	0–50	12–16
	Pulse rate	0–150 BPM	48 bps	4	12
Nonmedical	Motion sensor	–	4.8–35 kbps	0–500	12–16
	GPS position	–	96 bps	1	32
	High-quality video	–	14 Mbps	–	–
	Video	–	0.3–10 Mbps	–	–
	Voice	–	50–100 kbps	–	–

and collecting the contextual information along with e-health data is not putting overhead at the sensor layer. For that a single IC chip consisting of a 3-D accelerometer and 3-D gyroscope can generate these contextual data. Similarly, room temperature and humidity data can be collected using thermocouple and hygrometer sensors, respectively. These sensors communicate with nearby microcontroller devices (e.g., Raspberry Pi) with wire protocols such as UART, SPI, or I2C [37]. Microcontroller collects the data from different sensors and transmits that data to a wireless chip after doing some light computation tasks such as running Advanced Encryption Standard (AES) algorithm.

There is various wireless communication protocols buildup for connecting these microcontrollers with the cloud or fog devices through the Internet or direct local connection. Some examples are 4G, Wi-Fi, Bluetooth, BLE, and 6LoWPAN. BLE and 6LoWPAN are low-power protocols and used generally for low data rate applications like fall detection or heart rate monitoring. In addition, these protocols have limited bandwidth, and the maximum value is about 250 kbps [37].

5.2 Fog layer

In an IoT-based healthcare system, a smart gateway can act as a fog device, and these smart gateways may be connected to each other. It is simply an embedded device with basic hardware and software resources. It should have the following things:

i. wireless communication capabilities supporting various wireless protocols (BLE, 6LoWPAN, Wi-Fi, and Bluetooth) to connect with sensor nodes;

ii. ethernet, Wi-Fi, or 4G for connecting it to cloud via the Internet;
iii. an SD card for storing the data and for installing the OS;
iv. software: basic programs, fog services, and specifications;
v. operating system: lightweight, for example, Linux kernel.

This smart gateway can be static or mobile depending upon the application requirements. Both types of gateways have their own advantages and disadvantages. A static gateway is often a computationally rich device with a well-equipped power supply. Therefore it is suitable for those applications, which are computation intensive and need fast processing of high-quality data. However, the main drawback is the lack of mobility support. In the case of mobile gateways, mobility support and ubiquitous access can be availed, but they have limited hardware resources and power supply. In IoT-based healthcare systems, both types of gateways are used, but generally, fixed gateways are used for remote health monitoring and other applications. The software part in a fog device contains the fundamental function and features of gateways such as data transmission, gateway management, and security mechanisms. For example, iptable in a gateway is used for blocking unused communication parts of the gateways. MongoDB and MySQL are used to manage the database in its local repository.

5.3 Cloud layer

The cloud layer provides fundamental features and basic services like big data storage and data analysis. It is to be noted that the total load on the cloud system would be less in an IoT-Fog-Cloud-integrated system as compared with the system where there are only IoT devices and cloud servers. This is because some of the tasks are performed at fog layer like running ECG extraction algorithm, security checking, and data analysis. In a fog-based IoT healthcare system, cloud computing is used to support fog services [54].

Similar to other conventional systems, mobile applications and web browsers are the terminals through which the health monitoring system can be accessed. The end user can access in real time using these terminals in a readable format. In the system, cloud or fog nodes send back the data to end users. The end user may be a person, a system administrator, a medical doctor, a specialist, or a patient. In the case of a patient, the data can be used to instruct or actuate things or devices.

6 Research challenges and future trends

As discussed, the IoT-based healthcare system promises to have numerous benefits over a conventional healthcare system. It has huge potential benefits for society. However, IoT technologies are still in their early stage, and it faces many issues and challenges. In this section, various issues and challenges are given with possible future research directions.

6.1 Data analysis and management

Data analysis is the field in which good research has been conducted in the healthcare domain, but this area demands continuous good quality research in the future. The research should focus on developing fog and cloud-based algorithms that can process raw health data collected from sensors and convert that information into some useful knowledge. Machine learning and big data are the two main tools that can be used to attain this objective in the data processing field. Machine learning has many excellent capabilities and can be used in the healthcare domain for providing a diagnosis to patients, aiding in the development of treatment plans and making new discoveries about disease trends. Although several works have been proposed [57, 58] using machine learning, more improvement is needed in this field.

6.2 Security, privacy, and trust

Due to sensitive patient health data and openness of the three-layered network between health devices and cloud, security and privacy issues become important issues in the IoT healthcare system. Health devices such as wearable sensors are the soft targets and can be manipulated by external entities for sending the wrong data to fog and cloud nodes. This will result in the wrong analysis of the data and can lead to an incorrect diagnosis. It can even result in the blackmailing of medical users. Therefore there is an urgent need for protecting the health information from being tempered and illegally accessed from unauthorized persons. There are some security mechanisms such as valid authentication, data access control mechanisms, and ensuring data confidentiality through which security and privacy in the healthcare system can be achieved.

The fog makes IoT-based healthcare system more vulnerable as compared with the IoT-cloud system due to the more number of entity interaction and data exchange frequency. Although the fog layer provides security services in the system, these are not sufficient. Adversaries may attack a fog node itself. Therefore security and privacy mechanisms for the healthcare system should be proposed. The research should focus on designing some novel and efficient cryptographic algorithms that can run on fog devices. Generally, there are no ideal encryption schemes known that is suitable in all environments and situations. For example, ABE and FHE schemes are good enough with excellent characteristics and features, but these mechanisms are difficult to implement on resource-constraint sensors. Future research should focus on improving these schemes. The implementation of the security level would incur extra costs in terms of latency and energy consumption. So, this trade-off should be well studied in the future [46]. In addition, there is a need to work on trust concerns of IoT systems [59].

6.3 Interoperability and standardization

Today, most of the things in IoT-based healthcare systems are connected to devices or cloud servers via thin client or web browsers, which are a generic interface for IoT-based applications. For real-time data processing and efficient communications between things in healthcare domains, there is a demand for a standard interface that can reduce the network delay, overhead of load, and processing of data. As all layers, that is, sensor, fog, and cloud, in healthcare systems implement the nonheterogeneous interface, therefore, interoperability becomes a critical issue. Generally, there are various data formats (e.g., XML for ECG data and high-resolution images for skin disease) for fog devices and cloud servers. In addition, there is a lack of well-defined standards in the healthcare industry.

6.4 Legal aspects, policies, and regulation

Legal aspects are essential as patient data are very critical in healthcare applications. The IoT-based healthcare systems are able to handle different legal issues such as data jurisdiction, intellectual property rights, and contract law. For instance, the cloud service provider has to take care of various national and international laws when handling patient-related data.

6.5 Resource allocation and management of smart and digital healthcare system

In the smart and digital healthcare system, there are a large number of sensing, computational, network, and storage resources such as sensors, medical devices, actuators, fog devices, edge devices, smart gateways, and cloud servers. The presence of this large number of resources increases the overall complexity of the whole system in terms of managing, allocating, and orchestration of these resources. In addition, the sensors transmitted the sensed data to fog devices through multiple communication links. It results in huge network complexity. Therefore future research work should focus on understanding this complexity and designing some novel methods to tackle this complexity by proposing efficient and scalable resource allocation and provisioning mechanisms. In addition, the methods should consider the limited network bandwidth while communicating between fog and cloud servers through the Internet.

7 Conclusion

The healthcare industry is one of the many domains in which the impact of IoT technology has been increased very rapidly. Nevertheless, the implementation of IoT is very challenging with strict and well-defined requirements of a healthcare system. To fulfill a specific requirement in healthcare, for example, an alert generation for caretaker and doctor of a patient affected by a particular disease, the use of recent technologies such as fog, edge, and cloud computing is unavoidable and inevitable. These technologies are beneficial to serve localized healthcare applications because placing the IoT devices near to the user or in the proximity of the user decreases the network latency and response time. This chapter presents the role of these technologies in an IoT-based healthcare system by discussing every technical aspect of these technologies for the realization of a complete and efficient IoT-based healthcare system along with all the related services, applications, techniques, and challenges in healthcare. This study would help all the researchers and policymakers in integrating the IoT innovation into healthcare industries in practice.

References

[1] S.B. Baker, W. Xiang, I. Atkinson, Internet of things for smart healthcare: technologies, challenges, and opportunities, IEEE Access 5 (2017) 26521–26544.
[2] J.J.P.C. Rodrigues, et al., Enabling technologies for the internet of health things, IEEE Access 6 (2018) 13129–13141.
[3] A. Kumari, S. Tanwar, S. Tyagi, N. Kumar, Fog computing for Healthcare 4.0 environment: opportunities and challenges, Comput. Electr. Eng. 72 (Nov. 2018) 1–13.
[4] M. Sajid, Z. Raza, Cloud computing: Issues & challenges, in: International Conference on Cloud, Big Data and ..., 2013, vol. 20, 2013, pp. 34–41.
[5] A.J. Hassel, D. Danner, M. Schmitt, I. Nitschke, P. Rammelsberg, H.W. Wahl, Oral health-related quality of life is linked with subjective well-being and depression in early old age, Clin. Oral Investig. 15 (5) (2011) 691–697.
[6] Z. Obermeyer, E.J. Emanuel, Predicting the future-big data, machine learning, and clinical medicine, New Engl. J. Med. 375 (13) (2016) 1216–1219.
[7] S. Piai, Bigger data for better healthcare, in: Intel, 2013.

[8] G.J. Mandellos, G.V. Koutelakis, T.C. Panagiotakopoulos, M.N. Koukias, D. K. Lymberopoulos, Requirements and solutions for advanced telemedicine applications, in: Biomedical Engineering, 2009.

[9] N. Center for Health Statistics, Summary of Health Statistics: National Health Interview Survey 2017, U.S. Dep. Heal. Hum. Serv., 2017

[10] World Health Organization, 'Ageing Well' Must Be a Global Priority, Central European Journal of Public Health, 2014.

[11] B. Negash, et al., Leveraging fog computing for healthcare IoT, in: Fog Computing in the Internet of Things: Intelligence at the Edge, 2017, pp. 145–169.

[12] Numenta, Numenta; 2005, (2005).

[13] B. Farahani, F. Firouzi, V. Chang, M. Badaroglu, N. Constant, K. Mankodiya, Towards fog-driven IoT eHealth: promises and challenges of IoT in medicine and healthcare, Futur. Gener. Comput. Syst. 78 (2018) 659–676.

[14] S.H. Almotiri, M.A. Khan, M.A. Alghamdi, Mobile health (m-Health) system in the context of IoT, in: Proceedings—2016 4th International Conference on Future Internet of Things and Cloud Workshops, W-FiCloud 2016, 2016, pp. 39–42.

[15] N.K. Suryadevara, S. Kelly, S.C. Mukhopadhyay, Ambient assisted living environment towards Internet of Things using multifarious sensors integrated with XBee platform, in: Smart Sensors, Measurement and Instrumentation, vol. 9, 2014, pp. 217–231.

[16] G. Zhang, C. Li, Y. Zhang, C. Xing, J. Yang, SemanMedical: a kind of semantic medical monitoring system model based on the IoT sensors, in: 2012 IEEE 14th International Conference on e-Health Networking, Applications and Services, Healthcom 2012, 2012, pp. 238–243.

[17] M.P. Hurtado, E.K. Swift, J.M. Corrigan, Envisioning the National Health Care quality report, J. Healthc. Qual. 25 (2) (2003) 53.

[18] C. Perera, A. Zaslavsky, P. Christen, D. Georgakopoulos, Context aware computing for the Internet of Things: a survey, IEEE Commun. Surv. Tutorials 16 (1) (2014) 414–454.

[19] M. Singh, G. Baranwal, Quality of service (QoS) in Internet of Things, in: Proceedings—2018 3rd International Conference On Internet of Things: Smart Innovation and Usages, IoT-SIU 2018, 2018.

[20] M. Bertini, L. Marcantoni, T. Toselli, R. Ferrari, Remote monitoring of implantable devices: should we continue to ignore it? Int. J. Cardiol. 202 (2016) 368–377.

[21] C.F. Hill, B.W. Powers, S.H. Jain, J. Bennet, A. Vavasis, N.E. Oriol, Mobile health clinics in the era of reform, Am. J. Manag. Care 20 (3) (2014) 261–264.

[22] B. Snaith, M. Hardy, A. Walker, Emergency ultrasound in the prehospital setting: the impact of environment on examination outcomes, Emerg. Med. J. 28 (12) (2011) 1063–1065.

[23] A.K. Maurya, A.K. Tripathi, Deadline-constrained algorithms for scheduling of bag-of-tasks and workflows in cloud computing environments, in: ACM International Conference Proceeding Series, 2018, pp. 6–10.

[24] A. Botta, W. De Donato, V. Persico, A. Pescapé, Integration of Cloud computing and Internet of Things: a survey, Futur. Gener. Comput. Syst. 56 (2016) 684–700.

[25] L. Minh Dang, M.J. Piran, D. Han, K. Min, H. Moon, A survey on internet of things and cloud computing for healthcare, Electron 8 (7) (2019).

[26] M. Díaz, C. Martín, B. Rubio, State-of-the-art, challenges, and open issues in the integration of Internet of things and cloud computing, J. Netw. Comput. Applicat. 67 (2016) 99–117.

[27] A. Darwish, A.E. Hassanien, M. Elhoseny, A.K. Sangaiah, K. Muhammad, The impact of the hybrid platform of internet of things and cloud computing on healthcare systems: opportunities, challenges, and open problems, J. Ambient Intell. Humaniz. Comput. 10 (10) (2019) 4151–4166.

[28] M. Elhoseny, A. Abdelaziz, A.S. Salama, A.M. Riad, K. Muhammad, A.K. Sangaiah, A hybrid model of Internet of Things and cloud computing to manage big data in health services applications, Futur. Gener. Comput. Syst. 86 (2018) 1383–1394.

[29] B.B.P. Rao, P. Saluia, N. Sharma, A. Mittal, S.V. Sharma, Cloud computing for Internet of Things & sensing based applications, in: Proceedings of the International Conference on Sensing Technology, ICST, 2012, pp. 374–380.

[30] C. Stergiou, K.E. Psannis, B.G. Kim, B. Gupta, Secure integration of IoT and Cloud Computing, Futur. Gener. Comput. Syst. 78 (2018) 964–975.

[31] E. Marín-Tordera, X. Masip-Bruin, J. García-Almiñana, A. Jukan, G.J. Ren, J. Zhu, Do we all really know what a fog node is? Current trends towards an open definition, Comput. Commun. 109 (2017) 117–130.

[32] OpenFog Consortium Architecture Working Group, OpenFog Reference Architecture for Fog Computing; 2017, (2017).

[33] M. Paksuniemi, H. Sorvoja, E. Alasaarela, R. Myllylä, Wireless sensor and data transmission needs and technologies for patient monitoring in the operating room and intensive care unit, in: Annual International Conference of the IEEE Engineering in Medicine and Biology—Proceedings, vol. 7, 2005, pp. 5182–5185.

[34] Á. Alesanco, J. García, Clinical assessment of wireless ECG transmission in real-time cardiac telemonitoring, IEEE Trans. Inf. Technol. Biomed. 14 (5) (2010) 1144–1152.

[35] S.K. Sood, I. Mahajan, A fog-based healthcare framework for Chikungunya, IEEE Internet Things J. 5 (2) (2018) 794–801.

[36] K. Wang, Y. Shao, L. Xie, J. Wu, S. Guo, Adaptive and fault-tolerant data processing in healthcare IoT based on fog computing, IEEE Trans. Netw. Sci. Eng. (2018).

[37] T. Nguyen Gia, et al., Energy efficient fog-assisted IoT system for monitoring diabetic patients with cardiovascular disease, Futur. Gener. Comput. Syst. 93 (2019) 198–211.

[38] A.M. Rahmani, et al., Exploiting smart e-Health gateways at the edge of healthcare Internet-of-Things: a fog computing approach, Futur. Gener. Comput. Syst. 78 (2018) 641–658.

[39] S. Dash, S. Biswas, D. Banerjee, Atta-Ur-Rahman, Edge and fog computing in healthcare—a review, Scalable Comput. 20 (2) (2019) 191–206.

[40] T. Nguyen Gia, et al., Low-cost fog-assisted health-care IoT system with energy-efficient sensor nodes, in: 2017 13th International Wireless Communications and Mobile Computing Conference, IWCMC 2017, 2017, pp. 1765–1770.

[41] Health Level Seven International, About HL7, Heal. Lev. Seven Int., 2017

[42] F. Touati, R. Tabish, U-healthcare system: State-of-the-art review and challenges, J. Med. Syst. 37 (3) (2013).

[43] A. Anzanpour, A.M. Rahmani, P. Liljeberg, H. Tenhunen, Internet of things enabled in-home health monitoring system using early warning score, in: MOBIHEALTH 2015—5th EAI International Conference on Wireless Mobile Communication and Healthcare—Transforming Healthcare through Innovations in Mobile and Wireless Technologies, 2015.

[44] netfilter, netfilter/iptables; 2014, (2014).

[45] S.R. Moosavi, et al., SEA: a secure and efficient authentication and authorization architecture for IoT-based healthcare using smart gateways, in: Procedia Computer Science, vol. 25(1), 2015, pp. 452–459.

[46] S.R. Moosavi, E. Nigussie, M. Levorato, S. Virtanen, J. Isoaho, Performance analysis of end-to-end security schemes in healthcare IoT, in: Procedia Computer Science, vol. 130, 2018, pp. 432–439.

[47] S.R. Moosavi, et al., End-to-end security scheme for mobility enabled healthcare Internet of Things, Futur. Gener. Comput. Syst. 64 (2016) 108–124.

[48] R. Deng, R. Lu, C. Lai, T.H. Luan, Towards power consumption-delay tradeoff by workload allocation in cloud-fog computing, in: IEEE International Conference on Communications, vol. 2015-Septe, 2015, pp. 3909–3914.

[49] T.N. Gia, M. Jiang, A.M. Rahmani, T. Westerlund, P. Liljeberg, H. Tenhunen, Fog computing in healthcare Internet of Things: a case study on ECG feature extraction, in: Proceedings—15th IEEE International Conference on Computer and Information Technology, CIT 2015, 14th IEEE International Conference on

Ubiquitous Computing and Communications, IUCC 2015, 13th IEEE International Conference on Dependable, Autonomic and Se, 2015, pp. 356–363.

[50] L.M. Vaquero, L. Rodero-Merino, Finding your way in the fog: towards a comprehensive definition of fog computing, in: Computer Communication Review, vol. 44(5), 2014, pp. 27–32.

[51] Y. Cao, S. Chen, P. Hou, D. Brown, FAST: a fog computing assisted distributed analytics system to monitor fall for stroke mitigation, in: Proceedings of the 2015 IEEE International Conference on Networking, Architecture and Storage, NAS 2015, 2015, pp. 2–11.

[52] K. Xu, Y. Li, F. Ren, An energy-efficient compressive sensing framework incorporating online dictionary learning for long-term wireless health monitoring, in: ICASSP, IEEE International Conference on Acoustics, Speech and Signal Processing—Proceedings, vol. 2016-May, 2016, pp. 804–808.

[53] M. Tentori, J. Favela, Activity-aware computing in mobile collaborative working environments, in: Lecture Notes in Computer Science (Including Subseries Lecture Notes in Artificial Intelligence and Lecture Notes in Bioinformatics), vol. 4715 LNCS, 2007, pp. 337–353.

[54] X. Masip-Bruin, E. Marín-Tordera, G. Tashakor, A. Jukan, G.J. Ren, Foggy clouds and cloudy fogs: a real need for coordinated management of fog-to-cloud computing systems, IEEE Wirel. Commun. 23 (5) (2016) 120–128.

[55] R. Steele, A. Lo, Telehealth and ubiquitous computing for bandwidth-constrained rural and remote areas, in: Personal and Ubiquitous Computing, vol. 17(3), 2013, pp. 533–543.

[56] H. Malik, M.M. Alam, Y. Le Moullec, A. Kuusik, NarrowBand-IoT performance analysis for healthcare applications, in: Procedia Computer Science, vol. 130, 2018, pp. 1077–1083.

[57] P.M. Kumar, S. Lokesh, R. Varatharajan, G. Chandra Babu, P. Parthasarathy, Cloud and IoT based disease prediction and diagnosis system for healthcare using Fuzzy neural classifier, Futur. Gener. Comput. Syst. 86 (2018) 527–534.

[58] S. Mohan, C. Thirumalai, G. Srivastava, Effective heart disease prediction using hybrid machine learning techniques, IEEE Access 7 (2019) 81542–81554.

[59] M.N. Alraja, M.M.J. Farooque, B. Khashab, The effect of security, privacy, familiarity, and trust on users' attitudes toward the use of the IoT-based healthcare: the mediation role of risk perception, IEEE Access 7 (2019) 111341–111354.

Multicriteria decision-making in health informatics using IoT

Shefali Varshney, Rajinder Sandhu, and P.K. Gupta

Department of Computer Science and Engineering, Jaypee University of Information Technology, Solan, Himachal Pradesh, India

1 Introduction

With the rapid increase in the growth rate of IoT, it has become a part of Industry 4.0, which is called the Fourth Industrial Revolution (4IR). The adoption of IoT devices has helped society in many ways, one being the enabling of emergency notification systems for monitoring of health-related issues in real time. Various health monitoring devices available range from monitors of pulse to blood pressure to blood glucose and, in major cases, they monitor specialized implants like hearing aids and pacemakers as well. For general and health-related monitoring of the elderly, specialized sensors can also be adapted. The use of IoT in 4IR helps to ensures that appropriate therapy is provided to people to regain lost mobility and to provide accurate treatment. The IoT introduces many new possibilities, such as consumer devices like wearable heart monitors and connected scales that can inspire healthy living. Several IoT platforms are being explored that provide support for chronic disease and antenatal patients, recurring medication requirements, and management of patient vital signs.

The healthcare domain is now considered to be one of the major research fields in which the use of IoT and 4IR has completely changed existing practices. Thus, due to the IoT becoming a basic part of the architecture of 4IR, new components used for constructing medical devices, like actuators, sensors, and image-processing devices, are being developed and produced. Advancements in technology are providing various cost-effective and secure solutions for patients in the healthcare field, and this field is thus becoming one of the major points of consideration in use of the IoT to provide real-time solutions, in hospitals and doctor's offices, where patients still receive much of their healthcare, but also in homes and in remote locations using telehealth solutions, as depicted in Fig. 1. There are many types of services, applications, and distinct protocols available that are accepted widely in IoT and 4IR. Therefore, when building an intelligent and efficient healthcare system, all points of care and technologies must be considered carefully.

Embedded technologies play an important role in providing healthcare services to people in remote locations. The IoT continues to improve as a technology with regard to analyzing and transmitting data. The process of decision-making is essential to designing health policy and medical processes. From previous studies considering probabilistic health outcomes, most of the decisions are made under some uncertain conditions; thus it is sometimes more difficult for healthcare decision makers to make correct decisions in complex situations. Healthcare decision-making is different from decision making in

IoT Based Data Analytics for the Healthcare Industry. https://doi.org/10.1016/B978-0-12-821472-5.00014-4

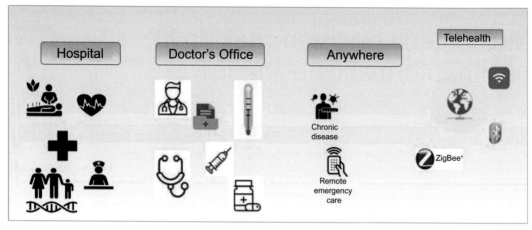

FIG. 1

IoT-based healthcare trends allow better care in hospitals and doctor's offices, but also anywhere, with telehealth and remote emergency care technologies.

various other domains, as it is a process that includes finite steps in a proper sequence, which makes this process easy to use. Several decision-making tools and methods are available for supporting healthcare decision-making using IoT for 4IR. Therefore, any advancements in healthcare decision-making are important, as they generate major benefits for both healthcare providers and patients [1].

This chapter presents a guide to the design of applications and an overview of the methods of multicriteria decision analysis (MCDA), also known as multicriteria decision-making (MCDM), for the IoT. Most of the decisions include multiple, often conflicting, healthcare objectives, which are subjected to some resource constraints. Above all, decision makers manage the situation by considering the available options under review. The standards of decision makers can also be applied explicitly or implicitly. MCDA is frequently applied to assist in healthcare decision-making [2]. Note that both the terms MCDA and MCDM are used in this chapter to refer to the same discipline.

2 IoT in healthcare

The IoT is powering 4IR, with smart devices able to make context-related decisions. Various smart devices in the IoT environment accept the data and perform processing to provide solutions in real-time. These IoT systems are further connected with the unlimited addressing ability of the net adaptation of IPv6 and cloud computing abilities. The processing and communication of data converge at innovation layers that include cloud, data, communication networks, and other devices [3]. The main key points that prominent platforms have to include need to be easily manageable by all devices; this device management must incorporate maximized throughput, minimized interruptions, low maintenance costs, and resource availability. With informative analytics, abundant IoT data analysis aggregates are required for appropriate decisions.

In real-time analytics, data quality is one of the major concerns: the cloud and accurate data must merge in a consolidated manner. Information is gathered from various data sources and platforms to be used in different fundamental services employed sophisticated analytics. As related to healthcare, the data comes from individual absolute biological, behavioral, and social characteristics and information. Appropriate use of this massive amount of data through the IoT is leading to more cost-effective healthcare and some remarkable outcomes, involving both clinical care and remote monitoring [4].

To obtain efficient health recommendations remotely, the data analysis task is performed in real time with the help of various algorithms. Clinical care systems also provide real-time solutions for patients using the IoT. These systems uses sensors for gathering data that is analyzed and stored through the cloud, which also provides the regular computerized information flow that enhances the quality of care at minimum expense. The basic structure of an IoT-based healthcare system consists of various building blocks. Some of these blocks are involved in collection of information from sensors using wireless sensor networks (WSNs) [5, 6]. The network provides a base for interfaces and displays and allows net connectivity for enabling base services. It must be durable, reliable, robust, and accurate. To perform healthcare-related actions, such as monitoring, remote surgeries, and diagnosing over the net, an IoT established healthcare system relates all the existing resources in the form of a network [7]. The entire structure is then committed to increasing the benefits of healthcare, from communities and hospitals to shelters. For example, to ease the problems of limited resources due to the rapid increase in the aging population, IoT-based smart rehabilitation has come onto the scene relatively recently [7, 8]. Wi-Fi has been extensively employed to merge the controlling devices. This network technology relates to all feasible healthcare assets, including doctors, hospitals, nurses, medical equipment, assistive devices, and rehabilitation centers, as well as the patients and clients..

A framework for IoT in healthcare, described in [9], includes three parts: Master, Server, and Things [10]. The Master includes the nurses, doctors, and patients who have credentials to access the system, using devices such as tablets, personal computers, or smartphones. The central part of the overall system is the Server, which is answerable for instruction generation, data analysis, database management, knowledge base management, and subsystem construction. The third part, Things, addresses every physical object related to the WAN, Short Message Service (SMS), and multimedia technology.

3 MCDA techniques in healthcare

With the rapid growth in technology, several MCDA approaches have been utilized by researchers in the computing field. This improvement and growth have expanded the accuracy of the results with the help of various computing approaches. The multicriteria decision analysis (MCDA) approach has recently been more commonly used. This approach is divided into two well-known techniques, multiple attribute decision-making (MADM) and multiple objective decision-making (MODM), the applications of which differ with respect to the type of problematic situations, according to whether the problem is a selection or a design problem. Moreover, MODM approaches have the number of decision variables fixed in an integer or constant domain. In comparison, the MADM methods are adjusted with an absolute number of particular alternatives. Use of MCDA is developing rapidly due to its ability to enhance the quality of decision-making. This is only possible by building a decision procedure that is much more efficient, explicit, and rational than standard decision-making procedures. Various private and public healthcare organizations and administrations have utilized MCDA applications, including the US Agency for Healthcare Research and Quality (AHRQ) [11]. Distinct surveys and studies have been addressed at identifying concerns, with the help of IoT adapted applications and equipment. Industrial players are also taking advantage of the extensive rise in market acceptance.

Healthcare decision-making is one of the major dynamic forces in the healthcare field. This means that utilizing an optimum decision-making procedure when handling patients is crucial in order to give appropriate care. Every patient's case involves some sort of decision-making process, which can lead either to successful or unsuccessful therapy. Even with the right diagnosis, choosing the appropriate treatment effectively and efficiently is essential. In the healthcare arena, planning and decision-making have become much more difficult in today's volatile market, even though these are the most important parts of the process in offering the required healthcare process, treatment, or information. For example, if a staff member in a hospital needs to decide on what treatment equipment or medication or therapy to prescribe to a patient, then an appropriate selection would be made more efficiently and effectively with the help of the MCDA technique. In many situations in healthcare, the best options, such as what kind of treatments to suggest or how to assign resources, are difficult choices to make and explain. MCDA has thus been presented to the healthcare domain as a suitable decision-assisting structure for resolving difficult situations with the help of available methods.

The healthcare field is categorized into multiple domains, including pharmaceutics, policymaking, hospitals, patient-level, health technology assessment (HTA), resource allocation, and others. Each of these domains is associated with massive amounts of data. To produce the necessary value in a secure manner, efficient, scalable, and accessible methods are necessary to handle the many details. Experimentation with the IoT in several other fields besides healthcare is establishing concerns about data security and privacy. Numerous studies have examined cybersecurity related to patient data.

Fig. 2 depicts some MCDA methods normally used for making decisions. These various approaches are AHP, VIKOR, ANP, PROMETHEE, and TOPSIS, defined in the figure. These techniques are particularly helpful in decision-making in the field of healthcare informatics and in medical decision-making. In [12], the AHP technique is applied for adjacent adaption of diagnostic tests for upper abdominal pain analysis. Various types of MCDA applications are accessible in the healthcare domain.

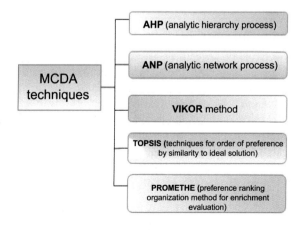

FIG. 2

Various types of MCDA techniques.

Table 1 Various MCDM techniques as used by others in the healthcare field.

S. no.	References	MCDM techniques used				
		AHP	**ANP**	**VIKOR**	**TOPSIS**	**PROMETHEE**
1.	Buyukozkan et al. [13]	✓				
2.	Sahin et al. [14]	✓				
3.	Aktas et al. [15]	✓				
4.	Huang et al. [16]	✓				
5.	Kriksciuniene and Sakalauskas [17]	✓				
6.	Alharbe [18]	✓				
7.	Singh and Prasher [19]	✓				
8.	Barrios et al. [20]		✓			
9.	Ozkan [21]		✓			
10.	Serkani et al. [22]		✓			
11.	Karadayi and Karsak [23]			✓		
12.	Bondor et al. [24]			✓		
13.	Zeng et al. [25]			✓		
14.	Chauhan and Singh [26]				✓	
15.	Buyukozkan and Cifci [27]				✓	
16.	Sadoughi et al. [28]				✓	
17.	Amaral and Costa [29]					✓
18.	Mishra et al. [30]					✓
19.	Chatterjee [31]					✓
20.	Farooq et al. [32]	✓				
21.	Kolios et al. [33]	✓			✓	✓
22.	Azam et al. [34]	✓				
24.	Opricovic and Tzeng [35]			✓	✓	✓
25.	Liu et al. [36]		✓		✓	✓

Table 1 signifies the usage of MCDM techniques in various healthcare fields and provides the guidelines for those discussed in this chapter.

3.1 **AHP**

Saaty developed the Analytic Hierarchy Process (AHP) technique in the 1970s [37], a structured methodology for making complex decisions, involving the modeling of the problem that needs to be decided as a hierarchy. The technique is related to consistency, its measure, and the interdependence of the set of elements of its structure. This decision-making process involves a large amount of data from various sources and uses both mathematics and psychology [38]. The ideas are normalized based on an individual's point of view in order to make better decisions. Many criteria and subcriteria are involved in order to rank the choices at each step. The choices are made on the basis of the criteria with high goal values. The criteria may be imaginary, with no measures which also helps in developing priorities for the criteria. But regardless of this some factors still exist which are difficult to measure. It would be so much easier to know how to measure these criteria's and it might lead to essential theories that depend on various factors for their clarifications [39]. From this, it could be understood that measurement in numbers can be calculated for benefit. Also, the same priority is not given in all the problematic situations and it has relative importance. Therefore, it is required to study how relative priorities are determined in decision making. This technique follows a systematic procedure:

(1) Describe the situation and identify the type of information in it.
(2) Create a hierarchy of the problem starting from the goal of the process, to the objective of the problem, going through middle levels to the bottom levels.
(3) Design a series of pairwise comparisons. Every element in the comparison matrix in the upper level is compared to an element in the level just below it.
(4) Now the preferences identified earlier are used to measure the priorities in the level just below it. This has to be done for every element.
(5) Now, for every element in the lower level, sum up their respective weighted values to acquire the complete priority for the elements.
(6) Repeat this process by measuring and summing until the last preferences of the substitutes at the lowermost level are attained.

A scale is required for making the comparisons, which indicates which element is more important or dominant than the other, concerning the criteria or the property with which they are compared. The scale is shown in Fig. 3.

3.1.1 AHP in healthcare
The AHP technique is used for making the best selection in any field, including in the healthcare sector for making critical choices. Some of the recent applications of AHP in healthcare are described in this paragraph. In [14], the factors and concepts of service quality were laid out, and a fuzzy AHP was developed to examine the proposed quality of service architecture. The main aim of this research was to

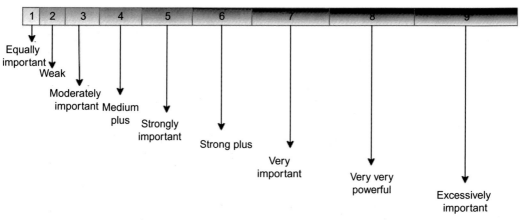

FIG. 3

Fundamental scales of numbers [14].

create a model that can examine the quality of the anticipated service in the healthcare region. Fuzzy AHP was utilized for estimating the proposed model. In [15], AHP was adopted as the approach used to choose the best site. This study was based on 19 subcriteria and 6 criteria, and all the districts were considered as alternatives. The alternatives were ranked with the help of the Saaty scale from 1 to 9. The adaptation of the hierarchy model was performed by using the software program Super Decisions 2.2.6. The experimental results depict demand to be an essential part of examining the optimum hospital site. Also, the study described in [16] determined that the service quality parameters and degree of importance were achieved with the help of AHP concerning patients and providers. Further, an adjustment was performed to retrieve the service quality index (SQI) for hospitals. The experimental results signify that the general hospital quality is related to staff attributes.

An innovative technique was designed for elderly citizens suffering from several health problems for checking their health symptoms [17]. The symptom checker was planted into an app called Help-to-You (H2U). The designs described used MCDA techniques such as AHP, with the help of fuzzy weights to deal with the indeterminacy of ambiguity. In [18] the work was based on accounting for the quality of healthcare by considering the absorption of indicators by implying an expert system based on AHP and the regression search of health data. The rankings based on the healthcare expenditure types and amount were evaluated and developed in EU countries. In [19] Fuzzy AHP was used in a web-based healthcare management system to measure the usable security. Every factor of usable security for the healthcare management system was provided. By designing and employing this type of web-based healthcare management system, the application usability was improved. Also, Fuzzy AHP helped in calculating the weights of each factor. In [20] the authors described the assimilation of the SERVQUAL methodology and fuzzy set approach to determine the service quality (SQ) of four hospitals in the Punjab state of India. In this study, AHP was used to figure out the preference of every subdimension and dimension of SQ attributes of healthcare. This preference was further utilized for ranking the hospitals from the patient's point of view. The experimental results indicate that the hospitals must focus on trustworthiness and reliability for offering good quality to patients.

3.2 ANP

The analytic network process (ANP) structures the decision process as a network, instead of a hierarchy as in AHP. It is a theory of multicriteria measures useful in obtaining the weighted scales from actual decisions that are a part of the primary range of numbers. These decision sentences display the impact of the sole element with one another in a pairwise comparison step [40]. ANP is a special case of making independent assumptions on a top to bottom level. It is also the autonomy of elements in a level of the AHP method. The decision is made in ANP on the basis of the number of scales in the AHP method. To consider all influences concerning the same criteria such that they would be important enough to synthesize, it is required to use the same criteria in creating all the comparisons. This criterion is called the control criterion; it is a meaningful way to aim the thinking to respond to the question of authority. Actual data and figures represent probability and likelihood, which can also be useful in relative form, rather than creating pairwise matrices as in AHP.

There are several worked-out models of ANP, most of them created by industrial engineers, executives, mature students, managers, and others who have examined and controlled the basic structure of this method. Therefore ANP has become an essential tool for predicting and presenting various competing approaches, with implicit interaction and explicit understanding and strength which is required in decision-making. It also can help in resolving conflicts in which there are several contrary powerful authorities. Hierarchy is defined as a cluster of base nodes or goals. It contains a sink node or a cluster in a possible hypothesis as an uncontrollable situation that depicts the decision alternatives. ANP structure is designed in the top to bottom manner where bottom levels do not support any estimation for top levels. Although it contains a cutoff at the lower level to display each alternative that this level is dependent on others also. This further indicates that elements are found to be distinct from each other. The ANP technique is quite useful in difficult situations among decision elements of a hierarchical structure over a network structure [41]. This technique contains the features of the AHP technique such as flexibility, simplicity, ability to review consistency in judgments, and concurrent use of qualitative and quantitative criteria [42]. In the ANP method, each issue is considered to be a network of criteria and the communication is done between each element in a network; communication between clusters, interconnection, and feedback are feasible in a network [43]. The following steps are followed in the ANP technique:

- A situation is first converted into a network for the simplicity of the problem. The network structure is thus obtained with the help of methods like nominal group or the Delphi method and perhaps through brainstorming as well. This point is the indication that the problem is now a network problem in which all the elements communicate with each other.
- Just as the pairwise comparison is performed in AHP, likewise every element in the cluster is compared pairwise. A comparison between the clusters is also performed based on their effect on reaching their goals along with the dependence between the criteria of the clusters. Eigenvectors help in providing the impact of the criteria on each other. Saaty nine-point scales help in measuring the relative importance of the elements.
- To accomplish all the priorities in the system with the internal significance of the vectors, interactions are required to be entered in a particular column of the matrix. This matrix is known as a Supermatrix, which is also known as a partition matrix, which depicts the connection among both clusters within a structure. The structure of the standard, goals, and substitute is shown in the three-level structure in the two forms (A) hierarchy and (B) network in Fig. 4.
- The last weight of an alternative is obtained from the alternatives column in the Supermatrix. The alternative with the larger value in the matrix is considered to be the top pick.

3.2.1 ANP in healthcare

The ANP approach has already been employed for choosing the best optimization projects, particularly Six Sigma, in areas that demand accurate decision-making, with the objective of accomplishing the selection of the project with maximum benefits at a minimum risk. In [21] the paper presents an ANP approach to determine and compute projects for healthcare authorities, enabling them to choose the project that ensures the desired financial achievement. The criteria weights are established with the help of ANP and then, finally, the results from ANP-DEMATEL (Decision Making Trial and Evaluation Laboratory) and ANP are related. Healthcare waste must be handled correctly due to pathological, infected, etc. contents that may result in an infection to the land surface and drinking water. Applied management systems could be the most appropriate answer from environmental, economic, technical, and social points of view. In [22] the major focus was to evaluate the present status of healthcare waste management in Turkey, in figuring out the best suitable disposal option with the help of various decision-making techniques. The actions taken for selecting an accurate healthcare waste disposal

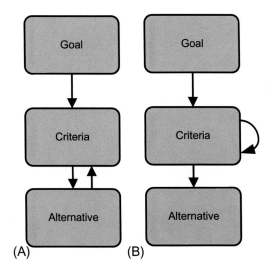

FIG. 4

Three-level structure of goals, criteria, and alternatives, where (A) represents a hierarchy and (B) shows a network.

scheme for advanced countries were calculated with the help of the ELECTRE III and ANP techniques. In [23] the authors used the ANP and AHP techniques based on experience acquired from hospitals for ranking the improvement projects. The main goal of this paper was to discuss the results of ANP and AHP methods for selecting a project in the field of healthcare in Iran.

3.3 VIKOR method

This method is used in multicriteria optimization and finding compromise solutions. It was introduced to resolve MCDM issues with conflicting and impossible-to-measure criteria situations, assuming that compromise is acceptable for conflict resolution and solution, the decision maker wants to find a solution closest to the optimal one (ideal), and the alternatives are examined based on the fixed criteria [36]. This technique emphasizes selecting and ranking among a group of alternatives with the existence of opposing criteria. This technique is an essential device in MCDM, especially under conditions wherein the decision maker is not capable of stating the options at the start of system planning. The compromise solution can be approved by decision makers in view of the fact that it offers the highest level of utility for the majority and the lowest level of regret for the losers. The terms of the solution are a base for agreement, including the decision maker's options by the weights of the criteria [44]. Supposing that every alternative is measured under every criterion concern, the ranking would be evaluated by comparing the strength of privacy to the perfect alternative. The VIKOR method is shown in Fig. 5 and traditional VIKOR algorithms include the following steps [46]:

- The alternatives are denoted with some specific terms. For each alternative, their respective components are defined. The number of alternatives and criteria are denoted, such as by m and n, respectively.

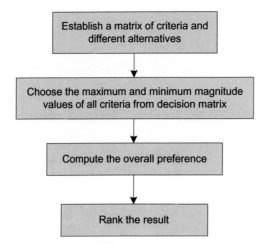

FIG. 5

Procedure of VIKOR methodologies [45].

- Obtain the highest and lowest rates of the criteria.
- Calculate the values of utility and the regret measures for the specific alternative; also, the relative importance of the criteria is represented.
- Evaluate the utility measure and regret measure, where the symbol v signifies the weight for the plan of the highest utility group.

3.3.1 VIKOR method in healthcare

This method emphasizes prioritizing and choosing from a group of choices and examines a solution, maintaining the maximum utility of the majority and lowest individual regret of the opponent. In [24] an MCDM approach based on fuzzy set theory and the VIKOR technique was applied to healthcare efficiency evaluation of some areas in Istanbul. This is a metropolis with 15 million people, making it one of the world's most populated cities. The focus of the paper [25] was to rank all the risk factors, with the help of the VIKOR approach, enforced on the database of patients with type 2 diabetes mellitus (T2DM). The data was taken from 53 T2DM patients and analyzed using the VIKOR method. In [26] the authors presented an enhanced version of the VIKOR approach with improved accuracy, sufficient for data in the medical field.

3.4 TOPSIS

The main purpose of the popular TOPSIS procedure is solving MCDM problems based on the principle that the chosen alternative must have the smallest distance from the positive ideal solution (PIS) and the largest distance from the negative ideal solution (NIS). Later the same principle was suggested by Zenely and Hall for resolving MCDM problems. In practice, the TOPSIS method has been employed

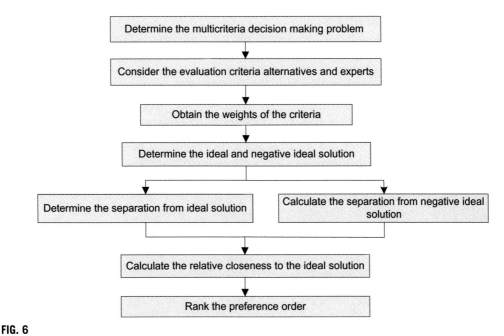

FIG. 6

Steps of TOPSIS methodology [47].

for MADM to resolve selection problems with a smaller number of alternatives. Standard TOPSIS methods are not applicable to problems with an unlimited number of alternatives, as occurs with the MODM technique, in contrast with MADM. With the conflict between the two distances between PIS and NIS, a compromise cannot be obtained with the smallest and the farthest distance from PIS and NIS, respectively. The unlimited alternatives situation of MODM makes it difficult to find the solution with the smallest distance from PIS and the farthest from NIS, so the criteria of the distances from PIS and NIS are replaced by "as close to PIS and as far to NIS as possible" [44]. To define these fuzzy terms, the integration function of the fuzzy set theory is applied. The TOPSIS process is shown in Fig. 6 and can be evaluated through the following steps:

- Evaluate the standard decision matrix.
- Determine the weighted normalized decision matrix.
- Measure the ideal and nonideal solutions.
- Using the n-dimensional Euclidean distance, calculate the separation measures.
- Calculate the relative closeness with respect to an ideal outcome.
- Rank the order of preference.

In [48] the authors claim that "the best alternative is the one that has the shortest distance to the ideal solution. Also, any alternative which has the smallest distance from the ideal solution is guaranteed to have the farthest distance from the negative ideal solution."

3.4.1 TOPSIS method in healthcare

The TOPSIS technique was proposed by Hwang and Yoon in 1981 [27] as one of the MCDM approaches that could help in determining solutions form a group of points. In [28] the criteria were determined in a study associated with the sustainability triple bottom line involving other standards, from a literature survey, and the location of a healthcare waste disposal center was chosen. Further, hybrid models of illustrative organic modeling, fuzzy AHP, and fuzzy TOPSIS methods were utilized to perform this survey. Thus this work creates a theoretical basis, in requirements of criteria, as well as hybrid methods and systematic activities, for selecting a location for healthcare waste disposal. In [29] the study evaluates an electronic service quality (e-sq) vision and identifies the main parts of e-sq. This e-sq framework ID is concerned with evaluating the benefits of service quality methodology. With the help of web service performance, the e-sq framework is demonstrated with a network service performance sample of the healthcare sector in Turkey, using a mixture of fuzzy AHP and fuzzy TOPSIS techniques.

3.5 PROMETHEE method

PROMETHEE (with its complement GAIA, geometrical analysis for interactive aid) is a decision-making method developed in the 1980s, which has undergone significant modifications and extensions since that time. From remarks on multicriteria problems, it is the consensus that this type of problem cannot be treated without the addition of information associated with the priorities and preferences of the decision makers [49]. The data requested by GAIA and PROMETHEE is generally clear and simple to specify for both analysts and decision makers. It contains a preference function related to each criterion, including weights indicating their relative importance. This technique is useful to attain the last rating of the alternatives, which requires a broad range of criteria. Further, an improved PROMETHEE II technique effectively builds its area of application and potentiality for solving these types of decision problems with various conflicting alternatives and criteria. PROMETHEE I is used for partial ranking of actions based on positive and negative flows, whereas PROMETHEE II is utilized for the complete ranking of the actions [50]. The basic process of the PROMETHEE technique is shown in Fig. 7.

3.5.1 PROMETHEE in healthcare

In [27] the author applied the PROMETHEE II technique to accelerate resource management and decision-making in an emergency department (ED). This MCDA technique is particularly useful in systems that require compound decision-making and includes various considerations. In [28] the PROMETHEE method was used to prioritize barriers for MADM, concerning the level of their harmful effects. Fuzzy AHP and PROMETHEE-GAIA are some of the most commonly used techniques by researchers in the field of location selection. However, these techniques also have their limitations; AHP handles the problem of size and building agreement, and PROMETHEE can provide better results with the help of the hierarchical format of AHP [29]. The objective is to avoid the limitations of the two methods by this integration.

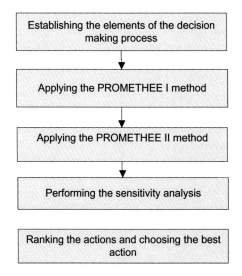

FIG. 7

Procedure of PROMETHEE technique [30].

4 Research directions

Numerous healthcare centers are being established around the globe providing better health-related facilities to their patients, with IoT playing a major part in some of their functionalities. Techniques such as remote monitoring of patients, observation of treatment progress from any location, and housing of vaccines are made possible with the help of the IoT. In [51–53], the authors have presented their innovative concepts for sustainable health centers, for monitoring various conditions, with the obtained data sent to healthcare professionals in real time. With this facility come various complex decisions that require the attention of healthcare personnel. This could be a decision related to the selection of the best healthcare facility, the best supplier for the hospitals, etc. The MCDM technique was introduced for the purpose of decision-making in difficult situations like these, in any field. This technique has provided solutions out of complex situations using several approaches, including AHP, TOPSIS, VIKOR, ANP, and PROMETHEE as described in this chapter. In healthcare decision-making, the MCDM technique has provided many benefits, with its property of giving the best solution. Whereas decision-making approaches are not a step-by-step process as typically employed in healthcare scenarios and as shown in this chapter, various aspects might be improved for a better outcome in the future, such as:

(1) The significance of the alternatives being presented in order.
(2) A defined set of scales for each criterion [38].
(3) The logic of the multiple attributes function.
(4) The usage of a single value score to determine a difficult problem.

The main idea of the MCDA methods provides the common solution as an answer to the complex problems. Apart from the healthcare field, the MCDA technique might be used in other fields for better outcomes.

5 Conclusion and future scope

The IoT is finding more and more applications in medical and health treatments, including fitness programs, chronic disease management, and health monitoring. Numerous types of services, applications, and distinct protocols have become accepted in this field. Therefore, in the design and implementation of an intelligent and efficient healthcare system, every decision point should be carefully considered. Healthcare decision-making is different from decision-making in many other domains. It is typically a process that includes finite steps in a certain sequence, which makes this process more stable. Various decision-making tools and methods can be used for supporting healthcare decision-making. In this chapter, practical guidelines for the proper application, designing, and reporting of MCDA methods have been provided. The MCDA/MCDM technique could be very useful in the future with the accelerated growth of technology for making complicated decisions. Hybrid MCDM techniques have already been used specifically for healthcare problems. However, a combination of two or more techniques could also lead to better decision-making in healthcare problems. Some of the main techniques of MCDM are elaborated in this chapter regarding the healthcare industry. It could not be endorsed that the tradeoffs in this to be based on a single criterion such as cost. Although if MCDM to be applied adequately could result in better outcomes, but the final result should not be the automated result of MCDM instead must be done by decision-makers. Advances in decision-making techniques could be very useful in the field of 4IR when employed appropriately. The feasibility of the studies focuses on the detailed researchers at the systematic applications that are the process that helps in reducing the mistakes instead of mathematical ones.

References

[1] M.J. Liberatore, R.L. Nydick, The analytic hierarchy process in medical and health care decision making: a literature review, Eur. J. Oper. Res. 189 (1) (2008) 194–207.

[2] V. Diaby, K. Campbell, R. Goeree, Multi-criteria decision analysis (MCDA) in health care: a bibliometric analysis, Oper. Res. Health Care 2 (1–2) (2013) 20–24.

[3] I. Frederix, Internet of things and radio frequency identification in care taking, facts and privacy challenges, in: Wireless Communication, Vehicular Technology, Information Theory and Aerospace and Electronic Systems Technology, IEEE 1st International Conference on Wireless VITAE, May, 2009, pp. 319–323.

[4] M. Simonov, R. Zich, F. Mazzitelli, Personalized healthcare communication in internet of things, in: Proc. of URSI GA08, vol. 7, 2008.

[5] P.K. Gupta, V. Tyagi, S.K. Singh, Internet of things based predictive computing, in: Predictive Computing and Information Security, Springer, Singapore, 2017.

[6] P. Gupta, T. Ören, M. Singh, Predictive Intelligence Using Big Data and the Internet of Things. IGI Global, 2019. https://doi.org/10.4018/978-1-5225-6210-8.

[7] L.M.R. Tarouco, L.M. Bertholdo, L.Z. Granville, L.M.R. Arbiza, F. Carbone, M. Marotta, J.J.C. De Santanna, Internet of Things in healthcare: interoperability and security issues, in: 2012 IEEE International Conference on Communications (ICC), IEEE, 2012, pp. 6121–6125.

[8] V.M. Rohokale, N.R. Prasad, R. Prasad, A cooperative Internet of Things (IoT) for rural healthcare monitoring and control, in: 2011 2nd International Conference on Wireless Communication, Vehicular Technology, Information Theory and Aerospace & Electronic Systems Technology (Wireless VITAE), IEEE, 2011, pp. 1–6.

[9] A. Piegat, W. Sałabun, Comparative analysis of MCDM methods for assessing the severity of chronic liver disease, in: International Conference on Artificial Intelligence and Soft Computing, Springer, Cham, 2015, pp. 228–238.

[10] Y.J. Fan, Y.H. Yin, L. Da Xu, Y. Zeng, F. Wu, IoT-based smart rehabilitation system, IEEE Trans. Industr. Inform. 10 (2) (2014) 1568–1577.

[11] R.V. Rao, Introduction to multiple attribute decision-making (MADM) methods, in: Decision Making in the Manufacturing Environment: Using Graph Theory and Fuzzy Multiple Attribute Decision Making Methods, 2007, pp. 27–41.

[12] U.S. Department of Health & Human Services, The Agency for Healthcare Research and Quality's (AHRQ), Webcast: Multi-Criteria Decision Analysis Techniques to Integrate Stakeholder Preferences in Comparative Effectiveness Research, Available from: http://effectivehealthcare.ahrq.gov/index.cfm/webcast-multi-criteria-decision-analysis-techniques-to-integrate-stakeholder-preferences-incomparative-effectiveness-research/, 2012. Accessed 1 January 2013.

[13] F. Castro, L.P. Caccamo, K.J. Carter, B.A. Erickson, W. Johnson, E. Kessler, N.P. Ritchey, C. A. Ruiz, Sequential test selection in the analysis of abdominal pain, Med. Decis. Mak. 16 (2) (1996) 178–183.

[14] G. Büyüközkan, G. Çifçi, S. Güleryüz, Strategic analysis of healthcare service quality using fuzzy AHP methodology, Expert Syst. Appl. 38 (8) (2011) 9407–9424.

[15] T. Sahin, S. Ocak, M. Top, Analytic hierarchy process for hospital site selection, Health Policy Technol. 8 (1) (2019) 42–50.

[16] A. Aktas, S. Cebi, I. Temiz, A new evaluation model for service quality of health care systems based on AHP and information axiom, J. Intell. Fuzzy Syst. 28 (3) (2015) 1009–1021.

[17] Y.-P. Huang, H. Basanta, H.-C. Kuo, A. Huang, Health symptom checking system for elderly people using fuzzy analytic hierarchy process, Appl. Syst. Innov. 1 (2) (2018) 10.

[18] D. Kriksciuniene, V. Sakalauskas, AHP model for quality evaluation of healthcare system, in: International Conference on Information and Software Technologies, Springer, Cham, 2017, pp. 129–141.

[19] N.R. Alharbe, Improving usable-security of web based healthcare management system through fuzzy AHP, Int. J. Adv. Comput. Sci. Appl. (2019) 1–5.

[20] A. Singh, A. Prasher, Measuring healthcare service quality from patients' perspective: using fuzzy AHP application, Total Qual. Manag. Bus. Excell. 30 (3–4) (2019) 284–300.

[21] M.O. Barrios, H.F. Jiménez, S.N. Isaza, Comparative analysis between ANP and ANP-DEMATEL for six sigma project selection process in a healthcare provider, in: International Workshop on Ambient Assisted Living, Springer, Cham, 2014, pp. 413–416.

[22] A. Özkan, Evaluation of healthcare waste treatment/disposal alternatives by using multi-criteria decision-making techniques, Waste Manag. Res. 31 (2) (2013) 141–149.

[23] E.S. Serkani, M. Mardi, E. Najafi, K. Jahanian, A.T. Herat, Using AHP and ANP approaches for selecting improvement projects of Iranian Excellence Model in healthcare sector, Afr. J. Bus. Manag. 7 (23) (2013) 2271.

[24] M.A. Karadayi, E.E. Karsak, Fuzzy MCDM approach for health-care performance assessment in Istanbul, in: Proceedings of The 18th World Multi-Conference on Systemics, Cybernetics and Informatics, 2014, pp. 228–233.

[25] C.I. Bondor, I.M. Kacso, A. Lenghel, D. Istrate, A. Muresan, VIKOR method for diabetic nephropathy risk factors analysis, Appl. Med. Inform. 32 (1) (2013) 43–52.

[26] Q.-L. Zeng, D.-D. Li, Y.-B. Yang, VIKOR method with enhanced accuracy for multiple criteria decision making in healthcare management, J. Med. Syst. 37 (2) (2013) 9908.

[27] A. Chauhan, A. Singh, A hybrid multi-criteria decision making method approach for selecting a sustainable location of healthcare waste disposal facility, J. Clean. Prod. 139 (2016) 1001–1010.

[28] G. Büyüközkan, G. Çifçi, A combined fuzzy AHP and fuzzy TOPSIS based strategic analysis of electronic service quality in healthcare industry, Expert Syst. Appl. 39 (3) (2012) 2341–2354.

[29] S. Sadoughi, R. Yarahmadi, M.H. Taghdisi, Y. Mehrabi, Evaluating and prioritizing of performance indicators of health, safety, and environment using fuzzy TOPSIS, Afr. J. Bus. Manag. 6 (5) (2012) 2026.

[30] T.M. Amaral, A.P.C. Costa, Improving decision-making and management of hospital resources: an application of the PROMETHEE II method in an Emergency Department, Oper. Res. Health Care 3 (1) (2014) 1–6.

[31] S.S. Mishra, K. Muduli, M. Dash, D.K. Yadav, PROMETHEE-based analysis of HCWM challenges in healthcare sector of Odisha, in: Smart Computing and Informatics, Springer, Singapore, 2018, pp. 163–170.

[32] D. Chatterjee, Delphi-FAHP and Promethee: An Integrated Approach in HEALTHCARE Facility Location Selection; 2015, (2015).

[33] A. Farooq, M. Xie, S. Stoilova, F. Ahmad, M. Guo, E.J. Williams, V.K. Gahlot, D. Yan, A. M. Issa, Transportation planning through GIS and multicriteria analysis: case study of Beijing and XiongAn, J. Adv. Transp. 2018 (2018) 2696037.

[34] A. Kolios, V. Mytilinou, E. Lozano-Minguez, K. Salonitis, A comparative study of multiple-criteria decision-making methods under stochastic inputs, Energies 9 (7) (2016) 566.

[35] M. Azam, M.N. Qureshi, F. Talib, AHP model for identifying best health care establishment, Int. J. Prod. Manag. Assess. Technol. 3 (2) (2015) 34–66.

[36] S. Opricovic, G.-H. Tzeng, Extended VIKOR method in comparison with outranking methods, Eur. J. Oper. Res. 178 (2) (2007) 514–529.

[37] Y. Liu, Y. Yang, Y. Liu, G.-H. Tzeng, Improving sustainable mobile health care promotion: a novel hybrid MCDM method, Sustainability 11 (3) (2019) 752.

[38] R.W. Saaty, The analytic hierarchy process—what it is and how it is used, Math. Model. 9 (3–5) (1987) 161–176.

[39] J. Figueira, S. Greco, M. Ehrgott (Eds.), Multiple Criteria Decision Analysis: State of the Art Surveys, In: vol. 78, Springer Science & Business Media, 2005.

[40] T.L. Saaty, Fundamentals of the analytic network process—dependence and feedback in decision-making with a single network, J. Syst. Sci. Syst. Eng. 13 (2) (2004) 129–157.

[41] S. Opricovic, Multi-criteria optimization of civil engineering systems, Faculty of Civil Engineering, Belgrade, in: Table II The Performance Matrix, 1998.

[42] M. García-Melón, J. Ferrís-Oñate, J. Aznar-Bellver, P. Aragonés-Beltrán, R. Poveda-Bautista, Farmland appraisal based on the analytic network process, J. Glob. Optim. 42 (2) (2008) 143–155.

[43] S. Kheybari, F.M. Rezaie, H. Farazmand, Analytic network process: an overview of applications, Appl. Math. Comput. 367 (2020) 124780.

[44] S. Opricovic, G.-H. Tzeng, Compromise solution by MCDM methods: a comparative analysis of VIKOR and TOPSIS, Eur. J. Oper. Res. 156 (2) (2004) 445–455.

[45] G. Villacreses, G. Gaona, J. Martínez-Gómez, D.J. Jijón, Wind farms suitability location using geographical information system (GIS), based on multi-criteria decision making (MCDM) methods: the case of continental Ecuador, Renew. Energy 109 (2017) 275–286.

[46] C.-L. Chang, A modified VIKOR method for multiple criteria analysis, Environ. Monit. Assess. 168 (1–4) (2010) 339–344.

[47] N. Zhang, G. Wei, Extension of VIKOR method for decision making problem based on hesitant fuzzy set, Appl. Math. Model. 37 (7) (2013) 4938–4947.

[48] E. Triantaphyllou, Multi-criteria decision making methods, in: Multi-Criteria Decision Making Methods: A Comparative Study, Springer, Boston, MA, 2000, pp. 5–21.

[49] J.-P. Brans, P. Vincke, Note—a preference ranking organisation method: (the PROMETHEE method for multiple criteria decision-making), Manag. Sci. 31 (6) (1985) 647–656.

[50] I. Giurca, I. Aschilean, C.O. Safirescu, D. Muresan, Choosing photovoltaic panels using the PROMETHEE method, in: Proceedings of the International Management Conference, vol. 8(1), Faculty of Management, Academy of Economic Studies, Bucharest, Romania, 2014, pp. 1087–1098.

[51] P.K. Gupta, V. Tyagi, S.K. Singh, Predictive Computing and Information Security. Springer, Singapore, 2017. https://doi.org/10.1007/978-981-10-5107-4.

[52] P.K. Gupta, B.T. Maharaj, R. Malekian, A novel and secure IoT based cloud centric architecture to perform predictive analysis of users activities in sustainable health centres, Multimed. Tools Appl. 76 (18) (2017) 18489–18512.

[53] D. Swain, P.K. Pattnaik, P.K. Gupta, Machine Learning and Information Processing. Springer, Singapore, 2020. https://doi.org/10.1007/978-981-15-1884-3.

[54] T.L. Saaty, Decision making with the analytic hierarchy process, Int. J. Serv. Sci. 1 (1) (2008) 83–98.

A research review on semantic interoperability issues in electronic health record systems in medical healthcare

8

Rimmy Yadav[a], Saniksha Murria[a], and Anil Sharma[b]
CT Institute of Management & Information Technology, Jalandhar, Punjab, India[a] Lovely Professional University, Phagwara, Punjab, India[b]

1 Introduction

With the rise in chronic diseases, people have become increasingly health aware and patients are now treated as "health consumers," striving for improved healthcare management. Predetection systems used in earlier days have become obsolete, due to their lack of efficiency and reliability. Hospitals are now adopting advanced systems that observe the health conditions of their patients, as sensitive and frequent estimations of physiological and behavioral patient information are vital for early recognition of physical and mental changes. Physical information checks normally include pulse rate, blood glucose level, blood pressure, weight checks, and others. For behavioral data gathering, continuous close monitoring is a prime requirement, which is difficult in a clinical environment. Chief solution to this problem is to make use of ubiquitous healthcare systems, involving the use of a large number of sensors and actuators to remotely monitor patients, which has brought a revolution to the medical industry by giving a quick disdain into the diseases by providing patients and their concerned doctors an alert to any possibility of a serious problem arising. With ubiquitous healthcare, electronic health records (EHRs), and other recent medical technology advancements, a vast amount of health-related data, such as XML (eXtensible Markup Language) data, relational databases, magnetic resonance imaging (MRI) images, scanned reports, tabular-based data and other relevant data types, are generated in the healthcare domain. These data have unstructured formats and semantics. Ubiquitous healthcare, for example, lacks a standard methodology for merging all these types of data. This chapter surveys the literature on semantic interoperability related to healthcare, including definitions, standards, vocabularies, obstacles, and future difficulties. Along with this, particular attention is placed on ontologies as a solution for the problems of semantic interoperability.

The EHR is the most extensively adopted and implemented electronic health application in the healthcare industry, serving as a framework for the movement from paper-based information to digital formats. An EHR acts as an interface between hospitals and other facilities for sharing health-related data, such as MRI images, tabular data, activities of daily living files, and many others. It digitally stores all the records of patients in databases, as the enormous number of patient records generated in the healthcare industry can be managed only by employing modern database technology. The EHR is responsible for merging this data, in heterogeneous formats, needed for patient data analysis for better healthcare management [1].

Interoperability of EHR systems is one of the most urgent requirements in the healthcare industry. With the help of interoperability, a patient's medical history or medical data can be shared with any hospital or clinic, which improves the quality of operations, workflow management, medical decision-making, and time savings. Semantic interoperability, which plays a main role in the EHR system environment, is the ability of systems to comprehend the semantics of data transferred by another system. In other words, the EHR system perceives and understands the patient data transmitted by another distributed EHR system, so that it is computer processable (readable by machines, not syntactically). But to achieve complete semantic interoperability is not an easy task, mainly due to the heterogeneity of patient data. Patient data may be structured, semistructured, or unstructured data. Many reasons are behind the semantic interoperability issue: different schemes used by different EHR systems, structure variations, standards, service level agreement policies, and different programs and database management systems implemented by distributed EHR systems are some of the causes [2].

Many different models, frameworks and methodologies have been proposed for EHR to improve semantic interoperability. The SIMB-IoT (Internet of Things) model [3] has been proposed to improve semantic interoperability among heterogeneous IoT devices in the healthcare sector. But drawbacks of the proposed mode are lack of security issues and syntactic interoperability. The OWL (Web Ontology Language)-based framework [4], as a semantic web, has been proposed to improve semantic interoperability among EHR systems. But the authors of this paper focused only on common data elements, not on archetypes (models of clinical content). Academia and industry access to EHR data and the involvement of hospitals is mandatory to improve research and development in the healthcare industry. The main goals of this chapter are to understand the importance of EHR and the vital prerequisite of interoperability of EHR to improve the grades of the healthcare domain; discuss interoperability standards; investigate various challenges related to interoperability; and finally suggest semantic web ontologies as a suitable elucidation of the semantic interoperability problems.

2 EHR and interoperability

2.1 Introduction and definitions

The electronic medical record (EMR) in a healthcare practice is a digital platform used to collect and store the patient's medical data, from their first appointment with concerned doctors and clinicians until their discharge date—i.e., a digital version of the patient's chart within that individual practice. The EHR system, on the other hand, provides access to individual health records saved in EMR systems among interacting organizations, such as for population health research and clinical research. In other words, the EHR serves as storage of many patients' medical histories in a digitized format to assist further innovation and growth, and committing to privacy for the patients. Interoperability means communication as well as sharing patient data between two or more software applications. Interoperability can be defined as the ability of two or more components to exchange a patient's health-related data and to understand and use this information accurately and effectively.

According to IEEE (Institute of Electrical and Electronics Engineers), interoperability can be defined as the capability of numerous EHR structures and their installed purposes to gather, transmit data, mine, and make use of that communicated data.

The benefits of interoperability can be described as follows:

1. Interoperability can frame the health-related records of patients so that, with the help of EHR, patient's health records can be updated easily.
2. EHR is basically implemented for merging the patient's entire medical history from distributed clinics, using possibly different computer systems and data formats.
3. Patient's data can be sent or received in standardized format to/from other providers.
4. EHR allows patients, clinicians, and doctors to access their own data or that of others from any location at any time.
5. Interoperable EHR systems cause paperwork to decrease drastically and help in avoiding unnecessary medical errors, tests, and medical imaging procedures.

2.2 Levels of interoperability

There are basically four dissimilar levels or categories of interchangeability or interoperability [4, 5]. These are categorized regarding whether the data can be followed by operating machines or by an intended user [5]. In the following sections, a concise introduction is provided.

1. No interoperability: Health information cannot be interpreted by human or machine [5].
2. Syntactic interoperability: Also termed technical interoperability. It can be defined in this context as the "ability of two or more communication technology-based applications to send and receive the patient's data from one machine," say EHR, to other machines or EHR, without any intervention. Its main target is to complete the intended tasks in a stipulated timeframe effectively [6].
3. Technical interoperability: This mainly focuses on machine components such as hardware, software, systems, and platform that enable system-to-system interaction [6].
4. Semantic interoperability: "It means integrating resources that were structured using different artifacts and different perspectives on the data. To achieve the semantic interoperability, systems must be able to exchange data in such a way that the precise meaning of the data is readily accessible and data itself can be translated by any system into a form that it understands."

3 Standards in E-health and interoperability

As already mentioned, the healthcare sector is currently facing the problem of interoperability of EHR systems. To meet this challenge, standards are playing a vital role. Standards serve as a common platform that coordinates the different applications and system software. In other words, standards can act as an interface between application programs and system software.

To provide full semantic interoperability, grades have to cover both the semantic and syntactic structure. To deal with the semantic interoperability, various standards have been developed by researchers and scientists. There are about 25 distinct standards that apply to the healthcare field, but four major standards are internationally accepted by the stakeholders: digital imaging and communication in medicine (DICOM), Health Level 7 (HL7), OpenEHR, Integrating the Healthcare Enterprise (IHE), and International Standard Organization.

3.1 Digital imaging and communication in medicine

DICOM is the most widely accepted protocol standard in the digital medical imaging field. It is endowed with all the important methods and techniques for identification, depiction, and processing of medical image data. In addition to this, DICOM will not be considering as a tool to for medical image processing and for communication in medical healthcare. It also has abilities such as transferring data with storage and display rules and regulations intended to include all operational characteristics of digital imaging.

DICOM companies also introduced the picture archiving and communication system (PACS), a medical imaging technology used to execute digital medical imaging. PACS is made up of various modules such as CT scanners, ultrasound, and archives for storing medical images [7].

3.2 OpenEHR

OpenEHR is a technology for e-health, including open specifications, ontology-based models, and software that can be implemented as design standards as well as to construct information and interoperability solutions for the healthcare sector. OpenEHR can be implemented as a specific program, incorporating numerous technical platforms, which in turn consist of information models and query languages. OpenEHR allows clinicians to save and reuse the patient's data from every connected EHR system. One of the interesting functionalities of OpenEHR is the launching of its product on a global scale, with complete lists of guidance in local languages within the healthcare and funding environment.

3.3 Health Level 7

HL7 is an international standards development organization. Its vision is to develop optimal and widely accepted standards, as well as implement standards, in the healthcare industry. HL7 provides standards for syntactic as well as semantic interoperability. It helps to improve delivery care and workflow management, reduce ambiguity, and sharpen the knowledge transfer of all of the stakeholders, including medical healthcare providers, corporate as well as government agencies, and last but not least, patient data. For clinical as well as administration purposes, the HL7 standard is helpful in patient's data transfer and the collaboration of electronic healthcare information. It employs a formal methodology. The HL7 provides a meeting place for various health specialists and IT experts to work together and with other organizations [8].

3.4 Integrating the Healthcare Enterprise

IHE defines a set of well-structured and defined profiles, called IHE health document sharing profiles, which are used in interoperable transfer of health-related documents. These profiles provide various mechanisms and methodologies for transferring documents within and across communities and make use of various important principles that cope with and resolve many of the common challenges in health data communication.

4 Model-driven architecture and its role in EHR

Recently, model-driven architecture (MDA) has developed rapidly as an organized, incorporated framework for designing as well as development. The Object Management Group (OMG) created MDA as a notional structure that splits the organizational or business process from technology-oriented platform decisions. Flexibility greatly improves and the functionality of MDA can deal with the complexity of any type of process. According to [9]: "One of the exciting features of models is that artifacts of software can be designed with giving focus on program code. Models basically use as a blueprint from which underlying features, such as programs and models, can be inherited and implemented either automated and semi-automated means." MDA approaches consist of two sets of models, platform-independent models and platform-specific models, that the organization chooses for implementation and support.

MDA is playing a vital role in the medical healthcare industry, as it provides a set of models for development and support. It is known that the healthcare sector is a multitenant, multidelivery model, involving different systems (e.g., multiple stakeholders, numerous standards, and various platforms, with varying multiple methodologies). Therefore, the different components and frameworks may not fit together architecturally, referred to as an architectural mismatch. In this case, interoperability is essential, but it can occur only when different EHR systems with different systems fit together. With the help of the MDA framework, stakeholders do not need to reengineer large-scale projects but rather only to implement the various paradigms such as grid and distributed computing, service-oriented architecture (SOA), and web services as a loosely coupled system.

Over the traditional approaches such as ontologies, MDA with semantic web has great potential to resolve the issues of semantic interoperability in the EHR domain.

5 Methods

5.1 Question of research

1. Why is semantic interoperability needed in an EHR environment?
2. What is the role of EHR standards in gaining full semantic interoperability?
3. What are the highlighted advantages resulting from EHR standards?
4. What is the importance of the semantic web with MDA in achieving EHR semantic interoperability?

5.2 Search plan

A review of the literature was conducted. For an effective literature review, journals were selected and studied from the years 2012–19. A set of journals and book chapters was selected from various search engine databases, such as Science Direct [791], THEIET (The Institute of Engineering and Technology) [20], IEEE [8], PubMed [611], and Taylor and Francis [175].

5.3 Search results

From the complete set of paper selections, interesting papers from the search databases were selected first. To further select and refine the collected papers, abstract analysis and empirical study of the papers followed as a refinement approach. The main keyword to begin the search of journals and book chapters was "EHR." The total number of papers screened was 1605 research articles on "EHR" AND "Semantic interoperability" OR "Semantic interoperability issues" AND "Electronic Health Record Systems." In this study, 1605 articles related to "Semantic interoperability issues in EHR systems" are represented. In Table 1, the collected papers are presented in tabular form. In Fig. 1, a chart representation is presented to pictorially show the number of articles found from years 2012 to 2019.

After eliminating the irrelevant papers, 185 papers or articles and book chapters were chosen. We combined the collected publications pertaining to challenges of semantic interoperability and the developed frameworks, methods, models, and approaches to cope with semantic interoperability represented.

Finally, the relevant papers were searched using keywords "Electronic Health Record" AND "Semantic Web" AND "Model Driven Architecture approach." Many fewer articles related to the MDA approach with semantic web in the EHR domain were found.

Table 2 shows the literature review pertaining to the standards, approaches, methods, framework, and ontology developed for semantic interoperability.

6 The challenges of EHR semantic interoperability

(a) The lack of well-acknowledged standards and good practices hinders the performance in managing the patients' health-related data as well as maintaining the privacy of patient data. Currently, there is no well-established model and standard that is legally valid.

Table 1 Number of papers found and collected from years 2012–19.

Year	Number of papers (journals and book chapters only)					Total number of papers found and reviewed
	Science Direct	Taylor & Francis	THEIET	IEEE Xplore Digital Library	PubMed	
2012	75	22	01	01	70	
2013	86	16	01	02	59	
2014	94	24	00	00	86	
2015	99	17	00	01	103	
2016	98	22	03	00	80	
2017	100	26	05	00	82	
2018	109	16	02	01	68	
2019	130	32	08	03	63	
Total	**791**	**175**	**20**	**8**	**611**	**1605**

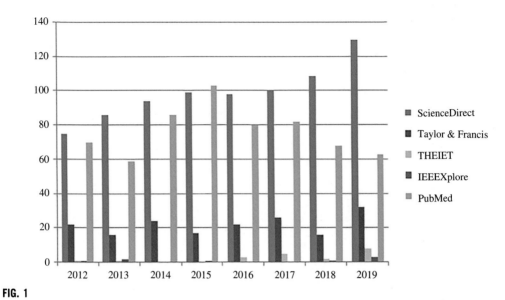

FIG. 1

Bar chart representation of collected papers from years 2012 to 2019.

(b) There is a lack of attentiveness among associations to the health information and knowledge representation interoperability standards that exist in the area of clinical research and clinical care. There is also a lack of mapping tools that can support semantic harmonization of heterogeneous health and life science data.

(c) There is a lack of appropriate documentation that clearly defines data content and formats of health information systems to be shared and interchanged across different system users.

(d) The privacy issue is one of the most challenging tasks in sharing the patient's data between EHRs.

(e) There have been major time delays in the adoption of standards-based health information systems by healthcare institutions. This is happening due to lack of incentives given to hospital staff members.

(f) There is a need for modeling technologies to enhance semantic interoperability to improve the quality of data transmission among stakeholders in the healthcare domain.

(g) Most of the available datasets in healthcare continue to be generated in named or registered formats. There are some serious initiatives to convert this type of data, but mapping local terms to concepts in a standard vocabulary is a complex and resource-intensive task. Resources, tools, and strategies are required to automate this process in order to improve the semantic interoperability.

(h) To achieve interoperability, there is a need to develop an EHR using standard modeling techniques so that the resulting EHR systems can easily communicate with EHR systems of other hospitals. The resulting systems can minimize administration costs, improve clinical performance, and enable better communication between stakeholders (such as patients and caregivers).

Table 2 Literature probe on semantic interoperability.

Ref. no	Proposed method/approach/technology	Advantages	Drawbacks
[10]	Authors proposed an infrastructure called Cancer Common Ontologic Representation (CaCORE) based on MDA approach CaCORE includes three modules: • Enterprise vocabulary services • Cancer data repository • API and model driven architecture-based information system	• Improves data interoperability	• Limited to one particular disease, i.e., cancer
[11]	Authors highlighted the issues of technical, semantic, and syntactic interoperability. An interoperability maturity model along with framework for ultralarge-scale system is proposed	• For evaluating the performance of proposed model and framework, authors take Malaysian healthcare systems as a case study • Improves the technical, semantic, and syntactic interoperability	NA
[12]	The problem of veracity dimensions when transferring patient data between EHR systems is highlighted They outlined the challenges faced in ASSESS-CT project by the European government	NA	• Authors point out that the current working clinical EHR systems are not natively interoperable • Semantic web will be helpful in improving the quality of patient's data in medical healthcare
[13]	Authors implemented HL7 with ontology as an approach to save the ontology into RDF triple store Saved ontology is then calculated with HL7 and Levenshtein distance and semantic fingerprints	• Improves the semantic and syntactic interoperability	NA
[14]	Semantic interoperability such as OWL (Web Ontology Language) is implemented as a framework for orthopedic implants (custom implant for bone cancer of tibia)	• It allows a single point of access for all common and individual information resources • Helps in surgery and implant manufacturing features	NA
[4]	A web based OWL (the ontology language) with secondary EHR datasets is used as a framework For implementation authors use Archetype Management System (ArchMS)	• Improves syntactic interoperability	NA

[15]	Authors designed a crisp ontology using a mediator such as DB2OWL or X2OWL. The proposed crisp ontology was then finally converted and implemented as a unified fuzzy ontology	• Heterogeneous datasets with different schemas, standards, terminologies, purposes, locations, and formats can be saved • It supports the idea of plug and play, which allows any EHR system with any structure to be integrated with the existing developed framework without affecting the working environment • Functionality of the proposed ontology can be extendable • The proposed ontology is more dynamic as it is multilingual with deep-learning medical queries	NA
[16]	Authors address the issue of process interoperability. For this, a framework is proposed For evaluating the performance of prototyped framework, a 2-year case study (palliative care information system) is used	• Helped in achieving interoperability, including patient care interoperability, clinical process interoperability, and administrative interoperability	• Time and space complexity increases • Only bound to single study (palliative care) of one system in one city
[17]	A clinical decision support system (CDSS) with OpenEHR archetypes, LIONC terminologies, and archetype query language (AQL) for automated systemic inflammatory response syndrome (SIRS) detection in pediatric intensive care is developed	• Helps in decision-making and analysis for complex task in health care • Early detection can be done easily	NA
[18]	Nationwide evaluating framework is developed The underlying framework consists of two parts: Health Information System (HIS) and PHIP (Patient Health Information Platform), which contain all electronic medical records of every patient • It supports statistical analysis of healthcare data • To deal with population health issues, big data analytics is performed at PHIP levels	• Improves quality of care • Improves decision-making • Improves flexibility at different regions in Sichuan, China	• The lack of incentive given hospital staff to implement a capable HIS is a loophole • Hospitals running outdated technology are also barriers in the way of achieving full semantic interoperability
[19]	Authors reviewed various ontology and data repositories including SNOMED CT, LOINCS, MedDRA, CTCAE, MEDICO, SEMIA, and REDLEX for healthcare	• The reviewed data repositories assist in annotation of produced data, i.e., links between the data and their relations	NA

Continued

Table 2 Literature probe on semantic interoperability—cont'd

Ref. no	Proposed method/approach/technology	Advantages	Drawbacks
[20]	Integrated Management System with fusing sensor to gather multiple diseases data (heart failure, diabetic data) is proposed. It stores these data to the cloud platform Data mining is applied to analyze the stored data	• Data repositories such as CardioSHARE, BIO2RDF, WHOs GHOs RDF, DailyMed, and Linked CT are used to publish health-related data in accordance with semantic web and with the linked datasets • Provides early predictions for critical health conditions	• Security issues in the cloud environment deteriorate the performance of the proposed system • Interoperability issues between the cloud platform and the proposed EHR system hampers the performance of the developed system • Semantic interoperability is seen throughout the process
[21]	Templates using SI model with SNOMED CT, which is ontology-based technique, are designed	• Syntactic interoperability is improved	• The proposed design methodology did not resolve the semantic interoperability
[22]	Authors reviewed various ontology and data repositories such as SNOMED CT, LOINCS, MedDRA, CTCAE, MEDICO, SEMIA, and REDLEX for healthcare	• The reviewed data repositories assist in annotation of produced data, i.e., links between the data and their relations	• Standard vocabulary leads to semantic interoperability
[23]	• Authors highlighted the data quality issue in the medical healthcare • Concept of informatics is applied to gather and analyze the data of patients suffering from malignant hyperthermia	• Improve data quality	• Standard modeling technique to improve the quality of data communication between the EHR systems is required • Semantic interoperability between the EHR systems is not managed by the proposed modeling technique
[24]	• An ontology-based clinical repository framework (OntoCRF) is proposed • Archetype models are implemented to design the repository of the clinical data	• The framework helps to maintain the data storage independent of the content specification • This framework aids in using available knowledge resources and can be linked with other archetype definitions	• Semantic interoperability occurred during storing the data

[25]	• The authors reviewed the role of informatics in four traditional medical systems: **(a)** Traditional Chinese medicine **(b)** Ayurveda **(c)** Islamic medicine **(d)** Traditional Malay medicine	• Various advantages are shared: **(a)** Data quality can be improved	Authors highlight the following issues: **(a)** There is need to harmonize the development process of standardization **(b)** There is a need to integrate national informatics infrastructure with well-established traditional medical systems. So a common standard modeling technique is required for transforming patient's health-related information to stakeholders
[26]	Tag based Q-UEL language is proposed for universal medical data exchange between the medical stakeholders	• It provides patient care between the physicians, specialists, and institutions through the internet • Vast quantity of data from the whole country can be collected and decisions can be made to improve the quality of data exchange between the stakeholders • Q-UEL is basically concerned with both content and management of data representation, including EHR records • It also improves the interoperability and interconversion, security, and disaggregation of the data	NA
[27]	Ontology for capturing semantics of circles of care	• Access-based privacy is employed in order to increase the privacy of patients • It captures patient's circle of care and helps to improve current access control systems to identify illegitimate access through access log	NA
[28]	A natural language-based knowledge representation method for medical diagnosis	• The developed ontology-based approach maps both the natural based user description and descriptive web knowledge into a common structure to increase interoperability	• The con of performance deterioration is still encountered

Continued

Table 2 Literature probe on semantic interoperability—cont'd

Ref. no	Proposed method/approach/technology	Advantages	Drawbacks
		• In addition to this, an automatic standardization (ICD-10 and SNOMED CT systems) method is adopted to normalize user inputs to ontology model • This model works like a human expert who is able to read and understand natural language inputs	
[29]	Designing reliable cohorts of cardiac patients across MIMIC and ICU	• It provides efforts to assess integration of Medical Information Mart for Intensive Care (MIMIC) and Philips eICU, two large-scale anonymized intensive care unit (ICU) databases using standard terminologies • Ontologies such as LOINC, ICD9-CM, and SNOMED CT are implemented to combine detailed care data from multiple ICUs worldwide	• Authors use alternative terms to describe the same concepts • The issue of mapping is incurred in integration of ICUs
[30]	Semantic interoperability in the OR: NET project on networking of medical devices and information system—a requirements analysis	• Several heterogenecus operating devices communicate with each other using embedded message communication techniques to improve semantic interoperability • Merging local healthcare unit to other healthcare unit globally with SOA	• Involves computational complexities in message transformation • Lack of context-based knowledge to map various operating device parameters
[31]	MPM system: A semantic middleware for a unified modeling representation of medical records	• Semantic mediation approach is introduced to solve the problem of heterogeneity and ensure interoperability between different medical systems • Proposed system explores similar EHR and compels to improve the quality of treatment by best shared experiences	• Mapping to conceptual knowledge is a clumsy task
[32]	Exploiting the FIWARE cloud platform to develop a Remote Patient Monitoring System	• New e-health system architecture provides healthcare services remotely (cloud-based framework) • Its approach optimizes the management of medical and paramedical personnel involved in the remote assistance of patients	NA

Ref	Title	Description	Findings/Recommendations
[33]	Ontological-based monitoring system for patients with bipolar I disorder	• It provided a PMS that integrates a clinical decision support system (CDSS) and electronic health record systems (EHRs) to help psychiatrists and primary care physicians tackle and manage bipolar I disorder • Interpretation of the semantic web ontology is used that helps physicians to detect disease at an early stage and helps to identify high-risk patients	NA
[33]	Internet of Things in healthcare: Interoperability and security issues	• REMOA targets home solutions for care/telemonitoring of patients with chronic disorders, developed with IoT sensing technologies • Sensing technologies are connected using wireless networking technologies	• Authors need to improve the standards for the interoperability of devices transferring patient's health related data
[34]	Effectiveness of web-based social sensing in health information dissemination—a review	• Authors highlight various challenging issues in IoT-based health monitoring systems	• There is a need to merge ontological models to have a complex medical healthcare domain • Requires changes in the E-health technology to collect patient-related data at instant of time • Needs to compel software modeling approach to separate data from the application logic
[35]	Choices for interaction with things on internet and underlying issues	NA	• To ensure interoperability, standards must be maintained between communicating applications • Various IoT middleware technologies were able to cope with the interoperability issue and had limited capacity

(i) Researchers must work on integrating the evolutionary process of standardization for health development and they must combine ontological models such as BN, SMILE, and Genie to have a high-quality representation of complex domains within the medical healthcare domain.

(j) Gathering patient data from remote locations is a very complex task.

(k) A modeling approach for separating the data from applications is still missing.

(l) The initial construction of ontology prior to any information sharing is challenging.

7 Future suggestions

(a) The semantic web can be implemented to improve data and to query data stored in huge heterogeneous databases.

(b) The MDA approach with interoperability in IoT-enabled devices in medical healthcare can improve the quality of data for better decision-making and analysis.

(c) To develop an effective EHR system, interoperability testing must be implemented amid the evolutionary process flow.

8 Conclusion

Medical healthcare is one of the most rapidly growing application areas for data analysis as well as data mining, aimed at further improvement in decision-making in health care. IoT-enabled medical healthcare-related devices are performing well in the field of medical healthcare. But semantic interoperability is a crucial requirement in medical healthcare. Semantic interoperability allows more than two EHR systems to communicate with each other in order to share their data, educate each other, use data for mining purposes and, most importantly, to improve the overall quality of healthcare. Furthermore, there is a lack of standards and frameworks to provide full semantic interoperability. We reviewed the articles and chapters from reputable search engine databases in search of quality research. In addition to this, the authors want to stipulate that developing research work by employing MDA including the semantic web to provide semantic interoperability will be a beneficial approach in future.

References

[1] A. Begoyan, An Overview of Interoperability Standards for Electronic Health Records, Society for Design and Process Science, USA, 2007.

[2] W. Hersh, Health care information technology: progress and barriers, JAMA 292 (18) (2004) 2273–2274.

[3] F. Ullah, et al., Semantic interoperability for big-data in heterogeneous IoT infrastructure for healthcare, Sustain. Cities Soc. 34 (2017) 90–96.

[4] M. del Carmen Legaz-García, et al., A semantic web based framework for the interoperability and exploitation of clinical models and EHR data, Knowl. Based Syst. 105 (2016) 175–189.

[5] V. Stroetman, et al., Semantic interoperability for better health and safer healthcare, (2009). pp. 1–34.

[6] H. Kubicek, R. Cimander, Three dimensions of organizational interoperability, Eur. J. ePract. 6 (2009) 1–12.

[7] O.S. Pianykh, Digital Imaging and Communications in Medicine (DICOM): A Practical Introduction and Survival Guide, Springer Science & Business Media, 2009.

[8] T. Benson, G. Grieve, Principles of Health Interoperability: SNOMED CT, HL7 and FHIR, Springer, 2016.

[9] W. Raghupathi, A. Umar, Exploring a model-driven architecture (MDA) approach to health care information systems development, Int. J. Med. Inform. 77 (5) (2008) 305–314.

[10] G.A. Komatsoulis, et al., caCORE version 3: implementation of a model driven, service-oriented architecture for semantic interoperability, J. Biomed. Inform. 41 (1) (2008) 106–123.

[11] R. Rezaei, T. Kian Chiew, S. Peck Lee, An interoperability model for ultra large scale systems, Adv. Eng. Softw. 67 (2014) 22–46.

[12] M.-C. Jaulent, et al., Semantic interoperability challenges to process large amount of data perspectives in forensic and legal medicine, J. Forensic Legal Med. 57 (2018) 19–23.

[13] A. Kiourtis, et al., Aggregating the syntactic and semantic similarity of healthcare data towards their transformation to HL7 FHIR through ontology matching, Int. J. Med. Inform. 132 (2019) 104002.

[14] M.M. Zdravković, et al., Towards semantic interoperability framework for custom orthopaedic implants manufacturing, IFAC Proc. Vol. 45 (6) (2012) 1327–1332.

[15] E. Adel, et al., A unified fuzzy ontology for distributed electronic health record semantic interoperability, in: U-Healthcare Monitoring Systems, Academic Press, 2019, pp. 353–395.

[16] C.E. Kuziemsky, L. Peyton, A framework for understanding process interoperability and health information technology, Health Policy Technol. 5 (2) (2016) 196–203.

[17] A. Wulff, et al., An interoperable clinical decision-support system for early detection of SIRS in pediatric intensive care using openEHR, Artif. Intell. Med. 89 (2018) 10–23.

[18] H. Zhang, B.T. Han, Z. Tang, Constructing a nationwide interoperable health information system in China: the case study of Sichuan Province, Health Policy Technol. 6 (2) (2017) 142–151.

[19] M. Zamfir, et al., Towards a platform for prototyping IoT health monitoring services, in: International Conference on Exploring Services Science, Springer, Cham, 2016.

[20] I.G. Chouvarda, et al., Connected health and integrated care: toward new models for chronic disease management, Maturitas 82 (1) (2015) 22–27.

[21] K.R. Gøeg, R. Cornet, S.K. Andersen, Clustering clinical models from local electronic health records based on semantic similarity, J. Biomed. Inform. 54 (2015) 294–304.

[22] X. Zenuni, et al., State of the art of semantic web for healthcare, Procedia Soc. Behav. Sci. 195 (2015) 1990–1998.

[23] B.G. Denholm, Using informatics to improve the care of patients susceptible to malignant hyperthermia, AORN J. 103 (4) (2016) 364–379.

[24] R. Lozano-Rubí, et al., OntoCR: a CEN/ISO-13606 clinical repository based on ontologies, J. Biomed. Inform. 60 (2016) 224–233.

[25] R.R.R. Ikram, M.K.A. Ghani, N. Abdullah, An analysis of application of health informatics in traditional medicine: a review of four traditional medicine systems, Int. J. Med. Inform. 84 (11) (2015) 988–996.

[26] B. Robson, T.P. Caruso, U.G.J. Balis, Suggestions for a Web based universal exchange and inference language for medicine, Comput. Biol. Med. 43 (12) (2013) 2297–2310.

[27] X. Dong, R. Samavi, T. Topaloglou, COC: an ontology for capturing semantics of circle of care, Proc. Comput. Sci. 63 (2015) 589–594.

[28] P. Scholten, et al., A natural language based knowledge representation method for medical diagnosis, in: 2016 SAI Computing Conference (SAI), IEEE, 2016.

[29] C. Chronaki, A. Shahin, R. Mark, Designing reliable cohorts of cardiac patients across MIMIC and eICU, in: 2015 Computing in Cardiology Conference (CinC), IEEE, 2015.

[30] B. Andersen, et al., Semantic interoperability in the OR. NET project on networking of medical devices and information systems—a requirements analysis, in: IEEE-EMBS International Conference on Biomedical and Health Informatics (BHI), IEEE, 2014.

[31] A. Dridi, A. Tissaoui, S. Sassi, The medical project management (MPM) system, in: 2015 Global Summit on Computer & Information Technology (GSCIT), IEEE, 2015.

[32] M. Fazio, et al., Exploiting the FIWARE cloud platform to develop a remote patient monitoring system, in: 2015 IEEE Symposium on Computers and Communication (ISCC), IEEE, 2015.

[33] C.H. Thermolia, et al., An ontological-based monitoring system for patients with bipolar I disorder, in: 2015 International Conference on Biomedical Engineering and Computational Technologies (SIBIRCON), IEEE, 2015.

[34] L.M.R. Tarouco, et al., Internet of Things in healthcare: interoperatibility and security issues, in: 2012 IEEE International Conference on Communications (ICC), IEEE, 2012.

[35] I. Mashal, et al., Choices for interaction with things on internet and underlying issues, Ad Hoc Netw. 28 (2015) 68–90.

Further reading

Wikipedia's Semantic Interoperability Definition 2017, Available from: https://en.wikipedia.org/wiki/Semantic_inter-operability, 2017. Accessed December 2019.

H. Leslie, International developments in open EHR archetypes and templates, Health Inform. Manage. J. 37 (1) (2008) 38–39.

I. Hwang, Y.-G. Kim, Analysis of security standardization for the Internet of Things, in: 2017 International Conference on Platform Technology and Service (PlatCon), IEEE, 2017.

W. Kuchinke, et al., CDISC standard-based electronic archiving of clinical trials, Methods Inform. Med. 48 (05) (2009) 408–413.

M.A. Brovelli, A. Maurino, ArcheoGIS: an interoperable model for archaeological data, in: International Archives of Photogrammetry and Remote Sensing, vol. 33, 2000, pp. 140–147. B4/1; PART 4.

T. Anzai, et al., Responses to the standard for exchange of nonclinical data (SEND) in non-US countries, J. Toxicol. Pathol. 28 (2) (2015) 57–64.

P. Mildenberger, M. Eichelberg, E. Martin, Introduction to the DICOM standard, Eur. Radiol. 12 (4) (2002) 920–927.

Core, CDISC CDASH, and Domain Teams, Clinical data acquisition standards harmonization (CDASH), Rep No CDASHSTD-10 Austin TX(2008).

D. Bender, K. Sartipi, HL7 FHIR: An Agile and RESTful approach to healthcare information exchange, in: Proceedings of the 26th IEEE International Symposium on Computer-Based Medical Systems, IEEE, 2013.

D. Kalra, T. Beale, S. Heard, The openEHR foundation, Stud. Health Technol. Inform. 115 (2005) 153–173.

IoT for health insurance companies

Vishakha Singh[a], Sandeep S. Udmale[b], Anil Kumar Pandey[c], and Sanjay Kumar Singh[d]

Independent Researcher[a] Department of Computer Engineering and IT, Veermata Jijabai Technological Institute (VJTI), Mumbai, Maharashtra, India[b] Banaras Hindu University (BHU), Varanasi, Uttar Pradesh, India[c] Department of Computer Science and Engineering, Indian Institute of Technology (BHU), Varanasi, Uttar Pradesh, India[d]

1 Introduction

The dawn of the Internet era has brought with it numerous challenges as well as opportunities, which seemed a little far-fetched some time ago. In this fast technologically revolutionizing world, new technologies are getting invented, evolved, and outdated within a matter of a few years. Thus, every enterprise needs to keep pace with state-of-the-art technology to remain relevant in the market. One such evolving technological paradigm is what is popularly called as Internet of things (IoT).

By IoT, we mean a group of internetworked devices working together, and requiring minimal human interference. It has started changing the business models drastically [1]. In 2017, the market of IoT was $235 billion, which is expected to grow up to $520 billion by 2021, as per Ref. [2]. There are three main types of IoT environments, namely, home, enterprise, and government. Out of these, enterprise IoT has the majority of market share [3]. The insurance industry can benefit from this by changing its service model. In this chapter, we focus on the health insurance companies and suggest how IoT can make their business models more serviceable.

Nowadays, people are becoming more and more aware of their health, their predisposition toward certain diseases, and the impoverishment caused due to high medical expenditure [4]. As a result, the market for health insurance is gradually expanding. This has created a huge competition among various health insurance companies, causing them to subsume IoT in their business model, to improve their quality of service. In this chapter, we discuss how IoT is being used to benefit both the insurer and the insured. For example, the biggest shortcoming that the insurers are aiming to overcome is the non-personalized nature of insurance products. This is being resolved by the collection of health-related data from their prospective customers, using IoT. Many insurers have come up with lucrative discounts on their products so that the policy buyers may agree to share their data. Apart from this, IoT is also being used to monitor the health of a policyholder. In this chapter, we discuss the challenges faced by health insurers and the IoT-based solutions employed by them. Some of these challenges are enumerated as follows [5–7]:

- increase the number of customer touchpoints (ways in which customers can get in touch with healthcare providers);
- pursuit of active risk prevention;

IoT Based Data Analytics for the Healthcare Industry. https://doi.org/10.1016/B978-0-12-821472-5.00008-9

- detect fraudulent claims; and
- offer individualized products and services.

A single industry does not need to accomplish all these tasks on its own. Rather, they are a part of a business ecosystem [8], where different industries work together for mutual benefit. For example, a smart wearable device (sensor) developed by some gadget manufacturer can be used to collect and pass the health-related data to an insurance company. This data can be used by that insurance company to determine personalized premiums or design a probabilistic model that determines the susceptibility of an individual toward certain diseases, etc. [5].

Some of the existing researches in this area are mentioned in the following section. After this, various ways to improve the business model of health insurance companies are elaborated in Section 3. The advantages of using IoT are discussed in Section 4. The issues related to the use of IoT, and their possible solutions are mentioned in Sections 5 and 6, respectively. Finally, the scope of future research along with the concluding remarks is in Section 7.

2 Related works

For a past few years, researchers have started taking interest in various aspects of IoT, including their applicability in improving the business model of insurance companies. Some of these aspects are listed as follows.

In [1], Lee et al. have thoroughly explained the five-layered architecture of an enterprise IoT. Vahdati et al. [9] have presented a model that tries to use IoT to personalize the premium rates for the insurers. In [10], Al Thawadi et al. have proposed a health insurance model in which health and lifestyle-related data are collected by the sensors, which is checked by a physician. He then predicts various treatments that might be taken up by an individual in the future. Based on this information, the insurance policy can be adapted. In [11], a proposal is made for using IoT only for the ancillary activities like automatic inspection of an insurance claim, etc. In [12], the concept of smart contracts is proposed in which IoT devices are used to automatically execute an insurance contract, thereby avoiding any delay in the final payment of a claim. Regarding transportation insurance, Li et al. [13] and Lertpunyavuttikul et al. [14] have presented a business model in which the premium rates are designed based on a person's driving behavior (which is determined by using the data collected by the sensors). They claim that this can encourage the drivers to drive safely.

3 IoT-based solutions for health insurance sector

The health insurance value chain consists of various processes such as underwriting, claim assessment, fraud management, billing, payments, etc., as shown in Fig. 1. Traditionally, these processes were done manually, which led to a huge amount of delay in the final payment of a claim. Not only does this create dissatisfaction among the existing policyholders, but it also instills skepticism among the new policy buyers. To fasten these processes, many insurance companies like Bajaj Allianz, ICICI Lombard, HDFC life, etc. have started using IoT [15].

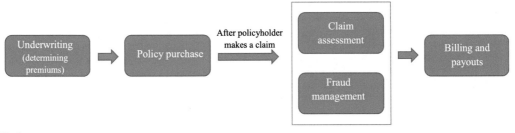

FIG. 1

Health insurance value chain.

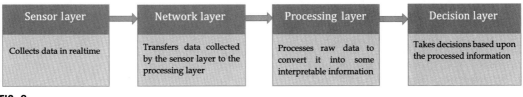

FIG. 2

Four-layered IoT architecture.

Let us have a look at a typical IoT architecture, before going into the details of how IoT has changed the business processes of such companies. The researchers typically use a multilayered architecture. We will be using a four-layered model, which has been briefly discussed as follows [1]:

1. *Sensor layer*: It consists of various devices like wireless/wired sensors, smartphones, and various types of embedded systems, like smartwatches. These devices collect health-related data like sugar level, blood pressure, sleep patterns, body temperature, etc. from their users.
2. *Network layer*: This comprises of various wired/wireless media, which are used to transfer the raw data from sensor layer to processing layer.
3. *Processing layer*: It is responsible for data management and analysis. For this purpose, it uses various tools of data analytics.
4. *Decision layer*: It takes decisions regarding the underwriting process, and claim settlement, based on the processed information. It is also known as the application layer.

Using the architecture shown in Fig. 2, let us discuss some of the IoT-based solutions that can change the traditional business model of a health insurance company.

3.1 IoT and touchpoints

Customer satisfaction is very important for service-based companies and the insurance sector is no exception. IoT can be used to increase the number of ways in which an insured person can get in touch with healthcare providers. Earlier, visiting the doctor/hospital was the only option. But nowadays, many insurers have collaborated with various hospitals as well as independent doctors to provide tele-consultation (using video conferencing, etc.) to its policyholders, in addition to the currently prevalent practices [5].

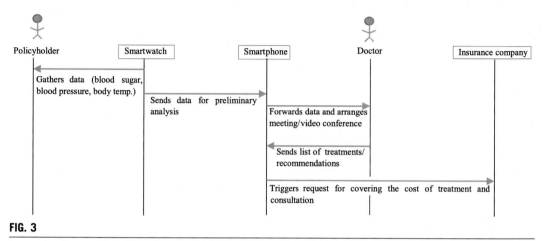

FIG. 3

Increasing the number of touchpoints using IoT.

Many applications based on artificial intelligence are being used for this purpose. They try to label the disease by which a person might be suffering, based on the symptoms specified by him/her and the data collected by the sensors. Thereafter, a doctor is consulted who sends a list of necessary treatments. If the doctor qualifies that the patient is suffering from any particular disease that is covered by the insurance policy, the smartphone application sends a claim request to the insurance company. This not only saves time but also increases customer's comfort as he/she does not need to go through the hassles of visiting the physician, in person. Another benefit of this application-based approach is that it may prevent the policyholder from making unnecessary and fraudulent insurance claims. Such teleconsultation services are already available in some countries like Germany and Britain. This scheme is illustrated in Fig. 3.

3.2 IoT and underwriting

Underwriting is the process of deciding what premium must be paid by the policyholder, periodically, during the term of his/her insurance policy. In a traditional scenario, the premium rate only depends on age, sex, and sometimes the medical history of the prospective policy-buyer. Although these rates were divided into broad categories, they were not individualized, per se. Also, there was a chance of fraud on the part of the policy-buyer. Thus, a need arose to get these rates individualized, as this would help insurance companies [16].

IoT can prove to be very useful in this context. Let us consider a scenario in which the policy-buyer is asked to wear smart devices that collect some health parameters, such as blood pressure, blood sugar level, number of steps taken per day, sleeping patterns, etc. After analyzing this raw input, the processing layer calculates the probability of him/her acquiring certain diseases. Then, based on these probabilities, the premium rate is decided for the given person. This method rationalizes the premium rates. Thus, it would be low enough to lure a person (into buying a policy) who is least likely to make an insurance claim. Also, it would be high enough to discourage a person (from buying a policy) who is most likely to claim the insurance money. This process is explained with the help of Fig. 4.

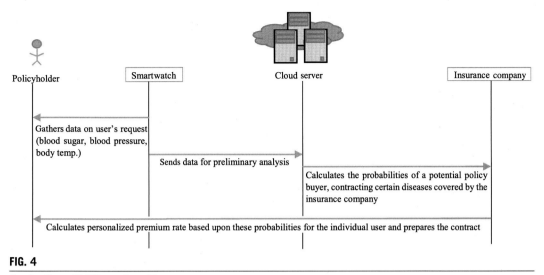

FIG. 4

Application of IoT in underwriting.

3.3 IoT and risk prevention

If the onset of a disease is detected early, the future damage to health can be prevented [5, 7]. But this requires continuously monitoring a person's health. In case he/she shows any symptoms of a certain disease, he/she can be immediately referred to a hospital, where he/she may be given treatment and recommendations (such as diet intake) that may prevent/diminish the impact of the ailment.

In this area, insurance companies can offer smartwatches, blood sugar patches, etc., and monitor the health of its policyholders remotely, by real-time data analysis either on a cloud or a fog platform [17]. If the disease is detected early and treated immediately, the amount of insurance money claimed is far lesser than when the disease gains more intensity. Thus, this technique can be very beneficial for insurance companies. This method has been illustrated in Fig. 5.

3.4 IoT and assisted living

In aging societies, people seek ingenious devices to manage their day-to-day activities, to attain an independent lifestyle. This is especially important for those elderly people whose mobility has diminished due to age-related factors and who have become dependent on others (people or machines) for carrying out even the most basic tasks. Such people have a positive outlook for IoT-related services [18, 19].

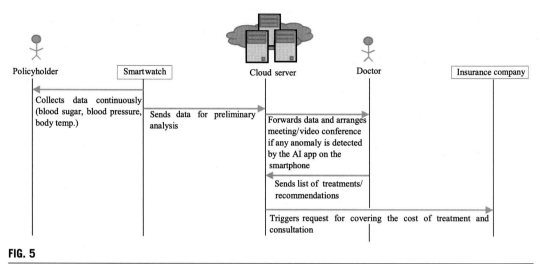

FIG. 5

Risk prevention using IoT.

In this context, once the policyholder loses his/her mobility and self-sufficiency, the insurer ensures the following:

- The provisioning of IoT-based smart devices that help the insured individual in his/her daily routine.
- Arranging the nursing service if and when triggered by the sensors or demanded explicitly by the policyholders [6].

3.5 IoT and fraud detection

If the policyholders allow the insurance companies to monitor their health parameters regularly, then it can help the latter to detect fraudulent claims made by the insured person. For example, consider a man who is getting regular insurance payouts due to his disability to walk properly. Assume that according to his/her insurance policy, he has to wear shoe force sensors every day for a few hours, and after monitoring for some time (days or months), the company finds out that the individual can walk and run like a normal person [20]. In such a situation, the insurance payouts can be immediately reduced or stopped. If such a model is adopted, it will act as a deterrent against any possibility of fraudulent claims made by the policyholders.

4 Advantages of IoT-based solutions

As discussed in the previous sections, IoT-based solutions are meant to benefit both the insurer and the insured. Some perks of using IoT in the insurance processes are mentioned as follows [21]:

- More precise risk assessment (the probability of a policyholder making an insurance claim) using various probabilistic models developed using the collected data.
- Costs can be reduced by actively preventing risks to the health of the policyholders.

- Better management and personalization of premium and other policy rates.
- Acceleration of growth in revenues, due to an increase in customer base and decline in fraudulent, unnecessary, or extravagant claims.
- Discounts and personal offers for the policy buyers (in lieu of sharing the personal health information via IoT sensors).
- Elimination of lengthy claim assessments resulting in faster claim settlement.

5 Issues related to IoT-based solutions

Although IoT has started to make inroads in the health insurance sector, the path ahead is rather bumpy. Its implementation is not as easy as it seems because many aspects should be kept in mind while doing so. The following are some of the problems related to the use of IoT in this field:

1. *Ethical issues*: The collection of health-related data with the help of smart devices creates a transgression in the privacy of the associated person. A person may not be comfortable in sharing his/her personal information with the insurance company. Also, the laws of some countries create a restrictive atmosphere for such companies that deal with their citizens' personal information, for example, the General Data Protection Regulation (GDPR) of the European Union (EU).
2. *Data security issues*: Even if the users agree to share their personal information with the company, there is an issue of a possible information leak/breach. Personal and sensitive data may attract the attention of maleficent hackers. If they get hold of such data, they might blackmail an innocent policyholder, especially if he is suffering from some socially stigmatized disease. Nowadays, many countries are considering to impose huge penalties on the corporates, in the event of a data leak/ breach (like the GDPR regime) [22].
3. *Possible deception on the user's end*: There is a chance that the user may not wear the sensors/ devices that are meant to collect his/her data. By doing so, the user may evade the monitoring mechanism of the insurance company and make many deceptive and unnecessary claims.
4. *Faulty sensors/hardware defects*: In some cases, due to some erroneous component(s), the sensor/ device may give incorrect readings, which may lead to wrong policy decisions. Although such cases would be less in number, they have the potential of diminishing the trust of policyholders.
5. *Dealing with the data volume*: We need enormous storage and analysis infrastructure for dealing with the huge volumes of data generated every day. The capabilities that we have today are a bit insufficient for this purpose.

6 Dealing with the issues related to IoT-based solutions

Most of the problems that occur while implementing IoT-based solutions can be rectified either partially or wholly. Some of these remedial steps are stated as follows:

1. *Informed consent*: The insurance companies should be transparent about the collection and monitoring of data of the policy-buyers. A transparency clause for data collection should be included in the contract itself. Ideally, if an individual does not wish to part away with his health-related information, he/she must not be forced to do so. Rather some other way to attain his/her approval can be thought of, for example, offering lucrative discounts.

2. *Ensure data safety and security*: The database where the policyholders' data is stored should be very secure. Besides, multiple layers of encryption should be employed on this data, so that the hackers may not be able to get hold of any information. Particularly, the security layer must be capable of resisting the ransom-based attacks.
3. *Prevent chances of user deception*: Preventing trickery on the part of the user is especially difficult when the sensors or wearables are easily removable. There must be a system to monitor whether the user is wearing the device or not. There is no one solution to this problem. But generally, biometrics may be used to ensure that only the designated user wears the device.
4. *Thorough hardware testing*: The sensors should be fail-safe and reliable. They should be passed to the policy-buyer only after rigorous testing. The insurer should also offer free of cost and quick maintenance and replacement to the insured.

7 Conclusion

IoT has the potential of bringing a revolution in the global economy. It is just a matter of time before the market for IoT will be skyrocketing. Thus, it is beneficial for the health insurance sector to utilize the research that is being done in this field. This chapter specifically dealt with the health insurance companies and their opportunities and dilemmas concerning the use of IoT. As discussed, health insurance policies can be individualized. Also, some processes like preparation of the contract, acceptance, and settlement of a claim, etc. can be automatized using IoT.

In the future, it would be interesting to see how the business models of health insurers change, following the invention of some compact and affordable devices that can analyze the urine, blood, sputum, and stool samples of a person and diagnose the diseases contracted by him/her [23]. This may expedite the process of disease diagnosis and also reduce the burden of unnecessary claims made against the insurance companies. The insurers must also keep updating their products and services, with the development of sophisticated smartphone-based applications, which can diagnose certain diseases, all by themselves, for example, Barnes et al. [24] have developed an application, which can diagnose urinary sepsis. Thus, the insurers need to be aware of the changing technology and keep updating their services accordingly.

References

[1] I. Lee, The Internet of things for enterprises: an ecosystem, architecture, and IoT service business model, Internet of Things 7 (2019) 100078. https://doi.org/10.1016/j.iot.2019.100078.
[2] A. Bosche, D. Crawford, D. Jackson, M. Schallehn, C. Schorling, Unlocking opportunities in the Internet of things, (2018) Available from: https://www.bain.com/insights/unlocking-opportunities-in-the-internet-of-things.
[3] J. Greenough, The corporate 'Internet of things' will encompass more devices than the smartphone and tablet markets combined, 2015, Available from: https://www.businessinsider.com/the-enterprise-internet-of-things-market-2014-12.
[4] R. Sengupta, D. Rooj, The effect of health insurance on hospitalization: identification of adverse selection, moral hazard and the vulnerable population in the Indian healthcare market, World Dev. 122 (2019) 110–129.

[5] F. Graillort, Digital Health: How Technology Could Reshape Health Insurance, 2018, Available from: https://coverager.com/digital-health-technology-reshape-health-insurance/.

[6] S. Behm, U. Deetjen, S. Kaniyar, N. Methner, B. Munstermann, Digital Ecosystems for insurers: opportunities through the Internet of things, 2019, Available from: https://www.mckinsey.com/industries/financial-services/our-insights/digital-ecosystems%20for%20insurers%20opportunities%20through%20the%20internet%20of%20things.

[7] R.P. Ellis, W.G. Manning, Optimal health insurance for prevention and treatment, J. Health Econ. 26 (6) (2007) 1128–1150.

[8] J.F. Moore, Predators and prey: a new ecology of competition, Harvard Business Rev. 71 (3) (1993) 75–86.

[9] M. Vahdati, K.G. Hamlabadi, A.M. Saghiri, H. Rashidi, A self-organized framework for insurance based on Internet of things and blockchain, in: 2018 IEEE 6th International Conference on Future Internet of Things and Cloud (FiCloud), IEEE, 2018, pp. 169–175.

[10] M. Al Thawadi, F. Sallabi, M. Awad, K. Shuaib, Disruptive IoT-based healthcare insurance business model, in: 2019 IEEE International Conference on Computational Science and Engineering (CSE) and IEEE International Conference on Embedded and Ubiquitous Computing (EUC), IEEE, 2019, pp. 397–403.

[11] L. Bader, J.C. Bürger, R. Matzutt, K. Wehrle, Smart contract-based car insurance policies, in: 2018 IEEE Globecom Workshops (GC Wkshps), IEEE, 2018, pp. 1–7.

[12] O. Mahmoud, H. Kopp, A.T. Abdelhamid, F. Kargl, Applications of smart-contracts: anonymous decentralized insurances with IoT sensors, in: 2018 IEEE International Conference on Internet of Things (iThings) and IEEE Green Computing and Communications (GreenCom) and IEEE Cyber, Physical and Social Computing (CPSCom) and IEEE Smart Data (SmartData), IEEE, 2018, pp. 1507–1512.

[13] Z. Li, Z. Xiao, Q. Xu, E. Sotthiwat, R.S.M. Goh, X. Liang, Blockchain and IoT data analytics for fine-grained transportation insurance, in: 2018 IEEE 24th International Conference on Parallel and Distributed Systems (ICPADS), IEEE, 2018, pp. 1022–1027.

[14] P. Lertpunyavuttikul, P. Chuenprasertsuk, S. Glomglome, Usage-based insurance using IoT platform, in: 2017 21st International Computer Science and Engineering Conference (ICSEC), IEEE, 2017, pp. 1–5.

[15] M. Mitra, IOT in insurance sector: home, auto and health insurance, Available from: https://www.mantralabsglobal.com/blog/iot-in-insurance-sector/.

[16] G. Hoermann, J. Ruß, Enhanced annuities and the impact of individual underwriting on an insurer's profit situation, Insur. Math. Econ. 43 (1) (2008) 150–157.

[17] T. Saheb, L. Izadi, Paradigm of IoT big data analytics in healthcare industry: a review of scientific literature and mapping of research trends, Telemat. Inf. 41 (2019) 70–85.

[18] L. Syed, S. Jabeen, S. Manimala, A. Alsaeedi, Smart healthcare framework for ambient assisted living using IoMT and big data analytics techniques, Futur. Gener. Comput. Syst. 101 (2019) 136–151.

[19] J. Offermann-van Heek, E.M. Schomakers, M. Ziefle, Bare necessities? How the need for care modulates the acceptance of ambient assisted living technologies, Int. J. Med. Inf. 127 (2019) 147–156.

[20] K. Corcoran, A woman lost her injury payout after she was caught posting runs on fitness app Endomondo, 2018, Available from: https://www.businessinsider.in/tech/a-woman-lost-her-injury-payout-after-she-was-caught-posting-runs-on-fitness-app-endomondo/articleshow/63331392.cms.

[21] N. Sakovich, How will IoT transform the insurance industry? 2018, Available from: https://www.sam-solutions.com/blog/how-will-iot-transform-the-insurance-industry/.

[22] A. Satariano, Google is fined $57 million under Europe's Data Privacy Law, 2019, Available from: https://www.nytimes.com/2019/01/21/technology/google-europe-gdpr-fine.html.

[23] K. Baggaley, Here's how smart toilets of the future could protect your health, 2019, Available from: https://www.nbcnews.com/mach/science/here-s-how-smart-toilets-future-could-protect-your-health-ncna961656.

[24] L. Barnes, D.M. Heithoff, S.P. Mahan, G.N. Fox, A. Zambrano, J. Choe, L.N. Fitzgibbons, J.D. Marth, J.C. Fried, H.T. Soh, et al., Smartphone-based pathogen diagnosis in urinary sepsis patients, EBioMedicine 36 (2018) 73–82.

Security and privacy challenges in healthcare using Internet of Things

10

Righa Tandon and P.K. Gupta

Department of Computer Science and Engineering, Jaypee University of Information Technology, Solan, Himachal Pradesh, India

1 Introduction

Healthcare is the domain in which the use of IoT has completely revolutionized the health scenario. By using IoT, one can keep the health records online and monitor the progress of the health of the desired one from any location. Therefore IoT is associated with everyone and can easily monitor the growth in population, rural urbanization, birth rate declination, economic growth, and unbalanced resource utilization. Making the use of IoT in healthcare systems has completely changed the healthcare field by providing many benefits to patients and doctors. By using various sensor-based devices in IoT for medical purposes, the healthcare industry has reached to a new horizon. However, with the increasing use of these devices, various concerns related to security are growing that in turn impose threats on the business model. Smart devices are connected to the global network for accessing information anytime and anywhere; therefore IoT-based healthcare systems can be a target for any assailant. For facilitating the endorsement of the IoT in the healthcare field, it is difficult to determine and classify different appearance of IoT security and privacy. In this chapter, we have focused on the various requirements related to security and privacy of information like confidentiality, integrity, authentication, availability, data freshness, data maintenance, nonrepudiation, interoperability, privacy, fault tolerance, and secure booting [1]. In an IoT environment, data security, data storage, securely communication, and controlling access have become serious challenges [2]. With the growth of IoT services in healthcare, many vulnerable nodes are deployed, and also, security architectures are of no use [3]. IoT in the healthcare domain is further divided into two main areas known as hospitals and private clinics. In both areas, IoT plays an important role by providing interconnection among patients, physicians, hospitals, and different cities. For collecting the data like blood pressure, heart rate, and sugar level, various IoT-enabled wearable devices play a significant role, and further, these data can be monitored to change of information. Using IoT in healthcare, security cameras are there for monitoring and capturing any invader.

Recently the use of sensor-based devices in hospitals and clinics has become common to keep the electronic records of patient's like personal details, patient's medical history, family medical details, and other diagnosis details of patients. All the details are filled and stored using electronic means. This information about a particular patient will be stored and kept confidential according to the security policies. Patients can access their reports online as well. In the healthcare field, patient's records are made available online for fast and easy access. While sharing data online with the patients, security can be the challenge. IoT is used these days in healthcare for making healthcare records available online. Using IoT the particular patient will get access to securely view their personal health records and get updated information too. Using IoT in healthcare can provide improved care to patients. If a patient met an accident and his family history and information about medical health are already present in the records using IoT, he will be treated quickly by getting his healthcare details. So, if the patient's healthcare record is present in electronic form using IoT, it means she/he will get treatment faster without any loss of time. Also, more than one doctor can have access to view the patient's record and discuss their current healthcare issue among themselves and treat them better. Hence, IoT in healthcare plays a very important part. In healthcare using IoT the repetition of healthcare records can be or major concern. So, healthcare records of the patients need to be updated on regular time interval. This will help in saving of time and effort of particular patient and doctor. If we are using IoT in healthcare, it will help in managing appointments with doctors and medication facilities and improving billing system, and human errors can also be reduced.

Without using complex integration technology, IoT makes use of different technologies in healthcare. IoT in healthcare helps in processing, analyzing, and manipulating the medical data effectively. To forecast the medical condition of the patient, IoT makes use of machine learning and big data analytic services [4]. IoT in healthcare helps in monitoring present and future health conditions of a patient. Patients can easily approach the concerned doctors without any loss of time because the patient's health status details will already be monitored by both doctor and patient [5]. Online assistance is available all the time where healthcare resources are managed easily and one can easily access the health specialists anytime. Various health professionals are also connected across the world using international collaboration [6]. There are many frameworks that have been proposed for healthcare based on IoT. The general framework is shown in Fig. 1. For different diseases, there exist different frameworks like for asthma management, a framework is proposed with different parties such as users or patients, doctors, medical service providers, and cloud server on which information is stored. All the activities of the patient like the respiratory flow of the patient and current activity level are recorded by using IoT sensors. Many machine learning models have been used for analyzing the current status of the patient, and any disbalance in the normal functioning will alarm for the quick medical intensive care [7].

For enabling different devices to work at the same time and work in collaboration, a gateway architecture-based framework is proposed for the healthcare system using IoT. Different tiers are used for healthcare system IoT with ubiquitous computing. This framework mainly consists of body area network, medical server, and hospital system. Sensors help in analyzing and monitoring the patient's record. Based on the monitoring, diagnosis is done, and results are updated accordingly. The cloud server is also used in this framework for storing the updated data [8]. A framework for chikungunya disease is proposed using fog computing and IoT. In this framework the data accumulation layer is used that helps in collecting data related to the initial stages of the chikungunya virus. Sensor devices are encapsulated into the patient's body for monitoring symptoms of this disease. Also, biosensors are present that help in monitoring maximum and minimum temperature, air quality, and other environmental conditions [9]. Another framework design is in such a way that it can handle huge data and lowers the latency. Health-related data and their processing are distributed between IoT devices and cloud resources. By distributing the load of data, power consumption will be lower. This helps in enhancing the processing speed of the data and increasing the overall performance of the healthcare-based IoT system. Wearable devices and smartphones are used on which the patient is able to get the updated

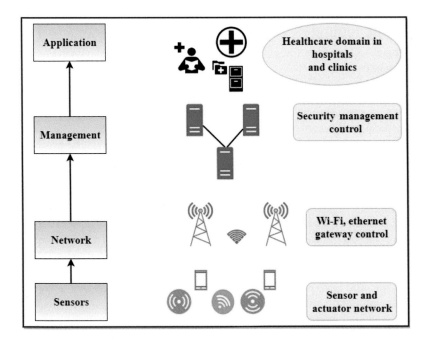

FIG. 1

IoT in healthcare.

notifications regarding the medical data of the patient [10]. One more existing framework in which the collaborative model is used includes the computing capabilities of the IoT environment. With the help of this framework, complex patterns of different diseases can also be captured. Also, other health-related information like swelling in the body, pain, itching, and rashes on the body will also be monitoring even if sensors are not functioning properly. For predicting complex pattern-related information, some machine learning techniques are used [11]. This framework mainly comprises doctors, patients, and database, which will be distributed. Health-related information will be sent to patients using wireless body area networks. Any patient can consult his doctor at any time and from anywhere [12–16].

IoT in the healthcare domain is monitoring and capturing the patient's data from several different places. Information is transferred through gateways, which will lead to leakage of data. For overcoming this problem a secure network should be used. IoT has reduced computation costs in the healthcare domain and also helped in processing the information related to healthcare in less time. In the framework, patient's information related to healthcare records like temperature of the human body and heart rate respiration rate is monitored using Raspberry Pi [17]. Ubiquitous sensor networks and wireless sensor networks are integrated with each other for sending the processed data to their destinations using a secure network. The integration of both ubiquitous sensor networks and wireless sensor networks helps in making a strong network through which we can send healthcare data to the patient's destination. This also helps with reliability. Gateways that are used help in analyzing data collected by sensors, and after processing, data are sent to cloud for storage [14,18–22].

2 IoT-based technologies for healthcare

Various technologies that can be enforced on healthcare using the IoT are shown in Fig. 2 and discussed in the succeeding text:

a. *Augmented reality*: This technology has brought a powerful change in healthcare using IoT. Using this technology, doctors can get assistance during different surgeries. Augmented reality helps in enhancing medical training, and also in upcoming years, this technology will have a bigger brunt in the healthcare field using IoT. With the help of augmented reality, existing procedures and mechanisms in the medical field can be contrived more adequate and decisive [1].

b. *Ambient intelligence*: Ambient intelligence is a vital part of the healthcare field using IoT. One of its areas is human-computer interaction as in healthcare field patients are humans [1,6].

c. *Cloud computing*: Cloud computing technology when incorporated with the healthcare field using IoT will provide resource-sharing facilities and also provide various services based on the requests. This technology also assassinates different processes and operations to meet the requirements of the patient [1,23].

d. *Grid computing*: When grid computing is enforced with IoT in healthcare, it will help in uninterrupted monitoring of patient's health status using sensors. Wearable devices are there that will monitor the saturation level of oxygen in patients, blood pressure of patients, ECG, EEG, etc. For monitoring and communication purposes, grid computing makes use of low-power consumption technology. Hence, this technology is more effective when in use [1,24].

e. *Big data*: Big data helps in enhancing the medical diagnosis and also helps in augmenting methods and procedures of the healthcare field. Healthcare data produced by sensors are first analyzed using different tools [1,6].

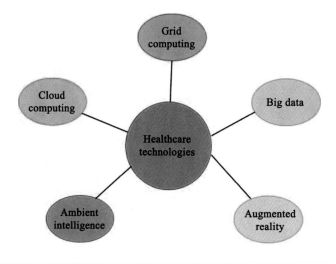

FIG. 2

Healthcare technologies.

3 Security and privacy challenges in healthcare

By using various sensor-based devices in IoT for medical purposes, the healthcare industry has reached to horizon. However, with the increasing use of these devices, various concerns related to security are growing, which in turn impose threats on the business model. Smart devices are connected to the global network for accessing information anytime and anywhere. So, IoT-based healthcare systems can be a target for any assailant. For facilitating the endorsement of the IoT in the healthcare field, it is difficult to determine and classify different appearance of IoT security and privacy. Various requirements that are related to security and privacy are confidentiality, integrity, authentication, availability, data freshness, data maintenance, nonrepudiation, interoperability, privacy, fault tolerance, and secure booting [6]. Fig. 3 represents the various security and privacy challenges in healthcare using IoT.

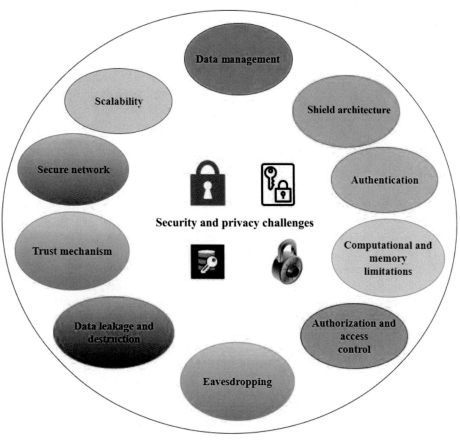

FIG. 3

Security and privacy challenges in healthcare using IoT.

- **Security and privacy challenges**: Sensitive data that are collected and analyzed by the sensors have contributed security large-scale challenges in the IoT environment.

  a. *Data management*: Medical data of patients can be captured using sensors in the IoT environment. Physical health data of different patients vary as the body of the patient is dynamic that will switch its state frequently. So, monitoring the data and managing the data securely is a major challenge [25]. Depending on the medical application we are using, the format of data may vary. Diseases related to skin are captured in image format; blood pressure and ECG are captured in other formats [26]. Handling different formats of data is quite difficult.

  b. *Shield architecture*: IoT architecture should be secure enough that there will be an end-to-end communication between resources that are used in the healthcare domain. In healthcare-based IoT architecture, secure communication between IoT devices should be there [27]. Healthcare-based IoT architectures should focus on software-defined networks (SDN) used by IoT devices. SDN security issues will also be considered for any secure IoT architecture. Disclosure of malignant traffic can be a challenging task for software-defined networks. Furthermore, intricacy involved in connecting the IoT network with cloud infrastructure would come out with security challenges [2,28].

  c. *Authentication*: In the healthcare IoT domain, authentication of data that are being transferred takes place. Also the original node from where data are sent is authenticated. Key distribution and key management can be a challenge for authenticating IoT devices. The authentication process also ensures that there should not be any large-scale overhead on IoT devices during key generation and exchange. Furthermore the authentication process assures key transfer integrity and also validates the keys when certificate authority is not there [2,29,30].

  d. *Computational and memory limitations*: Processors of low speed are encapsulated in healthcare-based IoT devices. The processing speed of these devices is not that powerful. Sensors and actuators are present in IoT devices that are not capable of handling computational costly operations. Hence a solution is required that will lower the consumption of resources and increases the performance of security and finding that solution can be a challenging task. Also, healthcare-based IoT devices are having lesser memory capacity. The fixed operating system is encapsulated in these devices, and their memory is not able to perform complex security protocols. That is another major challenging task [1, 31–33].

  e. *Authorization and access control*: In the authorization process, only authorized resources will get access rights [34]. For access to right verification, there are only a few mechanisms associated with IoT device nodes [35]. Hence, distributing and managing authorization mechanism and access control mechanism to various nodes that are present at IoT devices can be a challenging task for the IoT network [2,36].

  f. *Eavesdropping*: This can put patient's health-related data at risk. While communication, eavesdropping can affect the privacy of patient's data by detecting the data during transmission by sensors of IoT devices. If the data are not protected while sending it to other nodes through sensors, then an intruder can eavesdrop the patient's data using a sniffing tool. Hence, protecting data from eavesdropping is also a challenging task [37].

  g. *Data leakage and destruction*: This includes leaking and exposing confidential data like health-related records and personal details of patients to external users. This unintentional leakage is unauthorized. Protecting data from getting leaked and destroyed is a major challenge for the IoT environment [37].

h. *Trust mechanism*: While communication and transferring data between nodes of IoT devices, trust mechanisms should be applied. Also, intervention language and integrity mainframe systems must be there [1].

i. *Secure network*: The main aim of the IoT network is to send sensitive data to other nodes. IoT networks mostly use wireless links to send data. The availability of trusted network resources and secure wireless networks is a security challenge. Many possible attacks on the network are DoS attack, spoofing attack, routing information attack, etc. Hence, paying attention to these attacks can also be challenging [38].

j. *Scalability:* The number of devices connected to IoT is increasing in number and is gradually connected to the IoT network. In an IoT environment, it is necessary to increase sensors and IoT devices on a large scale to save the waiting time for processing and analyzing the data. Scalability can enhance efficiency, saves time, and also builds patient's trust in the IoT network. Hence a scalable security scheme must be designed without conciliating various security requirements and can become a major challenging task [1, 25].

The Internet of Things helped a lot in the healthcare domain. But there are few security challenges faced at each IoT layer using the healthcare domain [39]. Some security challenges at IoT layers are discussed in the succeeding text and also shown in Fig. 4.

i. *Perception layer*: This layer is also known as the sensing layer. Many sensors are present at this layer. Sensing technologies are also used like radio frequency identification (RFID) and wireless sensor networks (WSN).

- Security of objects: Microchips are used in radio frequency identification for wireless communication. These microchips are encapsulated into the objects for identification. An attacker can easily attack to these objects and can harm healthcare-related data. Hence the security of these objects is another challenge that needs to be considered [40].
- Data confidentiality: In the perception layer, wireless sensor networks are used in which sensor nodes play an important role in sending data to the data processing layer. So, health-related data that are sent to the processing layer should be kept secret, and integrity should also be maintained. Data confidentiality is a major challenge here [41].

ii. *Network layer*: In this layer, transferring healthcare data to the correct destination using secure networks takes place. Here, network technology like Wi-Fi and Bluetooth is used.

- Aggregating information: Sensor nodes are present at the network layer that helps in collecting healthcare data that should be a reliable one. Aggregating the collected data and security of that data is a big challenge [42].
- Protection of networks: After aggregating the data related to healthcare, that data will be sent to the destination by sensor nodes using particular network topology. Hence, protecting the network from any kind of attacker or intruder is an issue [43]

iii. *Data processing layer*: This layer helps in providing services like processed information of healthcare data and provides scalable data and abstraction. User authentication takes place here [39].

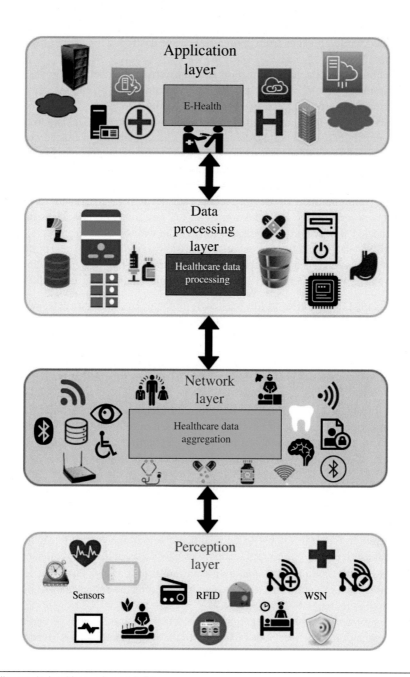

FIG. 4

Security challenges in healthcare-based IoT layers.

- Storing huge data: In this layer a large amount of data is processed and sent to further destinations. So, processing huge data can be the issue. Managing large data for processing can also be the other issue [39].
- Data anticipation: It allows users to view their information related to their healthcare field. It helps in interacting with the environment also. Providing all the information at the same time may lead to overhead and traffic in processing and updating the information, So, only relevant information should be given and provided at a particular time [39].

iv. *Application layer*: This layer is responsible for providing services to users. Here, we are talking about healthcare, so this layer will provide healthcare information to the patients. Patients can view their updated information related to their diseases.

- Authentication: Patient's authentication should take place for verifying the user's identity. Only authorized users should get access to these services such as viewing information, paying hospital bills, and using hospital facilities. Hence, more secure authentication procedures should be followed for giving access to the information and services of the users [44].

Hence, securing healthcare-related information from attackers and conflicting parties is a big challenge. Data integrity should be maintained throughout the whole process starting from collecting data to sending it to the destinations. For securing the healthcare data, it is necessary that the integrity and confidentiality of the data should be maintained. Encrypting the data is one of the ways to secure healthcare data from any intruder. Many encryption schemes are available for securing healthcare data. An encryption scheme can also affect the security mechanism of the data. So, that encryption scheme should be selected for encrypting the data that would not affect data integrity and confidentiality. Advanced encryption standard (AES) with a key length of 256 is one of the encryption schemes that is used for encrypting data and securing it from any unauthorized access and attack. One more encryption scheme that is very cost-effective is a homomorphic scheme that uses 1024 bits security key [7].

4 IoT-based healthcare applications

The following are the applications of healthcare using IoT (Fig. 5):

(a) *Diabetes prevention*: For preventing diabetes in any patient, IoT can be used in the healthcare domain. There is a term called diabetes mellitus that explains any kind of disorder in the glucose level of the patient. That disorder is caused by the disarrangement of fat, carbohydrate, and protein subsistence that will affect insulin action and secretion. IoT in healthcare can monitor the blood glucose level of the human body that will prohibit the risks caused by diabetes mellitus disorder. Also, it will help in planning the regular meals, activities, and medication time of each patient [1].

(b) *Electrocardiogram monitoring (ECG)*: Electrocardiography records the electrical activity of the patient's heart. It is important to monitor heart signals, heart rate, and heart rhythm so as to know whether the heart of a person is functioning properly or not. So, for monitoring those signals of heart, ECG is used. IoT-based applications in the healthcare domain have the potential to get the maximum information regarding the patient's heart and give that important information to respective patients and medical staff [1,45]. ECG in real-time applications is also monitored using electrodes when they are placed on the case of smartphones. All the processes of collecting ECG signals and processing them is done in real time [46].

(c) *Blood pressure monitoring*: Monitoring the blood pressure of a person is equally important so as to avoid problems related to circulatory systems. Using IoT in healthcare helps in monitoring blood pressure of patients timely that will prevent any kind of circulatory-related problems in a patient's body [1,6].

(d) *Body temperature monitoring*: Body temperature is a significant sign that helps us to know whether the whole body is adapting the environment temperature and conditions or not. Hence, measuring the body temperature for knowing the proper body functioning is also important [1]. IoT in healthcare uses body temperature sensors that will measure various variations in body temperature of a patient and the report of the same to the body temperature measurement system [6].

(e) *Oxygen saturation monitoring*: To measure the oxygen level of the patient, the pulse oximetry device is used. IoT with the combination of this device pulse oximetry monitors oxygen saturation level. This integration with IoT gives better results [1,47].

(f) *Rehabilitation system*: Medical issues like aging and issues caused due to cerebrovascular diseases are very common these days. Patients suffering from these diseases have to go to

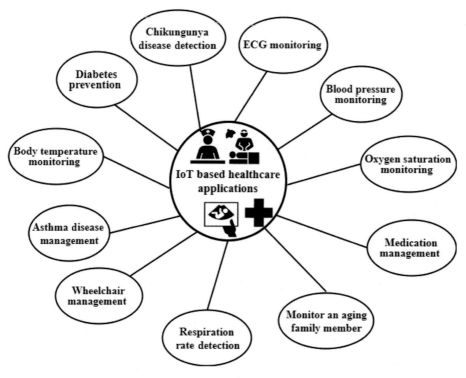

FIG. 5

IoT-based healthcare applications.

rehabilitation centers or clinics [1]. But in IoT a method is designed called ontology-based automation method that helps in blunting the complications caused by the earlier discussed medical issues [48].

(g) *Medication management*: For managing the medication burden of any patient, IoT in healthcare makes use of a tool that will help in managing all the medication of the different patients without making it complicated [6].

(h) *Wheelchair management*: For differently abled patients, fully automated smart wheelchair facilities are available using IoT in the healthcare domain. This will help the patient in doing their work without any interruption [6].

(i) *Asthma disease management*: For managing this disease, IoT sensors are there that help in monitoring patient's respiratory flow. Sensors are also responsible for monitoring the current status of the patient and can also report to the doctor for any misfunctioning in the normal activities of the patient. This will help in the fast medical treatment for that particular patient [7].

(j) *Chikungunya disease detection*: This disease can also be detected at the initial stage and can be controlled using biosensors and body sensors that are encapsulated into the patient body. Monitoring the symptoms of this disease, controlling the disease, and giving the correct treatment to the patient are done [9].

(k) *Respiration rate detection:* Respiration rate of patients will be monitored and updated to the database. Updated information on the patient's reports and disease records will be given to the patient. Also the time to time respiration rate will be monitored, processed, and stored [49].

(l) *Monitor an aging family member:* In the healthcare field, sensors are used for finding and tracking the activities of senior residents. Many systems are designed for monitoring the activities of the seniors at home only. Useful information is only sent using wireless networks [14].

The comparative chart for security challenges and various applications using IoT is shown in Table 1.

IoT devices analyze the healthcare application data for simplifying it. Hence the security of healthcare-related applications is important [50,51]. Various threats by security obligations are as follows:

➤ Energy escalation: In IoT, sensors are present that help in analyzing healthcare data. Wearable devices are also there in IoT healthcare systems. The main problem in wearable devices is the consumption of energy [52].

➤ Physical attacks: One of the serious threats is a physical attack. IoT devices collect information about patient's health from a vulnerable environment. IoT devices can easily be manipulated, and any attacker easily changes the information that is sent by IoT devices [52].

➤ Manipulating data: IoT devices receive and store data of healthcare applications [53]. Data manipulation and data changing can be a major threat for IoT applications [54–57]. Any attacker can damage the data that could lead to wrong treatment of a patient by doctors [58–60].

   • **Case study**: IoT plays an important role in the healthcare system by reducing the time required for the patient's care in various hospitals and clinics. IoT also helps in improvising real-time monitoring and tracking of medical equipment and staff. For example, an old man of

Table 1 Comparative chart for security challenges and various applications using IoT.

Security challenges/applications	Diabetes prevention	Electrocardiogram monitoring	Blood pressure monitoring	Body temperature monitoring	Oxygen saturation monitoring	Rehabilitation system	Medication management	Wheelchair management
Authentication [1]	Yes	Yes	Yes	Yes	Yes	Yes	Yes	Yes
Authorization and access control [4]	Yes	Yes	Yes	Yes	Yes	Yes	Yes	Yes
Data leakage and destruction [1]	No	No	No	No	No	No	No	No
Trust mechanism [6]	Yes	Yes	Yes	Yes	Yes	Yes	Yes	Yes
Secure network [8]	Yes	Yes	Yes	Yes	Yes	Yes	Yes	Yes
Computational and memory limitations [1]	No	No	No	No	No	No	No	No

65 years is suffering from blood pressure problem, and for checking the blood pressure of the body, he has to go to the hospital on a regular basis. This will consume much time and efforts. Also, this may lead to leakage of his medical reports and personal information as there are many hospitals where there is a manual process of registration till now. Nowadays, there are many applications available related to healthcare using IoT that provides a secure network for storing, transferring, and managing the patient's information. In terms of security, authentication and authorization of the patient are required. An old man if uses this application, this will help in reducing the time required for a checkup and also helps in monitoring his blood pressure day to day. As the network used in healthcare using IoT is secure, this will avoid the data leakage of patient's information. This will also help in alerting doctors and other staff members in case of any emergency. As shown in Fig. 6, the patient is sending a request for healthcare services using IoT. IoT application will connect to the doctor using a wireless connection. The request will securely be received by the doctors in the hospital. The doctor will start monitoring the blood pressure of the patient as per the request. All the information such as patient's personal details and lab reports will be securely uploaded to the database. The whole process is very secure, and there are no chances of information leakage and destruction. Hence, using IoT in healthcare results in efficient and secure data collection, processing, and storing in very less time.

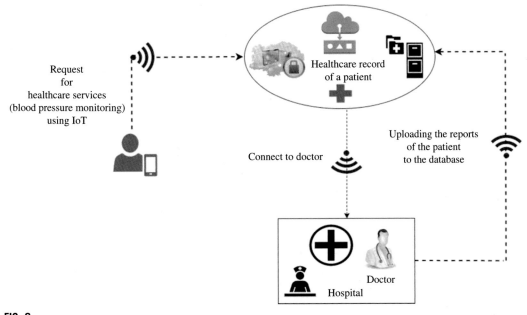

FIG. 6

Blood pressure monitoring of a patient using IoT.

5 Conclusion

The Internet of Things is used to enhance the healthcare field. The collection of data and storing of data take place using IoT devices. IoT in healthcare has made many things easy, but the security of data is important and is a matter of concern these days. To keep this thing in mind, many security and privacy issues are discussed in this paper. Many threats related to data are also discussed. Information about patients is sensitive, so it should be kept confidential. Health-related information can be damaged due to any attack. Any attacker can harm the person or healthcare data of the patient. So, some security solutions and mechanisms should be there for handling the various security-related issues of healthcare data using IoT. In future work, we will try to analyze various solutions related to security and privacy challenges in healthcare using IoT and also will try to implement those solutions for providing security to the healthcare-related applications using IoT.

References

[1] G.J.F. Carnaz, V. Nogueira, An overview of IoT and healthcare, in: S. Abreu, V.B. Nogueira (Eds.), Actas das 6as Jornadas de Informática de Universidade de Évora, Escola de Ciências e Tecnologia da Universidade de Évora, Évora, 2016 Retrieved from http://hdl.handle.net/10174/19998.

[2] M. Conti, A. Dehghantanha, K. Franke, S. Watson, Internet of Things security and forensics: challenges and opportunities, Futur. Gener. Comput. Syst. (2018) 544–546.

[3] A. Giaretta, S. Balasubramaniam, M. Conti, Security vulnerabilities and countermeasures for target localization in bio-nano things communication networks, IEEE Trans. Inf. Forensics Secur. 11 (4) (2016) 665–676.

[4] D. Xu, W.H. Li, S. Li, Internet of things in industries: a survey, IEEE Trans. Ind. Inform. 10 (4) (2014) 2233–2243.

[5] J. Gubbi, R. Buyya, S. Marusic, M. Palaniswami, Internet of Things (IoT): a vision, architectural elements, and future directions, Futur. Gener. Comput. Syst. 29 (7) (2013) 1645–1660.

[6] Islam, S.M. Riazul, D. Kwak, M.D. Humaun Kabir, M. Hossain, K.-S. Kwak, The internet of things for health care: a comprehensive survey, IEEE Access 3 (2015) 678–708.

[7] S. Sharma, K. Chen, A. Sheth, Toward practical privacy-preserving analytics for IoT and cloud-based healthcare systems, IEEE Internet Comput. 22 (2) (2018) 42–51.

[8] Y.E. Gelogo, H.J. Hwang, H.-K. Kim, Internet of things (IoT) framework for u-healthcare system, Int. J. Smart Home 9 (11) (2015) 323–330.

[9] S.K. Sood, I. Mahajan, A fog-based healthcare framework for chikungunya, IEEE Internet Things J. 5 (2) (2017) 794–801.

[10] H. Mora, M. Signes-Pont, D. Gil, M. Johnsson, Collaborative working architecture for IoT-based applications, Sensors 18 (6) (2018) 1676.

[11] J.F. Colom, H. Mora, D. Gil, M.T. Signes-Pont, Collaborative building of behavioural models based on internet of things, Comput. Electr. Eng. 58 (2017) 385–396.

[12] N. Bui, M. Zorzi, Health care applications: a solution based on the internet of things, in: Proceedings of the 4th International Symposium on Applied Sciences in Biomedical and Communication Technologies, ACM, 2011, p. 131.

[13] T. Wu, F. Wu, J.-M. Redouté, M.R. Yuce, An autonomous wireless body area network implementation towards IoT connected healthcare applications, IEEE Access 5 (2017) 11413–11422.

[14] A. Kulkarni, S. Sathe, Healthcare applications of the Internet of Things: a review, Int. J. Comput. Sci. Inform. Technol. 5 (5) (2014) 6229–6232.

[15] P.K. Mishra, M.R. Pradhan, M. Panda, Internet of things for remote healthcare, Int. J. Comput. Sci. Eng. 4 (4) (2016) 106–111.

[16] W. Lee, J.W. Park, A framework for building a collaborative environment in an Open IoT platform, in: Proceedings of International Workshop Ubiquitous Science and Engineering, 2015, pp. 19–22.

[17] R. Kumar, M.P. Rajasekaran, An IoT based patient monitoring system using Raspberry Pi, in: 2016 International Conference on Computing Technologies and Intelligent Data Engineering (ICCTIDE'16), IEEE, 2016, pp. 1–4.

[18] L.M.R. Tarouco, L.M. Bertholdo, L.Z. Granville, L.M.R. Arbiza, F. Carbone, M. Marotta, J.J.C. De Santanna, Internet of Things in healthcare: Interoperability and security issues, in: 2012 IEEE International Conference on Communications (ICC), IEEE, 2012, pp. 6121–6125.

[19] N.-C. Fang, Using Internet of Things (IoT) technique to improve the management of medical equipment, Eur. J. Eng. Res. Sci. 4 (5) (2019) 148–151.

[20] S.R. Moosavi, T.N. Gia, A.-M. Rahmani, E. Nigussie, S. Virtanen, J. Isoaho, H. Tenhunen, SEA: a secure and efficient authentication and authorization architecture for IoT-based healthcare using smart gateways, Procedia Comput. Sci. 52 (2015) 452–459.

[21] L. Catarinucci, D. De Donno, L. Mainetti, L. Palano, L. Patrono, M.L. Stefanizzi, L. Tarricone, An IoT-aware architecture for smart healthcare systems, IEEE Internet Things J. 2 (6) (2015) 515–526.

[22] M. Hassanalieragh, A. Page, T. Soyata, G. Sharma, M. Aktas, G. Mateos, B. Kantarci, S. Andreescu, Health monitoring and management using Internet-of-Things (IoT) sensing with cloud-based processing: Opportunities and challenges, in: 2015 IEEE International Conference on Services Computing, IEEE, 2015, pp. 285–292.

[23] Council, Cloud Standard Customer, Impact of Cloud Computing on Healthcare, Technical ReportCloud Standards Customer Council, 2012.

[24] E.K.L. Hariharasudhan Viswanathan, D. Pompili, Mobile Grid Computing for Data and Patientcentric Ubiquitous Healthcare, in: The First IEEE Workshop on Enabling Technologies for Smartphone and Internet of Things (ETSIoT), 2012.

[25] B. Farahani, F. Firouzi, V. Chang, M. Badaroglu, N. Constant, K. Mankodiya, Towards fog-driven IoT eHealth: promises and challenges of IoT in medicine and healthcare, Futur. Gener. Comput. Syst. 78 (2018) 659–676.

[26] W. Raghupathi, V. Raghupathi, Big data analytics in healthcare: promise and potential, Health Inform. Sci. Syst. 2 (1) (2014).

[27] S. Raza, T. Helgason, P. Papadimitratos, T. Voigt, SecureSense: end-to-end secure communication architecture for the cloud-connected Internet of Things, Futur. Gener. Comput. Syst. 77 (2017) 40–51.

[28] H.H. Pajouh, R. Javidan, R. Khayami, D. Ali, K.-K.R. Choo, A two-layer dimension reduction and two-tier classification model for anomaly-based intrusion detection in IoT backbone networks, IEEE Trans. Emerg. Top. Comput. (2016).

[29] Y. Yang, H. Cai, Z. Wei, H. Lu, K.-K.R. Choo, Towards lightweight anonymous entity authentication for IoT applications, in: Australasian Conference on Information Security and Privacy. Springer, Cham, 2016, pp. 265–280.

[30] F. Wu, L. Xu, S. Kumari, X. Li, J. Shen, K.-K.R. Choo, M. Wazid, A.K. Das, An efficient authentication and key agreement scheme for multi-gateway wireless sensor networks in IoT deployment, J. Netw. Comput. Appl. 89 (2017) 72–85.

[31] J. Vora, P. Italiya, S. Tanwar, S. Tyagi, N. Kumar, M.S. Obaidat, K.-F. Hsiao, Ensuring privacy and security in E-health records, in: 2018 International Conference on Computer, Information and Telecommunication Systems (CITS), IEEE, 2018, pp. 1–5.

[32] H.K. Patil, R. Seshadri, Big data security and privacy issues in healthcare, in: 2014 IEEE International Congress on Big Data, IEEE, 2014, pp. 762–765.

[33] P. Gope, T. Hwang, BSN-Care: a secure IoT-based modern healthcare system using body sensor network, IEEE Sens. J. 16 (5) (2015) 1368–1376.

[34] E. Bertino, K.-K.R. Choo, D. Georgakopolous, S. Nepal, Internet of things (IoT): smart and secure service delivery, ACM Trans. Internet Technol. (TOIT) 16 (4) (2016) 22.

[35] S.R. Moosavi, T.N. Gia, E. Nigussie, A.M. Rahmani, S. Virtanen, H. Tenhunen, J. Isoaho, End-to-end security scheme for mobility enabled healthcare Internet of Things, Futur. Gener. Comput. Syst. 64 (2016) 108–124.

[36] F. Li, J. Hong, A.A. Omala, Efficient certificateless access control for industrial internet of things, Futur. Gener. Comput. Syst. 76 (2017) 285–292.

[37] H. Tao, M.Z.A. Bhuiyan, A.N. Abdalla, M.M. Hassan, J.M. Zain, T. Hayajneh, Secured data collection with hardware-based ciphers for IoT-based healthcare, IEEE Internet Things J. 6 (1) (2018) 410–420.

[38] M.S. Virat, S.M. Bindu, B. Aishwarya, B.N. Dhanush, M.R. Kounte, Security and privacy challenges in internet of things, in: 2018 2nd International Conference on Trends in Electronics and Informatics (ICOEI), IEEE, 2018, pp. 454–460.

[39] G.S. Matharu, P. Upadhyay, L. Chaudhary, The internet of things: Challenges & security issues, in: 2014 International Conference on Emerging Technologies (ICET), IEEE, 2014, pp. 54–59.

[40] J. Gubbi, R. Buyya, S. Marusic, M. Palaniswami, Internet of Things (IoT): a vision, architectural elements, and future directions, Futur. Gener. Comput. Syst. 29 (7) (2013) 1645–1660 ISSN: 0167-739X, Elsevier Science, Amsterdam, The Netherlands.

[41] I.F. Akyildiz, W. Su, Y. Sankarasubramaniam, E. Cayirci, Wireless sensor networks: a survey, Comput. Netw. 38 (2002) 393–422.

[42] Y. Sang, H. Shen, Y. Inoguchi, Y. Tan, N. Xiong, Secure Data Aggregation in Wireless Sensor Networks: A Survey; 2006, (2006) pp. 315–320.

[43] M. Zorzi, A. Gluhak, S. Lange, A. Bassi, From today's Intranet of Things to a future internet of things: a wireless- and mobility-related view, IEEE Wirel. Commun. 17 (2010) 43–51.

[44] Q. Gou, L. Yan, Y. Liu, Y. Li, Construction and strategies in IoT security system, IEEE International Conference on Green Computing and Communications and IEEE Internet of Things and IEEE Cyber, August 20–23(2013) pp. 1129–1132.

[45] P.K. Dash, Electrocardiogram monitoring, Indian J. Anaesthesia 46 (2002) 251–260.

[46] M.S. Mahmud, H. Wang, A.M. Esfar-E-Alam, H. Fang, A wireless health monitoring system using mobile phone accessories, IEEE Internet Things J. 4 (6) (2017) 2009–2018.

[47] H.A. Khattak, M. Ruta, E.D. Sciascio, CoAP-based healthcare sensor networks: a survey, in: Proc. 11th Int. Bhurban Conf. Appl. Sci. Technol. (IBCAST), 2014, pp. 499–503.

[48] Y.J. Fan, Y.H. Yin, L. Da Xu, Y. Zeng, F. Wu, IoT-based smart rehabilitation system, IEEE Trans. Ind. Informat. (2014) 1568–1577.

[49] U. Satija, B. Ramkumar, M. Sabarimalai Manikandan, Real-time signal quality-aware ECG telemetry system for IoT-based health care monitoring, IEEE Internet Things J. 4 (3) (2017) 815–823.

[50] S.V. Zanjal, G.R. Talmale, Medicine reminder and monitoring system for secure health using IOT, Procedia Comput. Sci. 78 (2016) 471–476.

[51] K.A.N.G. Kai, P.A.N.G. Zhi-bo, W.A.N.G. Cong, Security and privacy mechanism for health internet of things, J. China Univer. Posts Telecommun. 20 (2013) 64–68.

[52] K.T. Nguyen, M. Laurent, N. Oualha, Survey on secure communication protocols for the Internet of Things, Ad Hoc Netw. 32 (2015) 17–31.

[53] C. Eken, H. Eken, Security threats and recommendation in IoT healthcare, in: Proceedings of The 9th EUROSIM Congress on Modelling and Simulation, EUROSIM 2016, The 57th SIMS Conference on Simulation and Modelling SIMS 2016, Linköping University Electronic Press, vol. 142, 2018, pp. 369–374.

[54] M. Elhoseny, G. Ramírez-González, O.M. Abu-Elnasr, S.A. Shawkat, N. Arunkumar, A. Farouk, Secure medical data transmission model for IoT-based healthcare systems, IEEE Access 6 (2018) 20596–20608.

[55] M. Almulhim, N. Zaman, Proposing a secure and lightweight authentication scheme for IoT based E-health applications, in: 2018 20th International Conference on Advanced Communication Technology (ICACT), IEEE, 2018, pp. 481–487.

[56] A. Bujari, M. Furini, F. Mandreoli, R. Martoglia, M. Montangero, D. Ronzani, Standards, security and business models: key challenges for the IoT scenario, Mobile Netw. Applicat. 23 (1) (2018) 147–154.

[57] M.A. Salahuddin, A. Al-Fuqaha, M. Guizani, K. Shuaib, F. Sallabi, Softwarization of the internet of things infrastructure for secure and smart healthcare, arXiv preprint arXiv:1805.11011(2018).

[58] H. Mora, D. Gil, R.M. Terol, J. Azorín, J. Szymanski, An IoT-based computational framework for healthcare monitoring in mobile environments, Sensors 17 (10) (2017) 2302.

[59] S. Tyagi, A. Agarwal, P. Maheshwari, A conceptual framework for IoT-based healthcare system using cloud computing, in: 2016 6th International Conference-Cloud System and Big Data Engineering (Confluence), IEEE, 2016, pp. 503–507.

[60] D.S. Rajput, R. Gour, An IoT framework for healthcare monitoring systems, Int. J. Comput. Sci. Inform. Secur. 14 (5) (2016) 451.

A secure blockchain-based solution for harnessing the future of smart healthcare

Sujit Bebortta and Dilip Senapati
Department of Computer Science, Ravenshaw University, Odisha, India

1 Introduction

In the recent past, enormous studies have been conducted on amiability of blockchain technology with several cross-disciplinary technologies, which have led to the emergence of some newly evolving standards. In this scenario the Internet of Things (IoT) acts as a distinctive paradigm having the capability to address a wide number of applications along with interconnected sensory devices capable of exchanging information on several heterogeneous platforms. The fundamental idea behind securing the transmission of data over heterogeneous platforms involves privacy preservation and security of highly sensitive underlying data. The modern day healthcare sectors involve continual exchange of healthcare data over digital platforms by exploiting the information communication technology (ICT). These data constitute of highly sensitive information regarding a person's underlying health conditions and personal information, making them susceptible to security breaches and thefts. Electronic health records (EHRs) are one such technology that facilitates the exchange of multiple health records over a distributed platform [1–4]. The healthcare data like many critical data require a high level of consistency to safeguard the patients' privacy and well-being at multiple levels of their transmission cycle. As per the 2019 reports of Forbes Magazine [5], the blockchain is considered to be one of the most popular tools in the United States among healthcare amenities for facilitating secure storage and transmission of medical data and also for management of redundant medical records. The rise of blockchain in the healthcare sector has observed much demand for assisting development of digitized healthcare platforms and to support the ubiquity of health services rendered toward patients.

The EHRs marked the most pivotal revolution in healthcare and medical sectors by facilitating digitization of the patients' medical files or health records. These advancements explicitly resulted in the transmission of EHRs over distributed platforms to favor remote monitoring and care for the patients. The EHRs usually encompass extensive implications regarding a persons' health that can provide sufficient background to the caregivers regarding the patients' health [2–4]. These EHRs can be used in consolidation with other supporting information to predominantly transform conventionally employed medical practices. All these facilities are aimed toward providing an on-time, cost-efficient, and precise diagnosis of patients to improve the quality of life and overall life expectancy. Several nations have

IoT Based Data Analytics for the Healthcare Industry. https://doi.org/10.1016/B978-0-12-821472-5.00004-1

taken proactive measures toward digitizing their healthcare services for extending appropriate medical attention to a wider population. The portability and accessibility of EHRs have made their adoption into the healthcare services much simpler. The Becker's Hospital's 2013 reviews [6] categorized Norway number one in the top 10 countries with adoption of EHRs. The large-scale adoption of these services has also led to the generation of massive health datasets. The primarily generated raw data usually constitutes of potentially uncorrelated and unstructured data acquired from patients and those produced by the physicians or caregivers. The EHRs may comprise textual data, image data, pathological test results, and other essential medical demographics. These data in their very raw form are potentially considered as by-products to EHRs. After applying certain sophisticated filtration and cleaning techniques, these data can be used in generating EHRs with qualitative, transactional, and quantitative medical data.

Several recent studies [4, 7–12] have considerably emphasized on the applicability and pressing requirements for using blockchain in healthcare sector and for the EHRs. By leveraging the capabilities of blockchain, a more persuasive analysis of some crucial issues can be obtained. As per the early December 2019 release from the Forbes, Anthem [5] one of the most popular US-based health insurer has introduced a pilot blockchain-based framework to secure their medical data. The company believes that in the upcoming years, all its 40 million members can have access to the blockchain platform for securely exchanging their health records and at the same time ensure privacy of its users. This can also provide the patients instant access to their medical records simultaneously facilitating the patients to manage the number of users accessing their medical data. Some of the commonly associated issues that can be dealt by employing blockchain in the healthcare sector are enlisted in the succeeding text:

- *Distribution of information*: Blockchain facilitates a decentralized platform for the distribution of health records across multiple platforms. The flow of health records plays an important role in facilitating some of the prime aspects of EHRs like remote patient monitoring, telemedicines, and telesurgery.
- *Transparency of information*: The immutability of blockchain makes it suitable to restrict the modifications made by multiple users to a record. This feature only provides the privilege to the authorized users to make modifications or to access the patients' health records.
- *Secrecy of transactions*: Blockchain-based platforms can either be public or private, based upon the requirements of users. This imposes certain constraints on the number of authorized users who can access patients' personal information and health demographics over a network of users.
- *Manageability*: It can add more robustness to the management of multiple health records over a largely populated network. The information contained in the records can be updated uniformly without the interference of any third party there by preventing any inconsistency and out-of-sync problems.

1.1 Frequently encountered issues in current healthcare facilities

Although the EHRs have emerged as a revolutionary technology in modern healthcare sectors, still it suffers from certain complications while working over large distributed platforms. Some of the commonly encountered issues with the conventional EHRs are discussed later.

1.1.1 Information sharing

Generally the conventional medical sectors follow three basic information sharing models that allow the exchange of medial information between different healthcare providers [6]. These models can be categorized as follows:

Push: This involves transmitting a payload of the healthcare information from one caregiver to another. It enables the large-scale sharing of data over multiple platforms, by sharing a common EHR for the patients.

Pull: In this scenario the healthcare provider can issue a pull request from another provider for querying any information regarding a patients' health.

View: Here the idea is that a healthcare provider can look into the records issued by another provider so that one can know the patients' previous medical history.

An example of these models can be illustrated by considering the patient-doctor scenario. A patient being treated in a care unit may require some advanced attention; hence the patient's health records and previous medical history may be pushed (or transmitted) to another care unit for progressive treatment or for any specialized treatment. Further, physicians at the second care unit may wish to seek more details regarding the patient's case history and may issue a pull request to the previous care unit seeking the desired records. Similarly, at a more progressive stage of the treatment, a physician may wish to view the pathological test results of the patient for further reference conducted by another doctor for drawing more idea regarding a patient's underlying conditions that comes under the view model. A high-level design of this scenario is presented in Fig. 1.

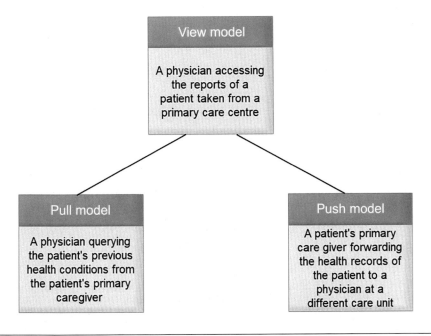

FIG. 1

An example of the push, pull, and view model for the exchange of health records.

1.1.2 Acquisition and storage of information

The information required for an EHR can be acquired using a variety of methods such as question-naires, population-based surveys, pathological tests, and magnetic resonance imaging (MRIs), entail-ing extensive health data [6]. The information should satisfy a number of criteria like the purpose of the EHR, the intended user, parties entitled to access the EHR, documentation of patient details (i.e., out-patients, pediatrics, inpatients, geriatrics, and speciality), patient's insurance documentation, system requirements (additional hardware/software requirements), and billing information. These require-ments are some of the most fundamental requirements that are essential to understand the goals that the EHR is meant to achieve.

The data stored in the EHRs should comply with uniform data storage standards to make data stored on EHRs more accessible and portable such that the intended organizations can facilitate quick re-sponse. This can further render more efficient and timely services to the patients. The convergence of some sophisticated computerized decision support systems (DSS) with EHRs can also assist in en-hancing the quality of diagnosis and essential predictions [13]. Typically the prescriptions and tests recommended by the physicians are to be stored electronically as the end results that can enhance the integrity of these records. The transmission of these records requires high degree of connectivity and communication efficiency to make their transfer secure and accessible.

1.1.3 Lack of consistency

The data in EHRs may not be consistent throughout its life cycle. This is the case that is commonly encountered with the push model, where the patient's complete pathological records may not be pushed into the EHR by the primary care provider, which gives rise to inconsistency at the new care provider's side. This may sometimes lead to misdiagnosis. The prevalent approaches in EHRs may be incapable in handling multiple health records generated by different healthcare units. Moreover, due to the presence of redundant patient information, being nonuniformly distributed over several platforms may lead to loss of essential features due to the lack of integrity. This also intercepts the sharing of data between different health organizations.

1.2 Enabling technologies of smart healthcare

Several state-of-the-art technologies have contributed in enhancing the standards of existing healthcare practices. These technologies that have assisted in revolutionizing the healthcare sector are discussed later.

1.2.1 IoT

The emergence of the IoT in healthcare sector has largely facilitated the capturing of real-time medical data perceived by various interconnected sensory devices as physiological signals. This has advanta-geously augmented the capabilities of healthcare services to more rewarding extents. The use of IoT in healthcare sector has been evidently observed in many areas like remote patient monitoring, assistive living, and telemedicine [14, 15]. The constant innovations introduced in these technologies have led to the evolution of more sophisticated infrastructures for capturing, processing, storing, and analyzing these data [16–18]. Hence, IoT provides a seamless mechanism to integrate most of these capabilities toward achieving a smart healthcare system.

1.2.2 WBAN

Wireless body area networks (WBANs) refer to the system of wearable sensory devices capable of wirelessly perceiving data from the environment and communicating among themselves. These technologies have favored the transformation of data acquired from the physical world to the cyber world. WBANs play an integral role in healthcare sectors for activity monitoring, fall detection, pulse detection, and acquisition of many more physiological signals that have benefited in the medical sector.

1.2.3 Blockchain technology

The blockchain technology provides a robust mechanism for handling largely growing networks. This technology provides a decentralized approach that adds more reliability and security to the information as compared with conventional distributed systems. Depending on the type of application, a blockchain may be either public or private.

- *Public blockchain*: This platform allows universal access to a user to join the blockchain network. It is decentralized and provides users to append changes to the records stored in the system. Here the network is permissionless, and no further change can be appended to a record once it is validated.
- *Private blockchain*: This is a permissioned platform, since the users participating in the network are restricted, and can be customized by the network administrator accordingly.

One of the most important aspects of blockchain technology is the smart contract. This is basically a set of protocols enforced between two negotiating parties. This advantages both the parties by adding more credibility to their negotiations without the interference of any third party. All the resources involved in the smart contract can be uniquely identified [19].

1.2.4 Dependability of EHRs on smart technologies

The dependability of EHRs on the wearable devices and IoT technology can be well observed by looking into the limitations of traditional EHR systems [20]. By leveraging the capabilities of IoT-based wearable sensory devices, the applications of EHRs can be drastically augmented to enable many issues like remote patient monitoring, remote recommendation of medications, and remote critical care solutions. The convergence of these technologies with EHRs facilitates seamless acquisition and integration of medical data over extensively heterogeneous platforms. A more stratified approach for managing and sharing of the patients' health demographics can be obtained. Further, more advanced monitoring devices and DSS can be integrated to render proactive services through EHRs. The prime goal behind this integration lies in improving the quality of healthcare services by configuring the IoT framework in context with EHRs for collection of more logical data to assist a context-centric medical prediction system. Fig. 2 provides an architecture of the smart EHRs enabled with IoT-based wearables and the services that can be accomplished by using this framework.

The conventional approaches employed in EHRs generally suffer through many limitations. The lack of a standardized audit upon the acquired data is one possible issue that intrinsically leads to many undesired problems. The lack of consistency and integrity in EHRs is a major issue that arises from the point of generation to the point of dissemination. This is more commonly encountered due to the change in platform between different organizations leading to inaccuracies in the generated payload. Further secrecy of the patient's personal information may be lost to some extent due to the lack of audit trials at each new level. It would be interesting to observe that these shortcomings in efficiencies are faced due to the variations in platform and institutional practices experienced due to the interorganization transfer of payloads.

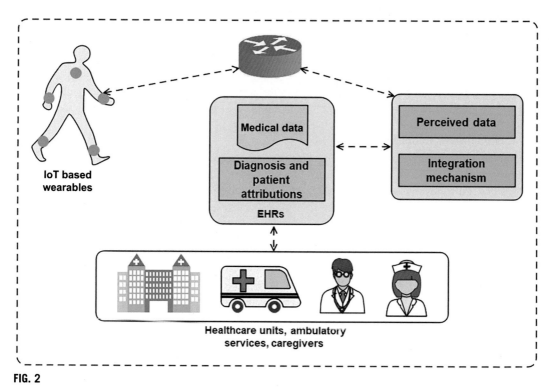

FIG. 2

An overview of the smart EHR infrastructure.

In context to the aforementioned scenarios, blockchain can be thought of as an essential construct for ensuring integrity, security, and standardization of data over a universal platform. The massive amount of payloads generated from healthcare sector count on the data continuously being generated from heterogeneously distributed platforms like clinics, pathological laboratories, hospitals, and insurance companies. The handling of these data is practically beyond the limits of conventional systems, which may distort the chronology of events that is to be enlisted in the EHRs. Further, variations in the workflow of data embedded into the EHRs may vary with different institutions [1–3, 20]. Thus the rationale behind using the blockchain technology for EHRs lies in efficient management of the distributed data in an integrated way by employing a decentralized ledger. This further provides an intelligent control upon the data stored and accessed from the EHRs. The resultant infrastructure can be used as a solution for many critical care units to provide timely diagnosis of diseases.

2 Emergence of blockchain-based EHRs

The healthcare sector has considerably benefited from large-scale advancements in information exchange and storage systems. Blockchain is one such explicit platform that enables a secure transmission between two peers without the need of any third-party interference and facilitates management of data over distributed platforms. In recent times, several industries have successfully implemented blockchain into their management framework and have immensely expanded their functionalities. One such recent example of blockchain applications can be observed in the energy distribution industries that involve the bulk generation and delivery of services over energy grids [21]. The use of blockchain here that can assist is managing the distribution of energy, checking the client access details, and billing information.

In the healthcare sector, EHRs are one such subsidiary, which frequently encounter various challenges like for acquisition and integration of medical data transmitted from diverse heterogeneous platforms or for preserving the privacy of patients' information and managing the number of people authorized to access these information. A notable example of such security breaches was when the United States' second largest health insurer Anthem was hacked and sensitive information of over 80 million employees and clients were filched [5]. The recently used security frameworks rely on secret key exchange mechanism between two parties that may not be sufficiently impervious due to the centralized architecture for storing and acquisition of data. Hence the blockchain-based framework is sufficiently capable of handling these issues due to its decentralized control mechanism and ubiquity of access. The prime motivations are to achieve a highly reliable, scalable, and integrative approach for acquisition, storage, and transmission of medical data over distributed platforms. The granularity observed in this mechanism is believed to be beneficial for healthcare systems to manage the number of users accessing the EHRs.

2.1 Types of blockchain

Blockchain has emerged as an essential technology behind many applications, since the time it was first introduced in 2008 by Satoshi Nakamoto to serve as a distributed database. Thus blockchain is now being pervasively used in many transactional applications that involve the management of multiparty transactions and also for applications involving substantial management of assets. Considering the growing popularity of blockchain-based services, a number of blockchain platforms that have been developed in recent years, we restrict our study to seven specific blockchain technologies [6, 22]. A classification for each platform is made, and some of the selection constraints that can assist organizations in choosing the right type of blockchain are discussed. These constraints basically involve the type of network infrastructure they follow (whether private or public), the pricing policies, and the appropriate programming languages required for implementing the platform's software development kit (SDK).

2.1.1 Ethereum

The Ethereum is an open source platform, which enables developers to build and use customizable decentralized applications that employ blockchain technology. It follows the public network infrastructure that provides a decentralized control. It is highly secure and immutable, where once the records are validated, no user can make further changes to it. It is permissionless, which privileges the users to participate in a network. They can be implemented using C++, Python, and Go.

2.1.2 Bitcoin

The Bitcoin is quite similar to Ethereum, since Ethereum was developed as an extension of the Bitcoin to achieve more flexibility and robustness. Bitcoin was originally developed by Satoshi Nakamoto by implementing C++ language to serve as a secure payment network. Later on, it has paved its way into several applications. Recently, it has been observed that some specifications of Bitcoin are also implemented using the JavaScript [22].

2.1.3 Hyperledger

Hyperledger is one of the most popular types of blockchain that has been extensively used in implementing many projects including smart grids, IoT applications, and supply chain management. Some of the popular projects include Hyperledger Fabric, Hyperledger Sawtooth, Hyperledger Transact, and many more that serve a multitude of applications [23]. This technology is also versatile in reducing the carbon footprinting, by facilitating code and distributed ledger reusability. It is a collaborative technology and may be both permissioned and permissionless.

2.1.4 Openchain

It is an open source scalable, real-time blockchain platform developed by Coinprism. It implements a distinct blockchain approach, unlike the Bitcoin, by considering a partitioned consensus framework. Here, every instance has a single authority responsible for validating the transactions. This framework is similar to the client-server architecture, which is a reliable and efficient centralized scheme specifically suitable for small organizations for facilitating management of digital assets.

2.1.5 IBM Bluemix Blockchain

The IBM Blockchain was originally released in March 2017, which deploys Hyperledger Fabric for enhancing the accountability and visibility for many tangible applications. It has evolved as an active catalyst in many diverse applications and business models. It is an active privately owned permissioned network. Presently, it has been widely used by most enterprises with additional infrastructural and security requirements.

2.1.6 HydraChain

The HydraChain is a blockchain platform that is meant to comply with specific organizational requirements. It is a joint contribution of the Ethereum project and brainbot technologies, which extends the idea of Ethereum protocol. It is an open source permissioned distributed ledger system with the idea for encouraging large-scale adoption of blockchain.

2.1.7 Multichain

It is generally used for implementing private blockchain platforms. These networks are permissioned, that is, the exchange of information may be done only between a restricted number of participants or may be specific to a location. It is more suitable for the financial sectors that require private peer-to-peer transactions.

All of these blockchain technologies play their own specific roles and are used in order of their compliance to the institutional and regulatory needs. A taxonomy of these technologies is presented in Table 1, along with some of their most important features and implementation platform for building the corresponding SDKs.

Table 1 A taxonomy of some popularly used blockchain platforms.

Type	Popularity	Network type	Pricing	Languages for implementation
Ethereum	High	Public	Open source	C++, Python, Go
Bitcoin	High	Public	Open source	C++, JavaScript
Hyperledger	High	Private/public	Open source	Python
Openchain	Medium	Private	Open source	JavaScript
IBM Bluemix Blockchain	Medium	Private	Free limited plan/payment required for upgrades	JavaScript, Go
HydraChain	Low	Private	Open source	Python
Multichain	Medium	Private	Open source	C#, Python, JavaScript, Ruby

2.2 Necessity for using blockchain

The implementation of blockchain-based EHRs can conquer some of the most challenging issues encountered in the conventionally used healthcare models. To meet the real-world requirements of healthcare industry like for the capture and analysis of physiological data, a highly scalable and reliable framework is required. To achieve this the system must allow data liquidity among several communicating interlinked platforms like healthcare units or hospitals, ambulatory services, physicians, and pathological laboratories to facilitate interoperability among them. The transmission of medical data over multiple platforms may require a secure distribution mechanism to ensure consistency and privacy of the transmitted data. Further the medical data captured from different IoT-based devices require a high level of accessibility to authorized users, thereby contributing to the enhancement in transparency of the data among users. All these issues discussed earlier can be achieved by employing a blockchain-based framework capable of meeting the dynamic growth in healthcare sectors. In the following subsections a detailed account of the solutions offered by the blockchain technology toward healthcare sector and EHRs are discussed. In Fig. 3, we provide the blockchain-based architecture for healthcare system involving EHRs as the fundamental information exchange platform. The data acquired from the sensory devices or wearables are transmitted to the IoT gateways from where they are transmitted to the cloud-based blockchain network.

2.2.1 Interoperability

In the healthcare sector, interoperability is often associated with the exchange of medical data between two different parties. These two parties may continuously communicate and facilitate the sharing of records until a specific goal is fulfilled. By leveraging the capabilities of blockchain, patients and physicians can have more control over the data being exchanged between different platforms, so as to ensure consistency and integrity of data [24–26]. By achieving interoperability among the communicating systems, several potential benefits can be obtained [24, 27, 28]. This may also result in improving the functional efficiency of many clinical care systems by providing access to a multitude of medical data.

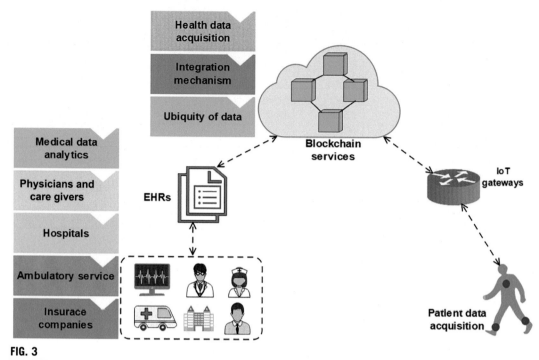

FIG. 3

The blockchain-based architecture for healthcare system.

The healthcare data generated by multiple IoT-based devices are generally channeled among different information exchange systems, which in turn moves through a number of platforms like healthcare units, pathological centers, and critical care centers to facilitate on-time care to the patients. The EHRs are often used to provide a more complete idea regarding a patient's underlying conditions encompassed within a single unit. The fundamental goal behind interoperability in the healthcare sector is to facilitate easy and on-time transmission of EHRs between different healthcare providers. It also facilitates proactive clinical care services to patients, provides personalized health recommendations, and reduces the cost and time constraints associated with physical transmission of EHRs over remotely distributed healthcare units. Hence, it may be said that by achieving interoperability between the different communicating healthcare information exchange systems, more accessibility, precision, timeliness, and cost-effectiveness in the healthcare services can be achieved.

2.2.2 Tamperproof

The distribution of EHRs and healthcare data over multiple platforms may make it susceptible to unwanted changes and therefore may give rise to inconsistency within the records. This can prove to be potentially hazardous resulting in imprecise health diagnosis and medication recommendations. Thus a highly reliable and consistent framework is required that empowers a controlled access to the users accessing the records and embedding changes into these records. The immutability and tamperproof mechanism facilitated by blockchain can largely advantage the healthcare sector for controlling number of users implementing changes into the records. This can also secure the consistency of information encompassed within the EHRs.

The convergence of blockchain technology into several critical infrastructures has produced more reliable, secure, and tractable solutions in managing the associated entities [29]. The transmission of EHRs over the Internet- and cloud-based platforms makes data vulnerable to thefts and tamper, which may cause loss of highly sensitive information. The work in [30] focused on blockchain-based tamperproof architecture for EHRs. A blockchain handshaker mechanism was provided, which behaves as a wrapper for integrating the EHR systems with public blockchain network (Ethereum) and the user applications. Thus, by employing blockchain technology for the exchange of EHRs and other healthcare information, we can obtain explicit interoperability and security.

2.2.3 Network size management

Most conventional infrastructures continue to face challenges as devices connected in the network gradually scale-up and continuously generate data. The healthcare sector comes under one such category that requires large number of connected devices capable of acquiring and processing medical data to function efficiently. Thus, to facilitate the smooth and uninterrupted functioning of these frameworks, highly reliable network architecture is required. Blockchain provides one such reliable and scalable platform for uninterrupted exchange of data between devices without the interference of intermediaries. The network size of blockchain has experienced tremendous growth since its initial arrival. In context to blockchain the network size refers to the number of devices or nodes that play an active role in the acquisition and transmission of data and information, thereby satisfying the scalability and reliability constraints of blockchain mechanism [31]. The use of blockchain in large networks also adds more traceability regarding the devices connected in the network and managing associated users and organizations.

The fundamental goal behind using blockchain in various cloud-based models and IoT-based dynamic systems is to mitigate the effects of latency and network failures for continuously growing networks. This has ensured the exchange of data generated from various medical systems with much efficiency and certainty. Hence, it would be essential to exploit the on-demand network scalability of blockchain for providing a secure and failure proof network [32].

2.2.4 Reliability

In the medical data exchange systems, reliability and consistency of the data are of foremost importance. To access healthcare data over distributed platforms requires reliability of data and the communication network. Most of the conventionally used distributed architecture offers a centralized control over patients' sensitive information that further poses security challenges. The use of blockchain in IoT-based environments using reliable communication protocols like the Zigbee or WiMax provide a highly reliable decentralized solution for the exchange of sensitive patients' information and EHRs over a distributed platform.

2.2.5 Transparency

For achieving transparency in any collaborative technology, a trust-based mechanism is highly inevitable between the participating entities. Prior to this the medical data or EHRs require to be highly secured and tamperproof. This will prevent the unauthorized users from accessing or writing into the records. Considering blockchain framework as the essential infrastructure for IoT systems to enable accessibility and seamless integration of health information, a substantially transparent framework can be obtained. Whenever any record requires being stored on the blockchain for sharing, the record may be readily available across the entire network. Here the control is not centralized, but is distributed throughout the blockchain network. This makes the healthcare information and EHRs more secure as no interventions from third parties or intermediaries have any effect of the information or quality of network, hence ensuring transparency of information throughout the network.

2.2.6 Cost efficiency

The healthcare services involve a huge number of data acquisition, processing, and analytics system, among which some are rented from service providers for processing massive datasets. Some instances may arise, when the user has to pay for some of the resources that are not being utilized to serve the current purpose. This may involve huge tenancy costs, thereby posing challenges on economy of the organizations or healthcare providers. Thus, by leveraging the capabilities of blockchain as a service (BaaS) framework, users can pay for the resources following the pay as you go model. This costs the users per unit of the transaction data, thus providing a data driven payment scheme. By availing blockchain services on the cloud platform, the users can avoid the upfront infrastructural costs. This can also greatly avoid the resources from being idle and can facilitate their better utilization.

3 Addressing security requirements of EHRs through blockchain

The growth in IoT-based healthcare applications has leveraged the efficient acquisition of a massive amount of physiological and sensory data. These data are mostly captured by various wearable sensory devices embodying the WBANs. The advent of IoT has drastically influenced services rendered toward healthcare community by facilitating the diagnosis of chronic ailments, detecting falls, autonomous care, and remote monitoring of patients. This requires the patients' health information to be heterogeneously distributed over multiple platforms at the same time so that caregivers and doctors can access these records to provide on-time care to the patients. To make these medical records or EHRs readily available at multiple locations simultaneously, the Internet serves as an appropriate medium [33]. This further gives rise to several adversaries mostly caused due to security and privacy threats imposed upon the medical records being transmitted from remote locations. Thus, to manage these vulnerabilities continually arising due to several security threats at different layers of information distribution, blockchain technology can prove to be the most fitting choice.

To better understand the compliance of blockchain for EHRs, we need to consider some commonly encountered security aspects. Although the security requirements of EHRs may be highly reliant on the specific applications it is supposed to serve, here some generic set of security requirements are provided. These requirements are specifically conjectured toward the design of advanced and secure healthcare systems [34–36]. Later are some crucial security aspects that are to be considered while designing these systems.

3.1 Authentication and authorization

Before establishing communication between different healthcare organizations, systems, and patients, a secure authentication of all the communicating parities is desired. A formal verification of their identities and roles is required to keep and account of each party involved in the system. Further, to ensure that the identified parities efficiently play their individual roles, some access rights are granted to each party so that they receive the stated information. These authorization and authentication checks are necessary at every layer of information flow cycle in healthcare systems to ensure integrity and privacy of the users' data.

3.2 Preserving confidentiality and permits

Preserving the confidentiality of medical data is crucial to IoT-based systems as the patients' personal data may be available on multiple devices heterogeneously distributed over a large geographical area. Hence the framework should be such that it can easily identify unauthorized users' and malicious activities taking place in the system and be able to robustly eliminate them from the network. The users' sensitive health information should only be made available to authorized users in the network subject to the roles they are required to perform. While processing the EHRs the patients should be entitled to track the flow of their health information, so as to understand who accesses their records and manage the persons who are authorized to update and manipulate their records.

3.3 Managing records' documentation

The patients and the caregivers and doctors are granted rights to manage and access the EHRs within certain organizational regulatory mandates and government policies. These policies are enforced to stipulate certain organized changes to EHRs, thereby preserving consistency and coherence of the underlying medical records. In EHRs and healthcare systems, it is essential to safeguard the chronology of events so as to assure better understanding of the health records. An account of all the modifications made to the records and users responsible for making modifications should be maintained as every stage. Apart from this the replication and reproduction of health records must be managed at each end so as to avoid inconsistency of information and redundant copies of same records.

Although the earlier discussed security and privacy issues of EHRs can be well-controlled when the EHRs are restricted only to an internal system or computer and is locally governed, problems may be severe when the records are heterogeneously scattered over multiple systems. To provide extensive services for smart IoT-based healthcare systems, data are required to be present over multiple systems at the same instance of time to facilitate near real-time services to the patients. Thus, toward this end, blockchain can prove to be a befitting choice as by exploiting the intrinsic properties of blockchain all these crucial issues can be addressed proactively and without the involvement of intermediaries. The requirement for employing blockchain in these applications is inevitable as it requires EHRs to be outsourced to the cloud storage so that the doctors and caregivers can readily access patients' health status from remote locations [37]. As the EHRs may be generated at different ends and may be modified by multiple users, this may sometimes give rise to different security and privacy threats, thereby affecting the consistency and integrity of EHRs. In the past several studies, [37–39] have focused on the adversarial effects of having a centralized storage system for health records and security breaches arising due to these systems. Thus the key motivation behind using a decentralized blockchain-based framework for the healthcare sector is to ensure safety, integrity, consistency, and confidentiality of patients' sensitive medical records at every stage of its' dissemination.

Several works have highlighted the use of some traditional and fundamental cryptographic schemes for preserving secrecy of healthcare information over the cloud-based servers in remote health monitoring applications [40, 41]. However, the disadvantages of these schemes can be more prominent than their advantages as for highly distributed and large-scale networks spanning over enormous geographic locations, these traditional cryptographic approaches may not work well in preserving all the users' rights and privacy. To this end, blockchain can provide more promising achievements as it can manage complex platforms with much ease and may eliminate any disputes arising between different medical organizations due to violation of rights. The blockchain technology implements smart contracts that are a much amiable choice in this scenario. This involves autonomously managing contracts between large organizations by eliminating the scope for intermediaries. As discussed in Section 1.2.3, for extending the healthcare services over larger geographic locations, public blockchain can be employed. This involves integration of patients' current pathological test results, EHRs, and other medical history into a public blockchain (e.g., Ethereum) so that the doctors can closely monitor the patients [37]. This also makes EHRs tamperproof as all the manipulations and changes made to EHRs by any of the authorized users are recorded and the chronology of each event is preserved. No changes can be implemented to any record without the knowledge of the administrator. The administrator keeps track of all the network activities and also identifies the user responsible for making any changes to the records. This also prevents any modifications into the EHRs at advanced stages of its development once all the users have validated the information contained in it. The patients and the healthcare organizations can be assured regarding the integrity and privacy of EHRs and exchange of any sensitive medical information in the blockchain network. Hence, by exploiting the capabilities of blockchain, a risk-free architecture for management of large-scale information over complex platforms can be achieved.

4 Case study for evaluating implementation of blockchain

To have a better understanding of the efficiency, reliability, scalability, and cost constraints of blockchain technology for real-life applications, we need to look into some of the important use cases where blockchain technology has been actually implemented as a working model. Here a case study of the Change Healthcare organization has been considered where the blockchain has emerged successfully [31, 42]. Change Healthcare [42], a blockchain-based healthcare provider launched in January 2018, is one of the largest health providers connecting 5500 hospitals and 9 million healthcare organizations. They claim to process up to 50 million transactions each day along with a throughput of 550 transactions per second. The prime motivations behind introducing Change Healthcare were to handle the plagued healthcare system in America, which is driven by frauds and inconsistencies. It is quite rare that the physicians, patients, and care providers find a complete up-to-date record, regarding patients' health. Further the transmission of these records over Internet-based platforms poses security challenges, and other issues like file format incompatibility, whenever it is distributed over multiple platforms; all these factors hugely affect the flow of medical information between different parties and quality of service (QoS) delivered to the patients.

As per the reports of McKinsey, a US-based consultation firm, it is observed that the healthcare industry could have saved up to $450 billion every year by facilitating up-to-date technologies and processes. Change Healthcare, along with its' intelligently spread network, contributed to the healthcare transformations by introducing blockchain technology. The Change Healthcare was initially a huge company capable of processing a large number of transactions each day. To meet the increasing popularity of devices capable of generating medical data from remotely situated locations, the company implemented blockchain technology into their existing systems. This framework was sufficient enough to handle the real-world traffic and meet most of the grueling requirements of healthcare industry. The framework adds more transparency to information accessed by users in the network using APIs, built specifically for the users to query their claims. For example, a hospital administrator or physician can query the blockchain via APIs for accessing the real-time status of patients. All the transactions in the system along with those that have appended some changes to their records are all recorded in the blockchain following the details of when and by whom were the changes made. Thus the traceability of clinical data and data transparency achieved by using the blockchain technology can revolutionize the healthcare sector drastically.

5 Observations and discussions

To better understand the potential of blockchain technology in addressing the large-scale requirements of healthcare sector, we have considered two specialized blockchain platforms that have been extensively used in numerous real-world applications. These frameworks have successfully paved their ways into our day-to-day lives. In this section the Bitcoin and Ethereum blockchain have been studied [43, 44]. The transaction value, hash rates, difficulty, and transactions communicated per day are some of the qualifying parameters that enable a user or an organization to choose an appropriate blockchain platform. An insight to each of the earlier constraints is provided, and the historical data from the past two years are analyzed.

5.1 Transaction value

The transaction value refers to transactions communicated by the blockchain. The transactions are quite similar to the transactions involved in conventional databases. The information stored in the transactions may vary as per the purpose of blockchain. For example, if the blockchain is to be used for facilitating the exchange of financial transactions, then the transactions may consist of payment details that the user may be liable to pay. Fig. 4A and B represents trend analysis performed over empirical data for transactions communicated by Bitcoin and Ethereum blockchains between 2018 and 2019, respectively.

5.2 Network difficulty

The network difficulty represents the complexity involved with finding a valid hash in largely spread blockchain network [45]. This is typically the time consumed by miner for including new transactions in the network. The importance of network difficulty can be realized in large blockchain networks by studying the interval at which the new blocks of transactions gets added to the network [45]. Fig. 5A and B represents the network difficulty of Bitcoin and Ethereum blockchains.

5.3 Hash rate

The hash rate is one of the most critical factors for blockchain networks. Following [46], hash rate may be said to be the power consumed by a blockchain to remain functional for generating and tracing the blocks at regular intervals. This is a computationally intensive task that cannot be satisfied by

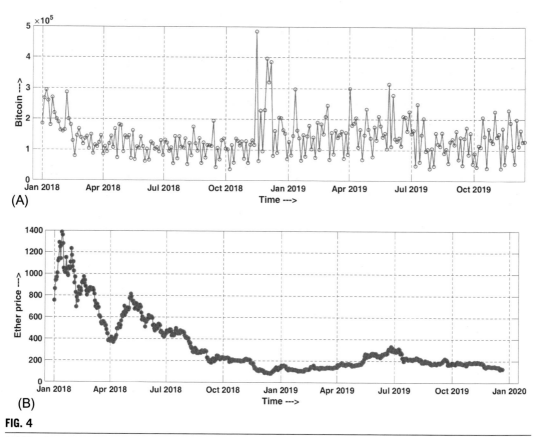

FIG. 4

(A) Total transactions communicated by Bitcoin blockchain between the years 2018 and 2019. (B) Total number of transactions sent by the Ethereum blockchain from the year 2018–19.

conventional databases as for retrieving and creating queries. Fig. 6A and B shows the trend analysis of the empirical data recorded for hash rates corresponding to Bitcoin and Ethereum.

5.4 Number of transactions communicated

This refers to the number of confirmed transactions transmitted per day. Once the mining process in blockchain is complete, transactions require to be validated. The transaction flow may constitute of confirmed and unconfirmed transaction. The confirmed transactions are those that have been verified for any possible changes in the record, and unconfirmed transactions are those that are still waiting to be verified in the blockchain network [47]. In Fig. 7A and B, the trade-offs between the number of transactions communicated per day in Bitcoin and Ethereum from 2018 to 2019 are provided.

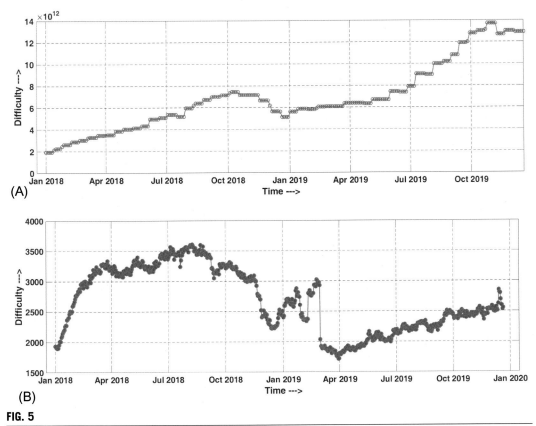

FIG. 5

Network difficulty in blockchain networks observed between 2018 and 2019: (A) The network difficulty for Bitcoin. (B) The network difficulty for an Ethereum blockchain.

6 Benefits of employing blockchain

The convergence of the capabilities of blockchain technology and IoT-based framework has led to the large-scale acquisition of physical world information including physiological and clinical data. The advent of these technologies has also facilitated the traceability of information flow, their authenticity among numerously spread devices, and their real-time accessibility to authorized users. This provides the reliability and consistency of transmitted data and information transparency and also enables a controlled mechanism for sharing of information and resources. The fundamental idea behind this solution is driven by the emergence of sensory devices capable of perceiving and transmitting data at various levels of healthcare sector. This ability of the sensory devices to capture data makes them an important component toward development of a highly sophisticated healthcare infrastructure.

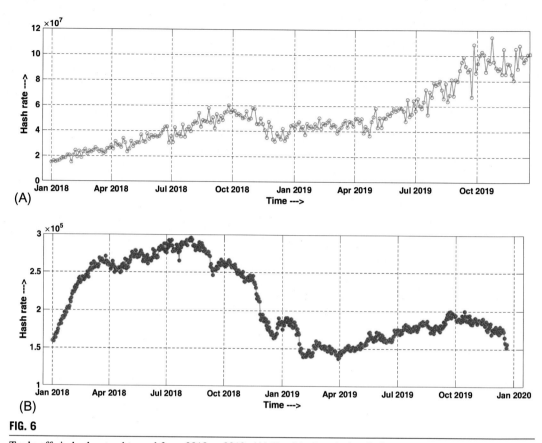

FIG. 6

Trade-offs in hash rate observed from 2018 to 2019: (A) Empirical analysis of Bitcoin's hash rate. (B) Empirical analysis of Ethereum's hash rate.

The use of blockchain is observed to be highly inevitable in the era of Internet for secure transmission of sensitive healthcare data over distributed platform [3]. Further the consistency of data contained in EHRs and their transparency at various cycles of healthcare system is also an essential trait that can be leveraged by using blockchain. Several blockchain platforms have now come into existence that provides options to each organization for facilitating more personalized services. The use of smart contracts has also brought in more control over the number of users responsible for accessing and manipulating a record at a given time. Using blockchain can also provide the administrators a detailed history of changes appended into the records along with the details of users appending the changes. Further the origin of data can also be identified by realizing the traceability aspects of blockchain.

The scalability of existing systems can also be extended to a much greater extent providing new scopes for delivering more specialized care to patients. The blockchain technology also claims the preservation of privacy and security of sensitive medical records by following robust authentication and certification mechanisms. Furthermore the physicians, caregivers, or administrators can also efficiently identify any issues causing mismatch within the records. The extensive nature of blockchain network

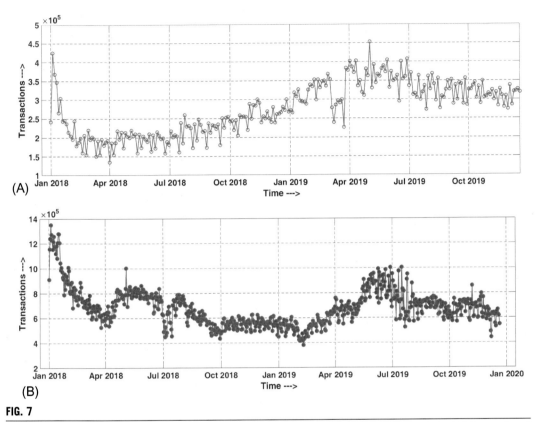

FIG. 7

Trade-offs between number of confirmed transactions communicated per data from 2018 to 2019. (A) The number of confirmed transactions communicated in Bitcoin. (B) The number of confirmed transactions communicated in Ethereum.

favors large-scale connectivity among healthcare service providers and patients all over the world. The healthcare organizations can also largely benefit by exploiting the capabilities of blockchain in effortlessly performing regulatory and institutional duties.

As evident in [42], like other industrial and supply chain communities, healthcare sectors also require audit trials that secure the transparency of medical records. EHRs may consist of a number of sensitive information like a patient's location, clinical data, insurance details, and billing information. This may make the EHRs susceptible to security threats. The work in [30] concentrated on a similar concern that proposed a secure blockchain-based architecture for safeguarding the information stored on EHRs employing blockchain handshaker scheme. Thus, by securing the data stored on EHRs, a number of issues can be addressed like malpractices in clinical care caused due to misdiagnosis, manipulation of medical insurance related information, delayed diagnosis, remote monitoring, and much more.

Apart from all these issues, some critical care issues like challenges faced due to delayed medical visits, inconsistency and discontinuity in information flow between different organizations (like hospitals, pathological test centers, and ambulatory services), granularity in information flow, etc. require active attention. All these issues can be handled much efficiently by using the decentralized blockchain framework. This framework attributes in adding checkpoints at every layer in the information flow process and in managing entities connected to the network.

7 Future research directions and challenges

The current blockchain-based smart healthcare systems are experiencing tremendous expansions and are rapidly evolving with the advent of several technological breakthroughs. In context to this, several future healthcare applications that can substantially benefit with the convergence of blockchain technology have been discussed in this section. Apart from this a comprehensive account of some potential challenges relevant to IoT-based healthcare systems is also provided.

7.1 Smart diagnostic systems

The fundamental idea behind designing these systems is to facilitate timely diagnosis and prediction of a large spectrum of diseases. This involves acquisition of clinical and physiological data from diverse sources. Although several studies have reported the considerable benefits involving role of IoT in these services [48, 49], however, the convergence of blockchain with these systems is still in its primary phase. The smart diagnostic systems mostly constitute of massive knowledge derived from diverse healthcare units and medical experts. Apart from these information, they also entail some sensitive patients' information like their names, medical records, location, and billing information; this may be prone to information theft, cause violation of privacy of the patients, and may also lead to financial loss. The blockchain technology can benefit these systems extensively by facilitating management of information acquired from diverse sources and securing their contents.

7.2 Smart wearable devices

Several wearable sensory devices and smart fitness devices are nowadays being used by a much larger population all over the world. Their ease of use and portability have made these low-power and low-cost devices much amiable across several healthcare platforms whether it may be for tracking the movement of elderly people or even for monitoring activity levels of individuals. These devices mostly comprise smart glucometers, oximeters, smart watches, etc. The location and health information carried by these devices makes them prone to several security and privacy threats, malicious intrusions, and so on. In [50, 51], different encryption schemes have been reported in this context along with their compliance for smart wearable devices. The use of blockchain technology can also prove to be quite efficient in this context in preserving the privacy and secrecy of information for WBANs. However, extensive monitoring activities and generation of keys can put some impact on energy constraints of the system and computational and communication overheads of blockchain-based framework [52].

7.3 Remote monitoring and critical care systems

The remote monitoring systems comprise a multisensor environment closely deployed in the patients' environment to facilitate continual and accurate monitoring of their activities or health conditions [49]. In Ref. [53] an IoT-based healthcare system was discussed for monitoring the movement of patients with Alzheimer's disease. In this specific scenario the deployment of highly interconnected sensory devices can efficiently assist in capturing information regarding patients' mobility patterns and prevalence of any conditions arising as a consequence of the disease. These systems for their efficient functioning may essentially require the transmission of patients' information between doctors and caregivers through EHRs. In this view the use of anonymization may show potential benefits for blockchain-based remote monitoring systems in preserving confidentiality and secrecy of EHRs [52]. Although by implementing blockchain in these highly distributed systems can have substantial benefits, however, this cannot prevent the patients' sensitive medical records and other underlying details against some specific attacks such as the linking attack that is quite common in these systems [54]. While the patients' information and EHRs are transmitted over different IoT-based systems, the attackers cause theft of some of the information being generated from these diverse sources and link them with the anonymized data to derive some private information from them, hence leading to linking attacks.

Although the advent of blockchain in the healthcare systems have brought upon drastic revolution to conventional healthcare practices, still the security and privacy issues require to be addressed with each advancements. Presently the implementation of blockchain has benefitted the patients and several healthcare organizations to improve the quality of life and well-being of its users. However, several scoping issues and applications are yet to be addressed. This has paved new platforms for the research community to explore and experiment the possible benefits and implication of blockchain with many cutting edge technologies.

8 Conclusion

The expansion of healthcare amenities driven by smart information retrieval and delivery mechanisms has increased the scope for inter- and intracare facilities among the healthcare industries. This involves exchanging of highly sensitive information regarding patients' health records like pathological test results, billing information, and insurance details. Thus a highly risk-free approach is required for facilitating the transmission of information over several remotely distributed platforms. In this chapter a secure architecture based on blockchain technology for acquisition and distribution of healthcare information was provided. The associated information is mostly acquired through IoT-based devices constituting of sensory devices over a large scenario to facilitate inter- and intracare health facilities to the patients. The EHR technology is discussed along with its compliance to blockchain platform. The pitfalls of conventional EHRs are discussed along with the benefits brought into their intrinsic traits by introducing blockchain technology that has been provided. It is observed that by exploiting the capabilities of blockchain framework, more traceability, transparency, scalability, reliability, and interoperability can be obtained in the existing healthcare services.

Some popular blockchain technologies are discussed along with their SDK implementations. This can greatly assist users and healthcare providers to select the appropriate type of blockchain that would suit their requirements. From the empirical studies, it is evident that blockchain platform can add more scalability, interoperability, and transparency in the exchange of EHRs and other medical information. The security aspects of IoT-based blockchain frameworks for healthcare sectors have been discussed. It is observed that by leveraging the intrinsic capabilities of blockchain, extensive functionalities can be achieved for the management of large-scale medical information and their secure transmission. To have a better understanding of the functionalities and services offered by blockchain, some popular real-world case studies have been investigated. The use of blockchain in most of the situations has been found to be more befitting to process millions of transaction within very less time. The empirical data obtained for two popular blockchain frameworks, namely, the Bitcoin and Ethereum, have been analyzed for some of the crucial parameters like network difficulty, transaction value, hash rate, and transactions communicated by blockchain. It is found that the Ethereum blockchain can handle increasing number of transactions as compared with Bitcoin. Thus it can be concluded that in the near future blockchain technology would be a remarkable choice for many industries to achieve interoperability, scalability, security, cost efficiency, and transparency of information flow.

References

[1] V. Dhillon, D. Metcalf, M. Hooper, Blockchain Enabled Applications: Understand the Blockchain Ecosystem and How to Make it Work for You, Apress, 2017.

[2] P. Zhang, M.N.K. Boulos, Blockchain solutions for healthcare, in: Precision Medicine for Investigators, Practitioners and Providers, Academic Press, 2020, pp. 519–524.

[3] A. Banerjee, Blockchain technology: supply chain insights from ERP, in: Advances in Computers, vol. 111, Elsevier, 2018, pp. 69–98.

[4] S.B. Caceres, Electronic health records: beyond the digitization of medical files, Clinics 68 (8) (2013) 1077–1078.

[5] L. Rosenbaum, Anthem Will Use Blockchain To Secure Medical Data For Its 40 Million Members In Three Years, Forbes, 2019. Available at:https://www.forbes.com/sites/leahrosenbaum/2019/12/12/anthem-says-its-40-million-members-will-be-using-blockchain-to-secure-patient-data-in-three-years/#4b822e816837.

[6] J.D. Halamka, A. Lippman, A. Ekblaw, The Potential for Blockchain to Transform Electronic Health Records, Harv. Bus. Rev., 2017. Available at:https://hbr.org/2017/03/the-potential-for-blockchain-to-transform-electronic-health-records.

[7] A. Shahnaz, U. Qamar, A. Khalid, Using blockchain for electronic health records, IEEE Access 7 (2019) 147782–147795.

[8] A. Roehrs, C.A. da Costa, R. da Rosa Righi, V.F. da Silva, J.R. Goldim, D.C. Schmidt, Analyzing the performance of a blockchain-based personal health record implementation, J. Biomed. Inform. 92 (2019) 103140.

[9] M. Kassab, J. DeFranco, T. Malas, V.V.G. Neto, G. Destefanis, Blockchain: a panacea for electronic health records? in: Proceedings of the 1st International Workshop on Software Engineering for Healthcare, IEEE Press, 2019, pp. 21–24.

[10] N. Kshetri, Blockchain and Electronic Healthcare Records [Cybertrust], Computer 51 (12) (2018) 59–63.

[11] B.L. Radhakrishnan, A.S. Joseph, S. Sudhakar, Securing blockchain based electronic health record using multilevel authentication, in: 2019 5th International Conference on Advanced Computing & Communication Systems (ICACCS), IEEE, 2019, pp. 699–703.

[12] E. Zaghloul, T. Li, J. Ren, Security and privacy of electronic health records: decentralized and hierarchical data sharing using smart contracts, in: 2019 International Conference on Computing, Networking and Communications (ICNC), IEEE, 2019, pp. 375–379.

[13] S. Bebortta, M. Panda, S. Panda, Classification of pathological disorders in children using random forest algorithm, in: 2020 International Conference on Emerging Trends in Information Technology and Engineering (ic-ETITE), IEEE, 2020, pp. 1–6.

[14] M. Talal, A.A. Zaidan, B.B. Zaidan, A.S. Albahri, A.H. Alamoodi, O.S. Albahri, et al., Smart home-based IoT for real-time and secure remote health monitoring of triage and priority system using body sensors: multi-driven systematic review, J. Med. Syst. 43 (3) (2019) 42.

[15] L.A. Durán-Vega, P.C. Santana-Mancilla, R. Buenrostro-Mariscal, J. Contreras-Castillo, L.E. Anido-Rifón, M.A. García-Ruiz, et al., An IoT system for remote health monitoring in elderly adults through a wearable device and mobile application, Geriatrics 4 (2) (2019) 34.

[16] D. Senapati, Generation of cubic power-law for high frequency intra-day returns: maximum Tsallis entropy framework, Digital Signal Process. 48 (2016) 276–284.

[17] S. Bebortta, D. Senapati, N.K. Rajput, A.K. Singh, V.K. Rathi, H.M. Pandey, et al., Evidence of power-law behavior in cognitive IoT applications, Neural Comput. Applicat. (2020) 1–13.

[18] S. Bebortta, A.K. Singh, S. Mohanty, D. Senapati, Characterization of range for smart home sensors using Tsallis' entropy framework, in: Advanced Computing and Intelligent Engineering, Springer, Singapore, 2020, pp. 265–276.

[19] K. Cagle, Rise of Smart Contract, Forbes, 2019. Available at:https://www.forbes.com/site/cognitiveworld/2019/03/10/rise-of-the-smart-contract/#380b21f479f4.

[20] A. Alamri, Ontology middleware for integration of IoT healthcare information systems in EHR Systems, Computer 7 (4) (2018) 51.

[21] C. Pop, T. Cioara, M. Antal, I. Anghel, I. Salomie, M. Bertoncini, Blockchain based decentralized management of demand response programs in smart energy grids, Sensors 18 (1) (2018) 162.

[22] S. Purkayastha, Compare Eight Blockchain Platform to Kick Start Your Next Project, RadioStudio, 2019. Available at:http://radiostud.io/eight-blockchain-platforms-comparison/.

[23] Hyperledger, Retrieved on November 2019 from:http://hyperledger.com.

[24] Y. Zhou, J.S. Ancker, M. Upahdye, N.M. McGeorge, T.K. Guarrera, S.C. Hedge, et al., The impact of interoperability of electronic health records on ambulatory physician practices: a discrete-event simulation study, J. Innovat. Health Inform. 21 (1) (2014) 21–29.

[25] W.J. Gordon, C. Catalini, Blockchain technology for healthcare: facilitating the transition to patient-driven interoperability, Computat. Struct. Biotechnol. J. 16 (2018) 224–230.

[26] K.R. Persons, J. Nagels, C. Carr, D.S. Mendelson, B. Fischer, M. Doyle, Interoperability and considerations for standards-based exchange of medical images: HIMSS-SIIM collaborative white paper, J. Digital Imaging (2019) 1–11.

[27] B.A. Stewart, S. Fernandes, E. Rodriguez-Huertas, M. Landzberg, A preliminary look at duplicate testing associated with lack of electronic health record interoperability for transferred patients, J. Am. Med. Inform. Assoc. 17 (3) (2010) 341–344.

[28] J.B. Perlin, D.B. Baker, D.J. Brailer, D.B. Fridsma, M.E. Frisse, J.D. Halamka, et al., Information Technology Interoperability and Use for Better Care and Evidence: A Vital Direction for Health and Health Care, (2016) NAM Perspectives.

[29] I. Koksal, The Benefits of Applying Blockchain Technology in Any Industry, Forbes, 2019. Available at: https://www.forbes.com/sites/ilkerkoksal/2019/10/23/the-benefits-of-applying-blockchain-technology-in-any-industry/#4c83a5eb49a5.

[30] M.S. Rahman, I. Khalil, P.C. Mahawaga Arachchige, A. Bouras, X. Yi, A novel architecture for tamper proof electronic health record management system using blockchain wrapper, in: Proceedings of the 2019 ACM International Symposium on Blockchain and Secure Critical Infrastructure, ACM, 2019, pp. 97–105.

[31] Q. Nasir, I.A. Qasse, M. Abu Talib, A.B. Nassif, Performance analysis of hyperledger fabric platforms, Secur. Commun. Netw. 2018 (2018).

[32] Biser Dimitrov, Major Improvements are Coming to Blockchain in 2020, Forbes, 2019. Available at:https://www.forbes.com/sites/biserdimitrov/2019/07/08/major-improvements-are-coming-to-blockchain-in-2020/#38cace0155b6.

[33] J.J. Hathaliya, S. Tanwar, S. Tyagi, N. Kumar, Securing electronics healthcare records in healthcare 4.0: a biometric-based approach, Comput. Electr. Eng. 76 (2019) 398–410.

[34] ANSI I, TS 18308 Health Informatics-Requirements for an Electronic Health Record Architecture, ISO, 2003.

[35] H. van der Linden, D. Kalra, A. Hasman, J. Talmon, Inter-organizational future proof EHR systems: a review of the security and privacy related issues, Int. J. Med. Inform. 78 (3) (2009) 141–160.

[36] P. Muñoz, J.D. Trigo, I. Martínez, A. Muñoz, J. Escayola, J. García, The ISO/EN 13606 standard for the interoperable exchange of electronic health records, J. Healthcare Eng. 2 (1) (2011) 1–24.

[37] S. Cao, G. Zhang, P. Liu, X. Zhang, F. Neri, Cloud-assisted secure eHealth systems for tamper-proofing EHR via blockchain, Inf. Sci. 485 (2019) 427–440.

[38] H.T. Wu, C.W. Tsai, Toward blockchains for health-care systems: applying the bilinear pairing technology to ensure privacy protection and accuracy in data sharing, IEEE Consumer Electron. Mag. 7 (4) (2018) 65–71.

[39] Y. Al-Issa, M.A. Ottom, A. Tamrawi, eHealth cloud security challenges: a survey, J. Healthcare Eng. 2019 (2019).

[40] N.A. Azeez, C. Van der Vyver, Security and privacy issues in e-health cloud-based system: a comprehensive content analysis, Egyptian Inform. J. 20 (2) (2019) 97–108.

[41] I. Masood, Y. Wang, A. Daud, N.R. Aljohani, H. Dawood, Towards smart healthcare: patient data privacy and security in sensor-cloud infrastructure, Wirel. Commun. Mob. Comput. 2018 (2018).

[42] Change Healthcare Using Hyperledger Fabric to Improve Claims Lifecycle Throughput and Transparency, Available at:https://www.hyperledger.org/resources/publications/changehealthcare-case-study.

[43] Bitcoin Explorer, Accessed on December 2019. Available at:https://www.blockchain.com/explorer.

[44] Etherscan, Accessed on December 2019. Available at:https://etherscan.io/.

[45] G. Walker, What is the difficulty?, Available at:https://learnmeabitcoin.com/beginners/difficulty.

[46] Explaining Hash Rate or Hash Power in Cryptocurrencies, https://coinsutra.com/hash-rate-or-hash-power/.

[47] Bitcoin Confirmations, Available at:https://www.buybitcoinworldwide.com/confirmations/.

[48] P. Amirian, F. van Loggerenberg, T. Lang, A. Thomas, R. Peeling, A. Basiri, S.N. Goodman, Using big data analytics to extract disease surveillance information from point of care diagnostic machines, Pervasive Mobile Comput. 42 (2017) 470–486.

[49] T. Saheb, L. Izadi, Paradigm of IoT big data analytics in healthcare industry: a review of scientific literature and mapping of research trends, Telematics Inform. (2019).

[50] H. Zhao, Y. Zhang, Y. Peng, R. Xu, Lightweight backup and efficient recovery scheme for health blockchain keys, in: 2017 IEEE 13th International Symposium on Autonomous Decentralized System (ISADS), IEEE, 2017, pp. 229–234.

[51] J. Liu, W. Sun, Smart attacks against intelligent wearables in people-centric internet of things, IEEE Commun. Mag. 54 (12) (2016) 44–49.

[52] M.U. Hassan, M.H. Rehmani, J. Chen, Privacy preservation in blockchain based IoT systems: integration issues, prospects, challenges, and future research directions, Futur. Gener. Comput. Syst. 97 (2019) 512–529.

[53] P.A. Laplante, N. Laplante, The internet of things in healthcare: potential applications and challenges, It Professional 18 (3) (2016) 2–4.

[54] G. Danezis, Statistical disclosure attacks, in: IFIP International Information Security Conference, Springer, Boston, MA, 2003, pp. 421–426.

Further reading

How Walmart Brought Unprecedented Transparency to the Food Supply Chain With Hyperledger Fabric, Available at:https://www.hyperledger.org/resources/publications/walmart-case-study.

How the National Association of REALTORS Improved Member Services With Hyperledger Fabric, Available at:https://www.hyperledger.org/wp-content/uploads/2018/05/Hyperledger_CaseStudy_NAR_V3.pdf.

Applications of IoT for human

Designing an effective e-healthcare system using Internet of Things

12

Tanveer Ahmed[a], Rishav Singh[b], and Rohit Verma[c]

Bennett University, Greater Noida, Uttar Pradesh, India[a] NIT-Delhi, Delhi, India[b]
Manipal Institute of Technology, Manipal, Karnataka, India[c]

1 Introduction

The development of healthcare system has been one of the major developments of academia and industry. However, this aspect of civilization has been varying across the world through different ages. In the present era of ICT driven economy, society healthcare also seems to witness a makeover in terms of applications that are not only technologically advanced but also controlled through codes of computation. This advancement in the working of healthcare is driven by the idea that our societies are undergoing a change to become a society running through networks of web [1]. Moreover, to achieve the highest accuracy in analyzing a disease and also its successful treatment, the latest inventions in IT seem to be lending the best hand at its development.

The web and many applications flowing through its network have become the essentiality of today's modern lifestyle. The societies are now curbing their urban (physical) limits by engaging themselves through connectivity created by the Internet. Hence, in healthcare, one major hurdle that needs to be passed is the reach to areas that have been left untouched by physical developments like smart hospitals. However, a solution to bridge this gap between technological advancement and marginalized areas and population is the remote monitoring of the patients living away from main centers of treatment. This personalization of the clinical care system emerges as a dream not far away from evading the differences between people with resources and people with less resources based on remote areas.

The rural areas, particularly in developing countries, have been the spaces where development in any form has been reported not to enter on time. In this case what remains unprecedented is the all-round implementation of the schemes of development, and the situation remains unnerved. The development of proper healthcare remains a challenge as the nonadaptability from the beneficiary side becomes the key lacunae in reaching this goal. In countries like India a major population lives in the rural areas that need complete training in the usage of ICT-based technologies mainly personal health systems for betterment of health conditions. Though there has been some level of change in using smart phones among the lower middle classes of Indian population, we still need a major campaign regarding digital literacy. The receptivity of changes and adaptability to new technologies must be motivated by schemes that make people understand the importance of IoT-based applications. These applications are aimed to revolutionize various aspects of technology on which human life has depended heavily lately. Hence, to design an effective healthcare system for implementation of best technologies in this field in

IoT Based Data Analytics for the Healthcare Industry. https://doi.org/10.1016/B978-0-12-821472-5.00019-3

general and for the rural population in particular, one needs to combine that element of ICT with healthcare technologies that may be handled not only by structural systems like hospitals but also personal. Hence, in this age, Internet of Things (IoT) comes as the only answer to the development of remote monitoring and personalized clinical care system [2].

IoT is a newly developed stage of the Internet that has become a reality. It is basically a crossover of "things" that are nothing but electronically run devices transmitting data through networks. The IoT is therefore a conjunction of several such devices connecting the flow of information and working of an enterprise or a system without any human interaction. Therefore, in the case of developing a healthcare system with the most advanced technology, IoT in this direction may help in reaching the aforementioned goals of remote monitoring and personalized clinical care system. Moreover, placing this idea of IoT in schematic development of healthcare systems in rural remote areas of developing countries seems to be the call of the day. Hence, this chapter aims to bring above the possibilities of better implementation of healthcare facilities through an understanding embedded in IoT.

The healthcare space may become extremely beneficial with the usage of IoT, with aspects that sustain and make possible IoT implementation of facilities earlier not present in the healthcare system. These features range across academically trained human resources mainly in ICT whether a doctor or a system engineer along with various developed machine-based technologies and gadgets that when combined together forms a system of advanced healthcare. Within this system the trained individuals both in medical and ICT expertise form an integral coordination with various machines through networks of web of various devices that run through various nodal centers combining together to form an effective healthcare system. This system will include the flow of data generated at each point be it the doctor or the patient or some intervener who is monitoring the subsidiary system monitored in the core system area. Moreover the patient (or family) will be made more technologically responsible and maintain their well-being on their own.

Hence the major objective of this chapter revolves around juxtaposing the present prevalent healthcare condition and bringing IoT to bridge the gap. This chapter thus aims toward developing a basic understanding of IoT as an answer to the two major gaps within the healthcare system today. These are majorly the requirements for remote monitoring and personalized healthcare system. This chapter aims to develop a basic understanding of how IoT comes up as the need of the hour for reaching the remote monitoring of healthcare implementation in remote conditions. This chapter then proceeds to bring above the possibility of creating a personal healthcare system in the hands of a person.

This chapter is divided into four major parts with each part acting as a precursor enabling the understanding for the need of the next stage, hence reaching the final determined destination of the healthcare system run through IoT. The first part brings before us a descriptive note on the general landscape of present day healthcare services through the lens of structural and policy level proceedings embedding IoT in its features. The second part however brings above the development of IoT as the demand of the day and solution to many issues of healthcare implementation. The third part then covers the two major areas of healthcare aimed to develop through IoT, which is remote monitoring and personalized clinical care system. The fourth part will put some light on the social and digital conjunction of these healthcare systems being developed under IoT theory, hence trying to highlight the various aspects of healthcare that will be affected both positively and negatively on the healthcare system in general. The fifth part marks the end of the chapter with a much needed discussion on the demand of an effective healthcare system rising in the changing social environment that is running very fast to become a society and economy connected through networks of web technology.

1.1 Situation of healthcare in India: IoT within the policies

The Indian healthcare scenario has been facing very slow growth, and the fight to overcome many chronic diseases is still on but at a very slow pace. The major victims of diseases remain generally unaware of the fact that there are certain symptoms and certainly certified preventive measures that may help them before landing in the soup. However, India shows a major holdup to catch up with international standards. Moreover, infant mortality and maternal mortality rate is quite high (http://www.optum.in/thought-leadership/library/internet-healthcare-things.html, Accessed on 4th March 2020) [3].

The cause to the healthcare system's slow growth rate may be understood in the overall growing population and also lack of proper medical facilities and moreover its overall accessibility by the most needy and remote rural population.

The IoT has made possible the newly developed congruence of various innovative devices that had already added to better treatment. However, with IoT, these ICT-driven devices are now conjoined to work in a cohesive manner producing much more accuracy and efficiency. With technology-driven healthcare the possibility to create a system that is more personal and centered around patient is high. With smart phones penetrating in households in remote areas, also, the idea of reaching the last patient might be reached. The limitations of tracking patients' daily symptoms, ailments, and alteration of medication accordingly may be met with IoT in the core healthcare aiming toward well-being of patients. Right treatment at the right time has always been the challenge among the Indian ailing patients as physical distance still marks the inability to reach the destined centers of treatment.

Indian Government has recognized this need and has been working to draft policies in this direction where healthcare and IoT are brought together. The schemes of government mainly regarding healthcare have been facing challenges in their implementation. There are several policy level and management-related constraints that keep the health facilities for rural or remote areas backward [4]. Kasthuri [5] very elaborately mentions the lag within the Indian healthcare system through the five A's. According to him the five A's are being aware of basic healthcare or lack of it; having access to healthcare facilities or lack of it; the absence of human resource; affordability or cost of healthcare; and, the last responsible A, accountability or lack of it toward the people and authority involved in providing healthcare facilities. This work by Kasthuri [5] hence marks the core lacking within the healthcare system right from the stage of basic knowledge about the symptoms and the government initiatives and their proceedings.

In one of the studies by Ramani and Mavalankar [6], the major challenges that this country is facing in the area of healthcare has been listed. The paper very broadly talks about the challenge and need of building health systems. It suggests scaling of financial hurdles by bringing the public private partnership concept to existence through various endeavors. It also talks about the tackling of nonfinancial obstacles in the delivery system through proper governance. Availability, affordability, and access to healthcare issues also remain a challenging aspect that needs to be addressed strongly.

In all these studies where loopholes of the healthcare system in India have been highlighted, the major issues remain very much managerial. Though there is still hope to reach the accurate levels of diagnosis in healthcare by developing more advanced technologies, the deliverance of the existing facilities remains a hard task to perform. The healthcare facilities that have been developed in the light of contextual understanding of diseases and lifestyle breeding illnesses need proper looking over. Hence, in this case, personalized healthcare systems become apparent to reach the goal of treating each one with the most effective way. In the cases of urban poor and rural remote area people, we also face the accessibility issue. Though there are several schemes available for the marginalized group, an efficient deliverance of these policies still remains in the backseat.

The whole system of healthcare hence needs a technological shift that would enable the proper reach of the facilities to the groups who have difficulties in reaching these centers. However, in the last few years, the government of India has decided to inculcate certain technological change through which they aspire to change the whole scenario of healthcare in India.

Point 5.1.4 in the Draft Policy of IoT by Government of India very strongly recommends a healthcare system engrossed with IoT and calling it smart health (remote). It describes two major agendas in this new endeavor. The first agenda is to set up projects for monitoring various vital parameters of patients like subtle changes in pulse, respiration, heart condition, temperature, and preventive warning on early onset of pneumonia (in small children) or other life-threatening problems, inside hospitals and at remote patient locations including old people's homes and ambulances, while the second point of agenda emphasizes to develop projects to detect and provide support to old age persons in case of fall [7] (p. 7, Government of India).

Hence, through these proceedings of government, we may see a wave of change in the approaches in which the health-related schemes are designed and implemented through the principles of IoT.

1.2 IoT: A new call of the day in healthcare

Healthcare all across the world has been revolutionized through inventions that have changed the diagnostic and the treating capabilities. However, major innovations like ICUs, labs, and operation theaters have been much more evitable and developed positively in the light of advancements rooted in ICT. In the present era when healthcare has successfully been integrating with these ICT-enabled systems, IoT emerges as an enhancing factor by connecting these devices of healthcare. IoT promises to help provide real-time data acting as an interface between the major points of the system that are doctors, patients, diagnosticians, etc. The dependence on the IoT in healthcare is increasing day by day considering many factors like easy access to healthcare and reduction in treatment costs [8].

Every individual has a unique sociocultural characteristic. Hence, it becomes very necessary that the whole ideology of medical treatment should be brought in cohesion with practice of well-being. Further, to achieve a more sustainable system of service providence to the patients, one needs to get through the early symptoms. Apart from this during a clinical time period for a patient treated with chronic disease, remaining in hospitals for a very long time becomes expensive. A person affected by this kind of medical condition can best be served with periodic diagnosis and early awareness for medication. But if this condition is fulfilled within the homecare, the whole medication becomes far more effective personally and economically. IoT, in this case, becomes the need of the hour as it enables a digital identity of a person within the healthcare domain [9].

Healthcare system has made several strides forward with technology giving much support in the form of low-cost sensors. Moreover, cloud computing and smartphones have taken the paradigm to new heights. Another major development is the wireless communication technology that helps keep up the healthcare of a person even from a distance. Hence the remote accountability of a patient has become a reality with real-time data transmission from patient to doctors and doctors to patients. Hence, appropriate action is now a possibility within the required timeframe [10].

In the history of development around ICT within the healthcare system, big data turned to be a revolutionary phase. It was with big data that healthcare was visualized in the light of new social changes particularly the lifestyle. The whole concept of developing a smartphone-based healthcare system began with the era of big data. Moving beyond this later came cloud computing, edge computing, IoT to frame devices, and their execution for the advancement of personalized healthcare systems giving way to remote monitoring. The fast and accurate diagnostic system was achieved through gadgets like smartwatches, smartphones, and headbands, which would help in detecting body temperature, blood pressure, and other indicators of changing health scenarios.

According to the work presented in Jin et al. [11], there are three main aspects for an IoT system: (1) network-centric IoT, (2) cloud-centric IoT, and (3) data-centric IoT. Each point has several attributes that form the structure of the whole system defining the IoT while giving support to each aspect making this relationship most vital for the technology to sustain.

Fengou et al. [12] proposed structural design of a telemonitoring system. This includes functions for data collection, data management, and analysis. It is based on the idea of IoT view that is centered on data. In another idea, data-centric IoT module, the flow of the data is described. Major work that comes up is by Lake et al. [13] who trace the flow of data reaching cloud platforms. In Lee and Palaniappan [14] a self-governing system was proposed that was designed to handle chronic diseases. This architecture can also be extended for remote monitoring.

2 Remote monitoring

A major problem that has been faced in medical history has been the effective health monitoring systems. It is due to this that many health risks remain unknown and are subsequently followed by the late and inefficient treatment. However, with growing technological alterations, we have witnessed at this juncture of ICT-enabled technologies that we have entered a phase where we may get an answer to the aforementioned problems. IoT with its strong wireless capabilities becomes the solution to ever known effective monitoring systems. The capture of data around a patient from time to time changing health-related symptoms may be recorded directly through this. Though a person and his/her doctor could be in a different space with distance becoming the smallest hindrance, the doctor will be able to analyze the ailments from far beyond physical touch. Hence, proper medical aid will be advised to the patient away from the doctor in the time when the ailment is not very aggressive [2] (p. 6229). The major problem of yesteryear has been that the people in remote areas die very early mostly with small and minute symptoms of a disease taking a big form due to negligence and late diagnosis.

Fig. 1 explains the working of a remote monitoring health system based on IoT. In this diagram a basic understanding of working remote monitoring health systems is represented. Sensors that are home based mostly are connected to the monitors in hospital through the interface of IoT. Through this interface, only real-time data recording and transmission happen, and command from hospital in real time about the quick health aid is made. The benefits and impact of the remote monitoring health system hence are listed in the diagram primarily as telecare management of chronic diseases and self-management of health indicators, and it also helps in reducing the hospital bills as the cost regarding stay in a hospital and also that of a nurse is minimized.

Remote monitoring hence allows the capture of patient data through wireless connections connecting the devices responsible for monitoring the patient and its condition from time to time. IoT is run through several complex algorithms and sensors that are there to share and analyze the data from the recipient to health authorities and vice versa. This may be seen as enabling that idea of healthcare where

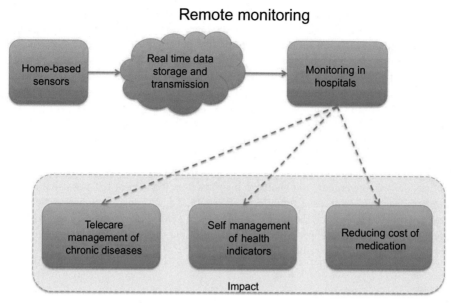

FIG. 1

Remote monitoring.

a remote understanding of the time to time health alterations of a person is recorded through machines using Internet [15] (p. 7). Hence, IoT becomes the guiding tool in reaching up to a level of inventions where distance is omitted through webs of connectivity of Internet. This connectivity is embedded in the ideological advancement of IoT that has now become the core tool of healthcare facilities to reach the marginalized.

In this process, remote monitoring, which may be also seen as telecare management of chronic diseases, becomes an easy process by giving space to self-management and reducing the cost of treatment. The self-management or homecare makes the cost of diseases bearable as now the daily living cost within the hospital is curtailed. The regular connectivity of patients and their families on smartphones and other devices have made it possible for the transmission of health-related data. Hence, to implement an IoT-enabled system of healthcare, one has to choose proper devices that will make the connectivity proper and achievable. In this case the devices with characteristics allowing interoperability and data transmission make the best applicability of this idea of healthcare innovation through IoT [16] (p. 29).

According to Woo et al. [17], the whole idea of connecting the IoT-based devices is through WBAN. However, this work goes a further step by proposing to connect these devices through a wearable sensor node run by solar energy and connected through IoT. These sensors may be attached to different parts of the body where they can monitor the body temperature and heartbeat. Through this, particularly for the node of the emergency accelerometer, the fallings in the body can be recorded, and proper action can be taken on time. This idea proposed in future may help in exchanging more significant data through wearable sensors that work on the principle of WBAN. This study then adds up to the technological development in the area of remote monitoring that is enabled and connected through the principle of IoT. The whole aim of achieving the remote monitoring is enabled through this proposed model that promises very patient friendly and wearable devices, the recordings of which will be displayed on a smartphone. Remote monitoring therefore becomes very easy with record keeping and transmission through wearable devices and smartphone connected through IoT.

With IoT-driven noninvasive devices, one may be able to analyze the collected data mainly physiological status of a patient who needs regular monitoring. Sensors are employed to connect to the body for collecting data around the symptoms and ailments. These sensors collect data from the body of patient recording their changing body temperature, blood pressure, and more so forth through gateways and clouds. This information is then sent through wireless connectivity to the people responsible for further action during treatment. Through this innovative change the role of a health professional that needed to be always in attendance with the ailing patient is curtailed. This change directly affects the cost of treatment and also devices replace human errors with much accurate data and analysis. The idea of the situational analysis of an emergency health condition without getting depended on human intervention hence can only be achieved through IoT-embedded devices.

IoT-driven healthcare system employs the technique of collecting data through cloud and gateways only to be analyzed by doctors and other trained medical professionals on time. The noninvasive nature of this technology based on ICT led to the innovation of IoT. IoT is emerging as a system revolutionary for healthcare all across the globe that aims to connect multiple sensors. IoT becomes very important in this era because in this world of growing population mainly in developing countries, this helps to balance the ratio of recipients of health assistance to that of health care givers. The world enters a new era with IoT in reference to medical facilities because this new development helps connect two categories of ubiquitous and uniquely identifiable devices: first ones having intelligence and second those that have web-based networking. This finally helps to establish a system that helps in reporting and decision-making [18].

The applicability of such innovative technological development can be seen in many particular situations in the healthcare system. One of the most vulnerable sites of healthcare giving is the monitoring involved with ICU patients. It is pertinent that this type of circumstance needs a full time onlooker to perform the right type of treatment at the required time. However, with dependency on human resources for this kind of full day job, the system becomes susceptible to errors. These include untimely reporting of a change in invasive and noninvasive aspects of the body. It is generally understood in medical studies that certain symptoms create patterns according to the diseases, and hence the medicines react on a prescribed timeline. However, the medical system also talks about special cases where anatomy of a human body differs from a fellow being. This may be understood as the idea that each human body has unique functioning. Keeping in consideration of this medical fact, the situation of an ICU patient becomes very risky as one's body reaction on the effect of a medicine is very different.

A continuous system to monitor patients was introduced by Prajapati et al. [19] in their work. In this proposed research model, it was highlighted that the basic idea of life saving activities that are involved in the treatment of ICU patients revolves around data collection of blood pressure, temperature, and several others. Patient who is bedridden is subjected to regular monitoring of oxygen level, brain pressure, heart rate, etc. These are very minute details that may indicate the need for alterations in medication at every changing time span. In this case devices if they are connected through IoT helps build a more accurate and timely healthcare systemic atmosphere. Hence the patients can be treated every minute through remote access of the body also by interconnection of networks to the server. In the case of the brain, certain devices can measure the intracranial pressure. Similarly, other body parts may be also connected to doctors through devices that can be worn. After collecting data, communication with doctors can be done via a personal digital assistant (PDA). After this the system can analyze the data, and preventative measures can be taken. This is the systematic way in which the IoT led healthcare system enables remote monitoring for patients.

Moving on to the other gaps that the healthcare system is facing in today's time is to gel up with changing societal norms and lifestyle. In a perspective to understand the recipient's point of view of various medical facilities, one may notice that the society is undergoing a big change in terms of lifestyles which highlights the shrinking of families to individuals leaving no relative available for healthcare. Today in this era of ICT-driven network-based economy and society, we as a country are definitely witnessing a makeover in our lifestyles. Today across classes, various gadgets have taken over the life processes. In various situations, we find ourselves getting economic gains from beyond a certain geographical boundary sitting at home only. The new wave of entrepreneurs sitting at home and running their business through the Internet is the era of life we are entering into. As discussed earlier, healthcare also has become a sector where one is getting used to certain devices that have made diagnosis more accurate and personal. IoT in this case acts as a web of interconnection that has made healthcare more personal.

2.1 Personalized health care system

The pertinent changes in societies have made the human relations in some cases very constrained to spaces leading to no reach and indifference. Old age homes are a very common outcome of this social change. India has particularly witnessed this change with much grave circumstances because of its old dependency of a society that was composed of joint families. The joint families gave space to each member of the family equally. The rights and responsibilities were equally distributed among all the members of the family. Hence, if a family member was ill and bedridden, it was not necessary that the close kin like parents, siblings, and spouses will only look after the person affected. It was the moral duty of each member to take responsibilities together. But with disintegration of the families and joint families being turned into extended families where children dissociate from the family geographically due to job conditions etc. to form a nuclear family in other cities. This makes it impossible for the children to reach their parents and other elderly kin members in time of health emergencies. The shrinking spaces in cities do not allow these kids to keep their elders in the same abundance that they have experienced all their life in small towns and villages. Moreover, even if the families live together, also the children who can look after the elders will have jobs and cannot be present in person to look after their close ones constantly. Hence the healthcare in the wake of changing social relations have also witnessed challenges of personal care and monitoring in case of patients with chronic diseases.

Moreover, in case of chronic diseases, it is the old age people who suffer the most. In this case recent developments in the coming of wearable devices and their connectivity to another device/monitor help to keep a tap on the every moment health conditions of a patient by their families or by themselves also very accurately and efficiently. This advancement in technology related to personalized healthcare devices may ease in keeping an account of the regular health conditions of the patients. This comes up as a very economical pursuit, and also it cuts down the expensive human resource expenditure on nurses.

The devices help the patient themselves and family members at far distance also to predict health conditions more accurately minimizing the risks of human error.

However, the role of IoT within a healthcare system may be understood in terms of its usage in creating connectivity within the various devices. The new developments in the form of wearable devices or gadgets of movable nature have made the possible implementation of IoT in creating the most effective personalized healthcare system.

Fig. 2 explains a fundamental diagram explaining the working of a personal health system enabled through IoT. This diagram shows that through wearable devices that are sensor based, storage and transmission of data are done in real time. The various features of the personal health system like health indicator, alarm system, regulating body activities like calorie intake or exercises, and its small pocket size make it easy for the user to carry it anywhere and maintain health at any point of time. The major impact of all these features is that this personal health care system helps the patient in minimizing the risk factors by monitoring and regulating various health indicators on regular intervals making the diagnosis easy and timely.

Moreover the probable conditions within which IoT may be applied in the healthcare system required certain preconditions. To ensure the performance of various factors within the IoT led healthcare system, the Consultants from Cambridge have used ANSYS software. The ANSYS software makes sure that the antennas within the wearable devices perform at their best. It is very necessary that antennas should be present within the implanted device. This is important because these antennas enable the devices to catch frequency from distant places often from more than one. The devices should also

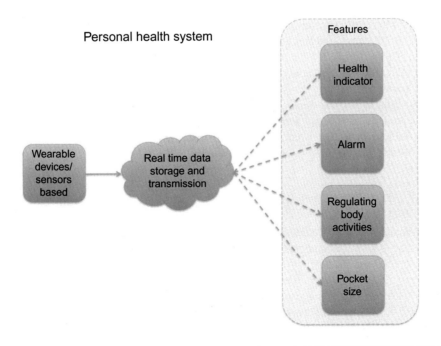

FIG. 2

Personal health system.

have the capability to be able to perform in various body types across gender and age (https://www. ansys.com/campaigns/internet-of-things/wearables-and-medical-devices, Accessed on 23rd March 2020 at 5.10 p.m.).

The various applicable aids that have been developed through IoT act as the elements of personalized healthcare systems; one of the various such devices is neurostimulators. The **neurostimulators** are implanted beneath the patient skin. This provides relief to the various types of pain by blocking pain messages reaching up to the brain. This is done by delivering mild electrical signals to the body. This new technique may be experienced by the patient's before opting for a long-term treatment. Another device that has been developed in creating a more personalized health system is the hearing aids. There are various people who are born or are subjective to inability to hear naturally. These types of people are categorized among the disabled. However, it has been understood through various works that apart from hearing inability, their minds are perfectly sound. The coming of Starkey hearing technologies 900 megahertz is actually a wireless hearing aid technology. These wearable hearing devices can now be connected with multimedia devices without using wires. The various accessible electronic devices like television and smartphones are used to monitor and get connected to the hearing aid equipment. Hence, these devices of wearable nature when connected through wireless interface may be used for more than one purpose. In some cases the left and right hearing devices part maybe used for both maintaining the volume and making a memory of everything received as messages heard while using those hearing aid devices (https://www.ansys.com/-/media/ansys/corporate/resourcelibrary/article/charged-up-aa-v9i1.pdf?la=en&hash=46173BFADE6595A8959937E154641A31F28419B3, Accessed on 23 February 2020 at 5.15 p.m.).

Body-worn devices have become increasingly popular among the various sectors of people. In the United States, soldiers are expected to wear these wearable devices so that while fighting in extreme conditions, their vital conditions and medical needs may be understood by the medical authorities sitting far behind. This will enable for a quick rescue to the injured soldiers more accurately and on time. In the case of athletes too, the measurement and recording of their performance has become very accurate and on time.

The various famous devices developed on these lines are as follows:

The device named "**Myo**" used to control the motion for patients who need to perform certain activities after a fracture. This device has come up as a revolution in orthopedic as now patients' progress and the angle of each movement of the bone can be controlled by the doctors monitoring through phone and computers.

Another device is the **Zio patch** device measuring heart rate and electrocardiogram (ECG) signals. This device is used to understand the various risk factors and its conditions defining the status of heart in case of heart diseases.

Dario is a small all size device that can be fit in a pocket and carrot along with a person anytime and anywhere. It easily determines diabetes metrics, which includes carb counting and the various accountability e-off exercises and meditation we do. It automatically records the blood glucose measurements and also so analyzes the effect of changing scenarios. It shares the information guarding the factors changing the matrix of diabetes and provides personalized reports with the patient and people concerned. It is also designed with GPS location tracker in case of hyperalert situations so that families can stay in peace. The best part of this device is that it gives accurate results in less than 6 s. Moreover, it does not require any batteries and cables to run, hence making the life of a diabetic patient and their family members easy and risk free.

There are various devices built to curb the problem of sleeping disorder. This was done via the combination of IoT with big data analytics [20]. The various applications like sleep cycle, sleepbot, beddit sleep tracker, and sleep time have become very common in usage by people who suffer from sleeping crises. A disorder in the sleep cycle is one of the major causes for stress building and causing problems related to heart and hormonal imbalance. In this case, applications work ok to relax the users and sleep comfortably and also analyze sleep cycles of the people using it. The apps offer per detailed statistics around sleep timings and ID its leverage efficiency e in percentage. These applications come with in-built features like ok relaxing sounds and smart alarms that ensure the working of these apps toward maintaining sleep cycle and hence fighting the stress issues.

The IoT has also been used to build application and wearable devices in case of a specific disease called Parkinson's disease. In case of this disease, regular data analysis regarding the patient is much needed. One of the studies has built a health data collection system where Fast Health Interoperability Resources have been used. In this system, it is very much possible to determine the abnormal indications in walking like a tilt or instability of stride length among patients with this particular disease. It is true storage and transmission of the foot pressure data that is analyzed to understand the regular condition of a Parkinson's disease patient to provide relevant aids at times of emergency [21]. While building a health system against Parkinson's disease, scholars have proposed that a model based on Unified Parkinson's Disease Rating Scale (UPDRS) assessment can be integrated with IoT [22]. This will be needed to build a system like a smart health system.

Another device using IoT for helping in monitoring the patients and their medication has also been developed. These devices are called ingestible sensors. They are used to bring precision in medication and to avoid any kind of side effects caused by wrong dosage. One of the examples of ingestible sensors is that they may use stomach juice to cure schizophrenia depressive disorders and also in case of bipolar diseases. In case of disorders due to depression, the sensors also maintain the record of the diet being taken by the patient and also show information about the wrong impact of those food items.

There are devices that have been developed to maintain moods among people suffering from diseases that are determined by a person's mental well-being. The complicated lifestyle today has triggered the issue of mood variations among the people dealing with emergencies at the workplace and creating work life balance. These devices can study the brain waves and release low intensity currents. This has been designed to study the impact of stress on human brains and also show the elevation of mood swings.

Another device that has been developed in this field is the smart glass. These glasses have a headset that is mounted on the lens along with and HD camera that transmits in real-time image on the LED screen. This device helps the people with hazy visibility to magnify the image and identify objects properly and with clarity. This device aims to cure visual disability in future.

The development of various devices of personalized healthcare system rationalizes the relation between social structural change with IoT becoming the solution for several questions being raised in the physically distanced society but merged through webs of Internet and electronic devices.

2.2 Impact and effects of an IoT-based healthcare system

After getting a view of the various practical applications in healthcare based on IoT, we also understand it's growing accomplishment in the same way. The various benefits that can be seen as an out shoot of IoT and its implementation in the healthcare system along with few limitations are the development of a smart hospital based on IoT.

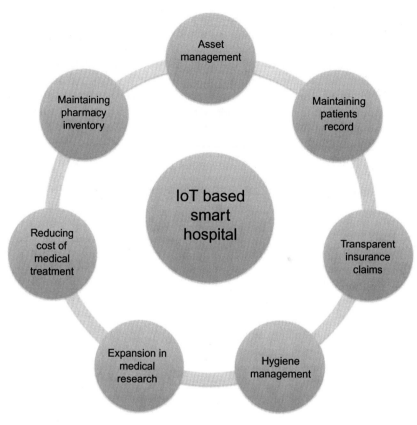

In Fig. 3 a listing of various features of a smart hospital based on IoT is done. However, a detailed description of all these features and its usage is also explained in the following part.

3 Uninterrupted health monitoring

Through all the personalized health-related gadgets and devices, the monitoring of health conditions among patients has become very prominent and easy. The various wearable devices like fitness bands and glucometers provide health status in the real time. It is also very much becoming convenient through these devices to keep up a check on the regular health status and lifestyle issues. These regular checkings may be keeping the count of calorie consumption, a check on the exercise routine, etc. In case of shrinking relationships and growing distances among family members both geographically and emotionally and in this situation, IoT has emerged as one of the boons in medical sphere to create a system where regular monitoring and personalized healthcare system is created within once home and connected to the medical experts very easily and efficiently.

3.1 Asset management through IoT

With the help of IoT, the hospitals today are able to implement various technical gadgets to maintain regular monitoring for patients curtailing human error. Another aspect of IoT development has completely changed the hospital scenario by making it more ICT-enabled smart hospital. This is primarily seen in the asset management activity managed through IoT. The various equipment like wheelchairs, oxygen pumps, nebulizers, and other monitoring equipment can now be traced with the help of their real-time locations by using the sensors connected though IoT in just a few minutes. Within the sphere of smart hospitals, it has also become possible to maintain hygiene monitoring through smart devices. The result of hygiene management through IoT is such that the hospital management has now overcome the age long issue of maintaining an eye on the hygiene workers, hence minimizing the possibility of any kind of infections to be spread within the hospital. Thus a smart hospital enabled through IoT will be able to maintain assets and maintain pharmacy inventory and cleanliness of the hospital making it a place less vulnerable to human errors and infections.

3.2 Keeping a track on patients records

With the help of wearable devices, a track of patient health can be kept by the medical experts regularly even if they are at their home. The diagnostic status of various risk factors that are sent to the physicians through the sensors embedded in the IoT interface helps in monitoring every small detail of a patient's health condition so that medical aid is sent to the patients on a regular basis and immediately. IoT also helps in maintaining a systematic record of patients so that the past medical history, current medical requirements, and also predicting future medical conditions along with the probable support needed are possible in the most efficient way. This record keeping helps in maintaining the crowd of patients that are recurring in nature. Medical records also help doctors to seek help from past records to implement the same medical treatment for new patients showing similar symptoms.

There are several survey reports that have covered the functioning of the system prevalent in the healthcare scenario. These surveys on a regular basis reveal that major issue that is at the managerial level for keeping detailed records of patients and their respective documents. The inability to record regular patient history has added to the failings of proper treatment by keeping an account of the medical history and so forth working in the proper direction. One of the surveys clearly mentions that this gap can be filled by using electronic health records (EHRs) based on the information technology. However, this system of record keeping through information technology (EHR), according to the report, is prone to mistake making. An alternate approach is employable electronic health record (EEHR). This is also known as WebEHR, which helps to connect the various health centers responsible for an all-round treatment of the patient. Thus, through this, quick sharing of proper information among these centers is enabled, which helps to take instant action against any problematic and emergency situation. Through this, we may observe how a small aspect of IoT-based application may help in the very old issue of record keeping. Medical treatment becomes very easy and comprehensible if the patient record history is available to the doctor. IoT in this case becomes the most reliable technological tool for an efficient data transmission through the long medical journey that the patient covers. This data transmission helps keeping the treatment more focused and accurate [23].

3.3 IoT enabled transparent insurance claims

The next level of achievement through IoT in the healthcare system can be seen in maintaining policies through IoT connectivity. The growing examples of health issues being tackled through insurance money have open post issues in the companies providing these economic benefits. There are several fraud cases of health emergencies in which money is retrieved. In these cases the data which are analyzed by the insurance companies are IoT enabled and are much more authentic and reliable. In other types of cases which are genuine the patients and family members have to suffer and they undergo various hassles to gain their claimed money. This creates a chaos within the family who supposed that with these insurance policies, they would not have to panic in case of medical emergencies but they suffer due to delay in information passing from hospitals to the insurance companies. With the help of IoT, the flow of information becomes very easy, and also so there is no laugh invert in the data transferred. This helps to create a transparent system between the insurers and customers (for setting financial issues). In many cases of an emergency, the decisions have to be made fast, and if there is connectivity through IoT, the process becomes quick ok as the healthcare system because of IoT becomes primarily data driven, which is transparent and devoid of any kinds of fraud.

3.4 Expansion in medical research through IoT

Healthcare system is halfway if there is no concrete research lab associated with it. IoT in this case becomes a very innovative and effective tool to overcome the gaps in medical researchers. It is true IoT that the collection of data of various medical conditions has become an easy task and its storage has become a huge resource to create a database for statistical analysis in the medical research field. This collection of data has seemed to be economical and hassle free, which is used for the evolution of better medical treatments for the diseases concerned. The smart monitoring of labs has also been possible due to IoT. This is because every wearable device when in the labs has to be checked of all the abnormalities before coming in practical usage.

3.5 Reducing the cost of medical treatment

One of the major benefits of a healthcare system developed out of IoT. Applications like smart thermostats help to keep the utility costs very less, while application of video surveillance can help us save on security measures with intelligent sensors. Moreover the data analytics tools can keep a pulse on both your patients and your financial health, reducing the cost of care and overall operations (https://www.cdw.com/content/cdw/en/solutions/digital-transformation/digital-transformation-in-healthcare.html, Accessed 23rd February 2020, at 5.30 p.m.).

3.6 Issues of concern within IoT led healthcare system

The IoT healthcare services should not be a subject of conflict between countries, and in case of attacks the authorized departments should be allowed to avail the services even under denial of service attacks. It is also a matter of concern that the data that are sent at a particular point of time should be refreshed and not data even a minute late. Freshness of data is a major concern because in case of delay in data delivery we would cause late posting of remedial instructions. The role of nodes involved in collecting and sending data should be placed with authorization and should follow nonrepudiation [23].

Privacy becomes an essential part of security within IoT enabled healthcare systems [24]. In this case many sensitive data of particular individuals and societies are transmitted through devices that are embedded in IoT. The storage of data within this system hence makes too much of private information vulnerable for creating social and political ramification. Say for example, if a country's head health information is derived and used against the interest of the country, it may create a situation of chaos. Also in case of an individual if the patient's personal records are open to the public, this may cause social backlashes to be experienced by the patient in case of diseases like AIDS, which in several conservative countries is considered a social taboo. Hence, maintaining integrity of data of patients becomes an essential task of IoT-based healthcare systems.

4 Conclusion

The challenges in the healthcare system have always been very intense and pertinent. IoT emerges as a boon to all the gaps that were faced by the system of hospitals. It is through IoT that wearable sensors are attached to the patients and crucial information is shared in high frequency to fight the menace of chronic diseases. The whole process of medical diagnosis and medication has become a very easy task because of the quick and accurate inflow of data between patient diagnosticians and doctors.

Earlier, there were various risks involved with the treatments including technology for key functions. However, with IoT, the functionality of these medical technological devices has become much more efficient when connected through an interface that is as reliable as IoT. Moreover, these devices help up to identify the symptoms from vital significance of diseases that are mostly chronic ok and have major chances of becoming fatal anytime. The data collection and analysis become possible in real time due to IoT and curving all the possibility of major moments where health of patients may be risked to untimely medication.

The two major aspects that have been highlighted in this chapter regarding the designing of healthcare systems based on IoT are remote monitoring and personalized healthcare systems. The two areas covered by IoT enabled health care systems feel a major gap that has always been a challenge to the medical frontier. Various aspects of both remote monitoring and personalized health systems have been listed in this chapter.

Handling the amount of data is one of the standing problems in the field. Moreover, for healthcare service providers, manually analyzing data is a tough bet. Despite the young age of IoT, it is inevitable that the idea takes healthcare to a new level. In addition, the issue of rural healthcare system may be fulfilled with devices enabled through IoT. This type of healthcare system tends to make the best of technology with unlimited potential.

References

[1] M. Castells, The Internet Galaxy: Reflections on the Internet, business, and society, Oxford University Press on Demand, 2002.

[2] A. Kulkarni, S. Sathe, Healthcare applications of the Internet of Things: a review, Int. J. Comput. Sci. Inform. Technol. 5 (5) (2014) 6229–6232.

[3] Internet of Healthcare Things' in India, Internet of Healthcare Things' in India: Changing the Landscape, Available at Optum site: http://www.optum.in/thought-leadership/library/internet-healthcare-things.html. (Accessed 5 January 2020).

[4] K.V. Ramani, S. Jane, D. Mavalankar, Management of RH Services in India and the Need for Health System Reform (No. WP2003-09-04), Indian Institute of Management Ahmedabad, Research and Publication Department, 2003.

[5] A. Kasthuri, Challenges to healthcare in India—the five A's, Indian J. Commun. Med. 43 (3) (2018) 141.

[6] (2) K.V. Ramani, D. Mavalankar, Health system in India: opportunities and challenges for improvements, J. Health Organ. Manag. 20 (6) (2006) 560–572.

[7] Draft Policy of Internet of Things, Department of Electronics & Information Technology (DeitY), Ministry of Communication and Information Technology Government of India GOI, 2015. https://meity.gov.in/writereaddata/files/Revised-Draft-IoT-Policy_0.pdf. (Accessed 5 January 2020).

[8] D. Niewolny, How the Internet of Things is Revolutionizing Healthcare, White Paper, 2013, pp. 1–8.

[9] M. Simonov, R. Zich, F. Mazzitelli, Personalized healthcare communication in internet of things, in: Proc. of URSI GA08, 2008, p. 7.

[10] V. Jagadeeswari, V. Subramaniyaswamy, R. Logesh, V. Vijayakumar, A study on medical Internet of Things and Big Data in personalized healthcare system, Health Inform. Sci. Syst. 6 (1) (2018) 14.

[11] Z. Jin, X. Wang, Q. Gui, B. Liu, S. Song, Improving diagnostic accuracy using multiparameter patient monitoring based on data fusion in the cloud, in: Future Information Technology, Springer, Berlin, Heidelberg, 2014, pp. 473–476.

[12] M.A. Fengou, G. Mantas, D. Lymberopoulos, N. Komninos, S. Fengos, N. Lazarou, A new framework architecture for next generation e-health services, IEEE J. Biomed. Health Inform. 17 (1) (2012) 9–18.

[13] D. Lake, R.M.R. Milito, M. Morrow, R. Vargheese, Internet of things: architectural framework for ehealth security, J. ICT Stand. 1 (3) (2014) 301–328.

[14] C.K.M. Lee, S. Palaniappan, Effective asset management for hospitals with RFID, in: 2014 IEEE International Technology Management Conference, IEEE., 2014, pp. 1–4.

[15] N. Dey, A.S. Ashour, C. Bhatt, Internet of things driven connected healthcare, in: Internet of Things and Big Data Technologies for Next Generation Healthcare, Springer, 2017, pp. 3–Cham.

[16] Y. Bhatt, C. Bhatt, Internet of things in healthcare, in: Internet of Things and Big Data Technologies for Next Generation HealthCare, Springer, Cham, 2017, pp. 13–33.

[17] M.W. Woo, J. Lee, K. Park, A reliable IoT system for personal healthcare devices, Futur. Gener. Comput. Syst. 78 (2018) 626–640.

[18] V. Karagiannis, P. Chatzimisios, F. Vazquez-Gallego, J. Alonso-Zarate, A survey on application layer protocols for the internet of things, Trans. IoT Cloud Comput. 3 (1) (2015) 11–17.

[19] B. Prajapati, S. Parikh, J. Patel, An intelligent real time IoT based system (IRTBS) for monitoring ICU patient, in: International Conference on Information and Communication Technology for Intelligent Systems, Springer, Cham, 2017, March, pp. 390–396.

[20] D.C. Yacchirema, D. Sarabia-Jácome, C.E. Palau, M. Esteve, A smart system for sleep monitoring by integrating IoT with big data analytics, IEEE Access 6 (2018) 35988–36001.

[21] D. Son, J. Lee, S. Qiao, R. Ghaffari, J. Kim, J.E. Lee, C. Song, S.J. Kim, D.J. Lee, S.W. Jun, S. Yang, Multifunctional wearable devices for diagnosis and therapy of movement disorders, Nat. Nanotechnol. 9 (5) (2014) 397.

[22] M. Nilashi, H. Ahmadi, A. Sheikhtaheri, R. Naemi, R. Alotaibi, A.A. Alarood, A. Munshi, T.A. Rashid, J. Zhao, Remote tracking of Parkinson's disease progression using ensembles of deep belief network and self-organizing map, Expert Syst. Appl. 113562 (2020), https://doi.org/10.1016/j.eswa.2020.113562.

[23] G. Carnaz, V.B. Nogueira, An overview of IoT and healthcare, in: S. Abreu, V.B. Nogueira (Eds.), Évora: Escola de Ciências e Tecnologia da Universidade de Évora, 2016. Retrieved from: http://hdl.handle.net/10174/19998.

[24] M. Dauwed, A. Meri, IoT service utilisation in healthcare, in: In IoT and Smart Home Automation, IntechOpen, 2019.

Heart rate monitoring system using Internet of Things

Rishav Singh[a], Tanveer Ahmed[b], Ritika Singh[c], and Shrikant Tiwari[d]

NIT-Delhi, Delhi, India[a] Bennett University, Greater Noida, Uttar Pradesh, India[b] CSIR-CSIO, Chandigarh, India[c] Department of Computer Science and Engineering, Shri Shankaracharya Technical Campus, Bhilai, Chhattisgarh, India[d]

Internet of Things (IoT) has been a mode of transformation in various sectors including healthcare. It has developed the healthcare sector in a way that has led to more appropriateness and effectiveness. Moreover, people have always followed the rule of "prevention is better than cure." Elaborating on this line of thought, we come toward several innovations at various junctures of medical history where research has been done to observe and analyze medical symptoms at an early stage. Hence the whole ideology of excelling in creating technology-laden medical treatment reaches its one of the best stage where IoT-supported devices become relevant in monitoring chronic diseases. Heart-related disease in today's world is one of the most prominent and most prevalent chronic diseases. Moreover a technologically sound healthcare system has helped fight this disease very efficiently. One thing that was missing in this whole engagement of heart-related treatment was all round monitoring of the body.

Heart monitoring becomes important because according to many studies, there are risk factors that need to be dealt with for timely recognition of the changes in heart condition. In the last few decades, it has been seen that several countries have undergone economic and social changes. The change in lifestyle and in eating habits has replaced our traditional diet with junk and processed foods. Moreover, growth in tertiary jobs has reduced the physical workout. This has triggered the cause of coronary heart disease very widely among the Asians [1]. Among Indians, deaths due to noncommunicable diseases have been 1.2 million among men and 0.9 among women. This statistic is the highest among the Indians only after China with the highest numbers of death [2]. According to a study that aimed to understand the cardiovascular disease risk factors among the National Capital Region population over 20 years of survey reports, several factors were pointed out. A longitudinal study was conducted with cross-sectional surveys as the main source of data collection from among women and men aged between 35 and 64 years. The first survey was done between the years 1991 and 1994, and the second was conducted between 2010 and 2012. The results obtained from this study listed out many factors for growing cardiovascular diseases. These indicators are overweight, alcoholism, high BP, diabetes, and several others [3].

In this case of growing risks around several bodily factors leading to heart-related diseases, the monitoring of these aforementioned factors becomes very important to tackle the cardio risks in wholesome manner. The new trend that has emerged is the creation of wearable devices that will be connected to other devices at the healthcare experts end. In the process of technologically driven, connectivity between the patient and doctor helps tackle the risks of distance and time and the change in any bodily factor. The body changes in terms of glucose, cholesterol, etc. determine the status of a heart, and hence the risk of this chronic disease affecting the heart is also analyzed, and any chance of stroke is evaded with proper medical attention.

IoT therefore could come in handy in this situation as it is a competent alternate to traditional systems. This new system helps in collecting important data from the patient related to his changing conditions. This information is then passed on to the doctors at any remote place only through Internet. Moreover, other medical aid like nurses or the doctors on duty available at the hospital can monitor the heart rate of the patient in the serial monitor through the real-time monitoring system. Through IoT, another milestone that has been achieved is the creation of a system in which warning is also transmitted from patients to doctors if the heartbeat goes abnormally low. This happens through a mobile application where a smartphone acts as a monitor where particulars of heart rate of the patient are displayed on the doctors' screen. Another technology of GPS is used in the new healthcare system run by IoT to determine the live location of the device. The exemplary model can therefore record the necessary details of a patient that could in turn be used to monitor the health of a person. This technology hence becomes vital in today's growing number of heart-related diseases because if we recognize beforehand the nature of the disease and condition of heart, then many future complications may be prevented [4].

This chapter thus aims to look into the details of a heart rate monitoring system run through IoT. The chapter is divided into certain sections that elaborate upon the elements of the chapter. The first part of the chapter lists out the statistics and description of the impact heart diseases are having on the humankind in the present era specially focusing on developing countries like India. In the second part, another major aspect that has been discussed is the heart diseases and the risk factors that trigger the heart problem. The third part deals with the explanation of IoT and its need in helping analyze the heart rate for effective monitoring. In this part the importance of IoT is highlighted toward curbing the growing menace of heart diseases. After this the fourth part then deals with explanation of particular studies that have been proposed to build a system run through IoT to monitor heart rates. Hence, this is one of the major parts of the chapter that makes us aware of the till date best technological advancement in this field. IoT-laden devices or a system run by IoT connectivity within the hospital and remote premises defines the new age heart monitoring. The various models have been discussed in this part of the chapter. The growing physical distance among people and also the dependence on social networking further demands the need of a certain technology that is remote sensing and also personal. Hence in this part, we aim to understand the operation of various healthcare systems that are remotely operated and are efficient as well in analyzing the risk factors leading to heart diseases. The last part of the chapter, which is the fifth part, brings above a picture of concluding remarks. Here, we look into the conditions of heart diseases as emerging out of the conjunction of the changing social scenarios adding to recognition of various risk factors, and the physical need for a heart rate monitoring device in the present day modern society is presented with the best of technologies available today.

1 Heart diseases and their risk factors

The lifestyle change mainly in India has given rise to degenerative diseases like coronary heart disease that is adding up to the high mortality rate among the Indian population. This constantly increasing number in the deaths because of heart attacks and heart-related diseases has prompted the medical science to understand the risk factors that are primarily responsible for these types of diseases.

Framingham Heart Study (FHS) was one of the most significant studies that began to question the most fatal disease and its factors. In this study the heart conditions of the people from Framingham, Massachusetts, USA were analyzed since 1948, and today the third-generation residents of the town are the participants in this longitudinal study. It has been claimed by many scholars that the pathophysiology of the heart disease was primarily understood through the detailed analysis of the data collected during this study (FHS) in all these years [5, 6].

Gradually the FHS became a very unique and successful platform that was not only adding to the general description of the causes and aftermath of a heart disease but also looking into the generational evolution of this disease. Under this study center, participants have been across generations through whom an outline of family patterns have been studied while adding the genetic evidence leading to the cause of the heart-borne diseases. The study became more relevant when it expanded its horizon geographically, and an observation into the diverse causes of heart diseases across cultures and regions was also known and understood. The contribution of this study has been such that after this a whole new perspective regarding heart-related diseases has been developed. The effect of the revelations made by this series of study is that it has been added to routine checkups that doctors are advised to check the blood pressures, abnormal weight, high cholesterol, unhealthy eating practices, physical inactivity, etc. (https://www.nhlbi.nih.gov/science/framingham-heart-study-fhs, Accessed on 23/2/2020 at 4.56 p.m.).

The lifestyle changes mainly in developing countries like India have given rise to chronic and degenerative diseases, including coronary heart disease, which is increasingly adding up to the number to mortality rate mainly among the urban dwellers. The various causes have been also seen emerging in the list produced by FHS in regard to risk factors leading to heart diseases. In this study, also the assessment of routine periodic ECG gave an idea of indications over time period as to the stages in which the heart disease develops.

However, with the prevalence of heart-related diseases, researches among the medical fraternity have led to the identification of various risk factors that affect the heart conditions leading to diseases. The preventive actions like curtailment of smoking, managing tension, control of sugar level, and reduction in taking cholesterol and saturated fats through both pharmacological and nonpharmacological have helped in reducing the chances of heart diseases. Moreover, to keep an eye on the fluctuation of these causal factors of a heart disease have also led to the timely action against any stroke, and hence the mortality due to coronary heart diseases have decreased to a significant level. However, this has not helped to a much grandeur level as to eradicate the deaths related to heart. It still remains one of the major cause of deaths [7].

This study by Chadha et al. [7] also identified the idea about the difference in rural and urban populations regarding risk factors for heart-related diseases. According to this study, there was a very minimum difference between dietary cholesterol pointing toward the merging food habits between the two populations. However, the sodium intake was more in both sexes mainly in urban areas hinting toward the eating habits of junk food items or preservative based ready-to-eat food, mainly packaged food items that are usually rich in sodium. The risk factor as pointed out by this study can be seen in the drinking habit among both the rural and urban population with urban population both men and women consumption being higher than their rural counterparts. Another list of risk factors found in this study included hypertension, obesity, diabetes, and family history of abnormal ECG. However, smoking habit was common in rural areas; more major risk factors were predominant in urban areas. The justification to this variation was seen in the basic life orientation of daily habits and routine. The lifestyle of urban population included very less physical activity as their economic activity was mostly in the tertiary sector, while their rural counterparts were still involved in the primary mode of production.

In a study by Harikrishnan et al. [8], the authors took data from Registrar General of India (RGI). The records of RGI on a constant basis in certain periods have reported mortality rates in India. These data give the details about deaths caused by CNV in the 1980s and 1990s. Investigating the data revealed that cardiovascular disease (CVD) has a major chunk on the deaths of people (15%–20%). Moreover, there was an increasing trend in the mortality report with 20.6% deaths in 1990, 21.4% in 1995, 24.3% in 2000, 27.5% in 2005, and 29.0% in 2013.

Study conducted by Prabhakaran et al. [9] for the young male population in India also revealed several factors related to heart disease. CVD have emerged as the most prevalent cause of death among people in both urban and rural areas. As per the statistics presented in this report, India witness 10.5 million deaths every year, and out of these 20.3% of death among men, 16.9% death among women was caused due to CVD. It has been observed that more than 80% of deaths happen in low- and middle-income countries.

The major causal factors of CVD are embedded in the societies we live in. The highest number of CVD is found in areas where there is consumption of unhealthy food and smoking of tobacco. There is no physical activity involved, and also there are sources of stress mainly mental operating on the diseased body. The aforementioned risk factors regarding CVD can also be seen as diseases in their own right, which in themselves require proper medical attention. These diseases act as preliminary stages of CVD, hence if these diseases will be cured, then the early stages of CVD may also be fought efficiently. This becomes very necessary because these are the diseases that add momentum to the pathological processes in vascular cerebral and myocardial aspects of the human body. This, if not checked over the years, may lead to stroke myocardial infarction and kidney failure. All these broadly come as cause due to coronary heart disease [8].

Reports regarding the health conditions have observed that the major risk factors (like obesity and abdominal obesity) are increasing. Bad habits in one's lifestyle are the major cause for heart-related diseases. However, the reports are regional. There is a need for a national level mass reporting on this issue.

A very prominent change in the rural and slum lifestyle has also added to the CVD risk factors in prevalence of no education or less of it. These populations seem to have very little awareness against smoking itself that causes one of the basic risk factors for heart-related diseases. The lifestyle everywhere is also hit by the curse of inactivity. Even in rural households with rapidly increasing usage of technologies which has replaced labor with technology has added to the risks in lack of physical activity leading to heart risks. There is a trend to opt for junk food in place of healthy food items. In this process, fat consumption like saturated and transfats have increased through increase in intake of processed food. Calorie-filled fast foods have become the comfort zone and aim of consumption, through which prestige is leveled among the consumers. The western food chains have become the new aspects of eating habit that is completely opposite to our traditional food items. This is hence taking a toll on our body. Another factor that has become the most prominent risk factor for heart condition is hypertension. Hypertension is recorded to be very pervasive phenomena among the Indian youth and has affected the health conditions in general and heart-related diseases in particular. Another growing abnormality is obesity that has become a reality for Indian population. Obesity stops the body from running efficiently and affects each part of the body like blood pressure and pulse rates. Hence through these major risk factors listed earlier, we may get a broad view of how the heart condition in India is changing and what are the relevant factors leading to this pervasiveness of heart-borne diseases.

In the study conducted by Gupta et al. [10], a number of use cases have been presented to report the abnormal lipids among Indians going beyond the low-density lipoprotein cholesterol. These lipoprotein lipids have very less low-density cholesterol but high triglycerides. Another major issue that has been pointed out in this study is the genetic factor that is passed through generations. Similarly, thrombotic risk factors have also been reported to be the causal factor for heart-related disease [10].

Through all these studies, we can make a list of the major risk factors for heart diseases. The list thus includes hypercholesterolemia, hypertension, smoking, diabetes mellitus, and dyslipidemia among both men and women [11].

However, bringing an understanding into the various risk factors regarding the heart-related diseases among population all over the world, we certainly identify with a requirement of technological interventions. The inculcation of technologies mainly ICT driven would be the major step toward resolving the core issue of indispensability of a technique that would help predict the causes of heart diseases prior to its becoming chronic. In this regard, artificial intelligence becomes inevitable within the healthcare system to become the core ideology to analyze the data collected around the heart patients. Moreover the major challenge in this regard is the collection and exchange of data in real time between the patient and the medical experts. This is the gap that in present day time of innovations may be filled by the gadgets developed on the basic idea of Internet of Things. The heart diseases have to be scrutinized on a regular basis that becomes mostly difficult. The inaccessibility of physical premises for the patients makes IoT very important and useful in regard to fighting heart diseases and its complicated risk factors. This is because the daily analyzing of the various risk factors like blood pressure, pulse rates, abnormal weight, and cholesterol may certainly give a very clear picture of symptoms of the heart diseases. These symptoms through IoT will be shared by the patient and doctors with both acting on the emergencies more profoundly and quickly.

2 IoT and monitoring of factors affecting heart

A number of IoT-based devices and systems have been developed and executed in the healthcare industry. IoT has transformed the healthcare system, and its major impact has been on the treatment of heart-related diseases. It gives health services by analyzing the physical status of a person suffering from heart condition without taking feelings into consideration.

Quite prominently IoT has introduced changes in the healthcare system, and now, ECG signals are collected from a patient at remote distance to conduct the diagnosis. A knowledge base has been created that has helped in processing rear ECG signals without any interruption of noises otherwise.

In the present scenario, there is an increase in the heart-related diseases. IoT has acted as a revolutionary base for a healthcare system in which heart is monitored regularly and remotely. There are sensors embedded in this new system that are attached to the person at home. These sensors are then expected to act as a connecting interface to a microcontroller that permit the checking of heart rate, hence reading and broadcast the collected data over the Internet. It is user friendly as the patient himself in consultation with the doctor can set the low and high heartbeat limit. It is after this setting that the system starts analyzing and the moment heartbeat rises above the prescribed limit, the system sends an alarm to the controller. Similarly the system remarks heartbeat for it going below the minimum limit. This system allows the user to login from anywhere, and through this, monitoring of the patient's heart rate has become possible in any given time and space. This definitely insures a timely medical attention for any alert that is made by this heart rate monitoring system conducted through IoT.

According to a study by Li et al. [12], IoT technique has made the monitoring of fundamental functions of a human body at any given point of time. This study marks the proposal of an IoT-based monitoring system for an all-round healthcare system against heart diseases. This system takes into account the patient's physical signs such as blood pressure, ECG, and SpO2 on a continuous basis. The authors further explain and elaborate their proposed system in following lines, "…it can also be combined with real time analysis with algorithms to access patients health conditions and give warnings to potential attacks in advance, which can make the pervasive healthcare more intelligent" [12].

The Internet of Things (IoT) makes it very feasible for the medical system to monitor various important functions of the human body in any condition and in any situational circumstance. Moreover, this regular monitoring helps the medical experts living in different physical premises to analyze the data transferred through patients in real time. This process is actually very economical in usage an application with minimalizing the human resources. There are various studies that prescribe the IoT-based heart rate monitoring healthcare system against the proliferation of pervasive heart diseases. In most of the systems, the monitoring of patients is done through analysis of various bodily signs such as blood pressure, ECG, SpO2, etc. In some studies the data are also connected to the environmental change indicating an effect on the patients' body. These genre of studies have mainly proposed IoT-based healthcare system with four different modes of transmission of data that takes care of a wholesome release as well as computation of data while also giving proper instructions for healthcare needs of the patient [13]. These models claim to be pocket friendly with service of the most advanced technology. The application of these models has thus justified the role of IoT in dealing with the heart-related issues. With IoT the age long issues of remote monitoring of a heart patient become possible in real-time analysis of the data. Moreover the old-age people and differently abled people have been the major population facing the risks regarding heart-borne diseases. The access to hospitals has been one of the major challenges for many populations including the two vulnerable groups mentioned earlier. Hence a technological advancement like sensors and wearable devices through an interface of IoT helps connect these vulnerable groups to the medical experts on time. In this regard, IoT becomes very useful condition in which a healthcare system is developed with properties like real-time monitoring, dynamic intervention, and self-care.

Coming back to the complicated nature of various risk factors that lead to the heart-related diseases, we get awareness toward various aspects that need to be analyzed in real time. Though some of the studies regarding the risk factors of the heart diseases have concentrated on the analysis of ECG, blood pressure, and heart rate, the causes enlisted have gone beyond these factors. There are various nonphysiological factors also that have frequently led to the causing factors of heart diseases. The particular cases of heart attacks caused by these symptoms are in medical terms known as "silent myocardial infarction (SMI)." In these type of cases, it is not the obvious signs of extreme chest pain or breathing issue or dizziness, but the symptoms are so "silent" that they are mistaken for regular discomfort. The symptoms like fatigue or mild chest pain may be confused with gastric pain, indigestion, or age-related pain. The only way out to diagnose SMI is electrocardiogram (ECG) that gives the status of heart muscle damage. However, the symptoms of SMI are not loud enough to immediately rush for ECG. In this case, doctors have listed with few major warnings related to SMI that may be listed as follows (https://www.health.harvard.edu/heart-health/the-danger-of-silent-heart-attacks, Accessed on 23.02.2020 at 9.02 p.m.):

- Discomfort in chest lasting a long time,
- Feeling of discomfort in the upper part of the body,
- Shortness of breath,
- Nausea or lightheadedness.

This nonphysiological list of symptoms causing SMI hints toward the need of a healthcare system that acts as a multitransmitter making way for both physiological and nonphysiological conditions. This becomes very important because SMI cannot be understood on face value of symptoms. Moreover, with mild strains in the body as hints of heart attacks, SMI actually leaves the body scarred with a heart attack that may not be loud in its coming but very impactful in the long run. Hence, in this case, IoT becomes an indispensable interface giving space to various transmitters collecting patient's information at various levels but at real and same time. These warnings are mainly context based and hence require an all-round analysis of the condition in which the patient is facing the issues. Hence in comparison with single parameter devices, a multirange parameter-based device will help many issues concerning the heart attacks. However, these many faced transmitting machines can be best run through IoT. A system of multilayered transmitting device connected to the medical experts through IoT is the best way out build a healthcare system for accurate prediction of heart risk factors for timely prevention of heart attacks both major and SMI (Fig. 1).

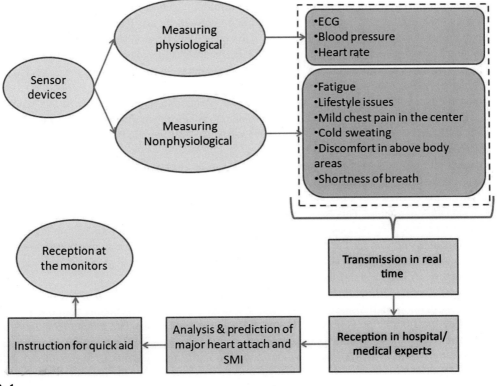

FIG. 1

Prediction for prevention through IoT.

3 Recent trend analysis

There has been a rising trend of using IoT and making devices embedded in it for monitoring purposes mainly for heart diseases. There have been models and projects that have focused on sleeping issues and its effect on the heart-related problems [14]. Another work developed a system that aimed to monitor people's brain bioelectrical activities [15]. Other systems on this trend have been designed especially for old-age people to understand their body [16]. The aims of these models and systems have been widely same. However, the architecture of each model remains diverse in terms of innovation. Another dimension of diversity that occurs among these systems based on IoT is the analysis and monitoring of various different aspects that affects the heart conditions. Few pay attention to the more obvious postural and physical science. On the other hand the others have been designed to monitor the more pathological aspects like heart rate blood pressure ECG. The systems also vary on analysis of physiological and nonphysiological parameters, through which a more contextual data are passed on the medical experts sitting in remote locations. However, we need systems that can collect data at different frequencies in consideration with proper medical requirements and avoid congestion of data to remote server or the medical expert recipient [12].

Another system model was initiated and executed by Rosli and Olanrewaju [17] that intends to identify heart issues among patients involved. The main goal of this model is to provide for such devices that will act as a monitor to changing heart conditions and they are handy and mobile in nature. Through this model a sensor for identifying pulse is developed through which ECG signals are obtained. These ECG signals then established heart rate processed and analyzed by an ARDUINO. This system can send an alarming text message if it is being used in the prescribed range GSM was employed to establish communication between phone and ARDUINO. In the initial stage of execution of this system, the patient gives specifics of a normal range (minimum to maximum). In the next stage, these specifics are broadcasted through a text encompassing the earlier normal range and being sent to ARDUINO to set a threshold. In this whole process of this system, a sensor attached to the body of the patient reads the required data until a message of stop is sent. This system thus helps to alarm the medical aid providers whenever the heart rate goes beyond the prescribed range and allow them to take the required action for the betterment of the patient.

A smartphone system was proposed in Wolgast [18] for the effective and quick monitoring of heart conditions in real time. In this model a sensor was used to send and receive the changes in heart condition. This whole system was based on sending and receiving body data through Android application developed on the smartphone. This model becomes unique in a way as it was able to minimize the noise congestion while receiving data by using a smartphone. In this model the Android application possessed two modes: one was for the display of ECG signal, and the other for showing the strength of signal. This work hence was able to establish a portable device that transmits a clear set of data with higher accuracy levels on a smartphone.

Dewan and Sharma [19] worked on a system proposing a hybrid algorithm to dig out patterns and relevant data regarding heart disease. This system, however, worked onto predicting patterns of a body change basing itself on the previous data. Hence, this system aims to use data mining techniques and analyze the records in hospital only to make predictions and take decisions on this basis. Another proposed model by Koshti et al. [20] uses IoT devices to monitor heart condition. In this model, body sensors are employed from wherever the data have to be extracted. After the data are collected, this system transmits these data using ZigBee modules. These data are then transmitted to a faraway location where they are received and monitored on a system. However, the issue with such a system emerges in its difficulty in tracking real-time data because computers are not portable like smartphones [21].

Another model proposed in Medhekar [22] uses Bayesian classification to forecast heart attacks. This proposed model executed at two different stages. The first stage was when the authors used data from a clinic to make models and classes. The particular dataset used in this model ranges from 3003 observations to 14 parameters collecting information regarding heart disease diagnosis. Moving further in the next stage, a testing was done in which a prediction probabilistic algorithm is employed to categorize the data collected as of high average and low risk.

Jambhulkar and Baporikar [23] also proposed a different model to monitor heart rate. In this model, naive Bayes was used to analyze heart condition. In the next phase of this model, WSN was added to the system to collect data regarding various patients in real time and then broadcasting that multiple user data to the main system that is central for all. After this, MATLAB was employed to pull out required aspects from the earlier datasets and hence predicting using DM technique. However, this model was only meaningful for hospital and not useful for users.

Moving away from the limitations of the aforementioned model, Raihan et al. [24] developed a model to predict heart rates and hence the possibility of an attack through data mining techniques basing itself on clinical data. In this model an Android application was developed using smartphones to collect patient information. These data are then analyzed by the application to calculate the risks involved in the happening of a stroke. In this process, multiple statistical techniques are used to predict the risk of a heart attack, and the prediction is immediately shown on the smartphone.

In another research done by Mamun [25], heart rate signs were brought together from fingers and ears using infrared transmitter and receiver pair module. In this model a filter was also used to minimize the noise, and a microcontroller was used to transmit the signals received and then display the same on the screen. In this system the heart rate data collected were found accurate, and this system works good for hospitals minimizing involvement of human errors.

Another work by Parihar [26] proposes a system in which heartbeat and temperature are displayed using microcontroller Arduino Uno; it claims to be one step ahead in this world of systems working in offline mode. This model consists of sensors that are attached to a body and through which heartbeat and temperature is analyzed continuously. This system works in real time, hence making possible the monitoring of body remotely.

A model based on GSM has been proposed in Sudhindra [27]. In this model the aim is to create a system to analyze anomalies in a normal heart rhythm. In this system on the occurrence of an unusual change in the body parameters, the signal is sent to nearby people through GSM; moreover in nonavailability of proper help from extreme nearby people, an SMS is sent to medical authorities and other people concerned.

On a similar pattern a work by Kumar [28] has been developed in the light of growing heart disease and risks around it; the sensor used in this system also is attached to a microcontroller that transmits the changing heart rate through Internet. In this model the user may set heartbeat levels. Any fluctuations in the heartbeat going above or below the prescribed limit are sent as an alarm to the doctor, and a live heart rate of the patient is displayed in front of the doctor. This thus helps in a timely action against any heart-related risks.

In the next research module proposed by Alam [29], the system is based on microcontroller that involves optical technology. In this system, heartbeat and body temperature are monitored through sensors collecting relevant information from the fingertip. Optical technology is used to spot the blood flow across the fingers. This system comes up as a very portable device in comparison with earlier models of recording data. In this model, WSN is also used, but unlike other remote monitoring systems, this device seems to collect changing body signs with much more accuracy even in adverse situations. Further the data collected through sensors are sent through GSM module to a place far away and are displayed on a screen of a phone.

In the study conducted by Goel et al. [30], specially designed equipment like electrocardiograph (ECG) and portable wrist strap watch was used to monitor a patient's pulse. As expected, using equipment of this sort not only is expensive but also adds additional complications. Moreover a medical expert must be present to monitor a person on a regular interval. Basically, using the so-called photoplethysmograph (PPG) technique, a patient's heartbeat was monitored. If there is an anomaly, the system sends an SMS to the family members and to the medical experts Goel et al. [30].

In a study by Majumder et al. [13], a model has been proposed that is based on IoT for creating a personalized e-healthcare device for prediction of cardiac arrest. The uniqueness of this model lies in its features of combining both detection and communication techniques. The major features of this model may be seen primarily as developing a *multisensory device* based on IoT for prediction of heart risk factors. Another feature of this model is to develop a smartphone-based healthcare system that may be connected through a sensor, a wearable body area sensor (BAS) for making the devices more user friendly. As people today are mostly well versed in a receiving data on smartphone monitors, the reception of medical instructions from the experts after analysis of data transmitted through BAS would be much more efficient and an easy procedure in dealing with risks related to heart. In this model a low-power communication module is proposed to be developed to send data to the smartphone. Hence, this model aims to develop easy-to-use devices connected through IoT. However, the model helps in dealing with cardiac issues as it serves to transmit data on various levels that may be physiological and nonphysiological. This thus emerges as a very helpful personalized system of healthcare giving a model for fighting against the uncertainty of heart diseases related risk factors.

While explaining the proposed model, this paper elaborates on the basic architecture on which the model is based upon. The architecture used in framing this model is IReHMo as enlisted by the authors. The authors in this model very cogently explain the characteristics of IReHMo and its applicability in developing a remote monitoring system in healthcare. This proposed model runs on IReHMo that when combined with CoAP makes it very easy for transmission of large number of data in a remote system of healthcare. The authors then further explain the importance of IReHMo by explaining its uniqueness in helping several types of home-based sensors operating in a remote healthcare system run through IoT by sensing layers. This reduces the bandwidth and the number of packets being sent, hence proving to be more efficient in sending data related to health in comparison with traditional remote health monitoring systems [31].

Through all these models proposed by various scientists, we get an overview of the latest contributions of IoT in fighting the menace of heart diseases. These models may differ in their approaches and methodological endeavors, but their sole aim is to employ the usage of IoT in understanding the various risk factors causing heart attacks. Another common idea that flowed through all the works listed earlier was the concern to develop a healthcare system that will be personal in nature. Hence through IoT acting as a connection between the personal pieces of gadget like smartphones as suggested by most of the studies, the monitoring of risk factors leading to heart issue may be evaded on time.

4 Conclusion

In this chapter, we tried to list out various technological advances for monitoring the conditions of a heart in real time. We discussed the history of heart diseases and highlighted the root causes. The effort of this chapter in the beginning was to develop historical understanding of the beginning of research into the causal factors of heart-related problems. It was observed that it was very late, that is, in 1945 that the country like America was able to diagnose the cause of heart diseases and recognize the symptoms like high blood pressure, stress, and sedentary lifestyle as the cause of heart disease that was by this time a cause of natural epidemic among the Americans. Hence a serious medical intervention was initiated soon after. But the series of researches in the reaching up to the various risk factors proclaimed that there are two major types of risk factors that is physiological and nonphysiological. The effect of the both types of risk factor is such that they may cause both major and SMI or mild heart attacks with both giving scar mark on the natural functionality of heart. This situation hence demanded for another level of technological inclusion that will be able to monitor and analyze the time-to-time indications caused by various risk factors. However, this new technology was considered to be more personal in form of gadgets and devices attached to the body of patient mostly wearable.

The coming of wearable devices was seen as a welcome innovation among the heart patients. However, in this case, the wearable devises that were supposed to act as sensors were not complete without an interface that would connect these gadgets to the machines in hospitals or near to medical and diagnostic experts. Here was the introduction of IoT to fill this gap in the technological intervention in building personalized healthcare system among the heart patients.

We in this chapter then pointed out the work done especially considering heart rate monitoring using the paradigm of IoT. Advantages and disadvantages of multiple solutions were discussed in detail. Through the discussion in this chapter, it was found that using the paradigm of IoT, it is indeed beneficial and, moreover, lucrative to focus attention on diagnosing heart-related conditions.

An observation and analysis of various models of IoT-based monitoring system were made in the last part of this chapter. In this part then the focus was to bring above the characteristics elements of IoT acting as the most accurate and viable and reliable innovation through ICT on which the treatment of heart-related diseases may be based upon.

References

[1] M. Chandalia, P.C. Deedwania, Coronary heart disease and risk factors in Asian Indians, in: Diabetes and Cardiovascular Disease, Springer, Boston, MA, 2001, pp. 27–34.

[2] World Health Organization, Global Status Report on Noncommunicable Diseases 2014 (No. WHO/NMH/NVI/15.1), World Health Organization, 2014.

[3] D. Prabhakaran, A. Roy, P.A. Praveen, L. Ramakrishnan, R. Gupta, R. Amarchand, D. Kondal, K. Singh, M. Sharma, D.K. Shukla, N. Tandon, 20-Year trend of CVD risk factors: urban and rural national capital region of India, Glob. Heart 12 (3) (2017) 209–217.

[4] T.V. Sethuraman, K.S. Rathore, G. Amritha, G. Kanimozhi, IoT based system for heart rate monitoring and heart attack detection, Int. J. Eng. Adv. Technol. 8 (5) (2019).

[5] R. Hajar, Framingham contribution to cardiovascular disease, Heart Views 17 (2) (2016) 78.

[6] S.S. Mahmood, D. Levy, R.S. Vasan, T.J. Wang, The Framingham heart study and the epidemiology of cardiovascular disease: a historical perspective, Lancet 383 (9921) (2014) 999–1008.

[7] S.L. Chadha, N. Gopinath, S. Shekhawat, Urban-rural differences in the prevalence of coronary heart disease and its risk factors in Delhi, Bull. World Health Organ. 75 (1) (1997) 31.

[8] S. Harikrishnan, S. Leeder, M. Huffman, P. Jeemon, D. Prabhakaran, A Race against Time: The Challenge of Cardiovascular Disease in Developing Economies, Centre for Chronic Disease Control, New Delhi, India, 2014.

[9] D. Prabhakaran, P. Jeemon, A. Roy, Cardiovascular diseases in India: current epidemiology and future directions, Circulation 133 (16) (2016) 1605–1620.

[10] R. Gupta, I. Mohan, J. Narula, Trends in coronary heart disease epidemiology in India, Ann. Glob. Health 82 (2) (2016) 307–315.

[11] M. Rao, D. Xavier, P. Devi, A. Sigamani, A. Faruqui, R. Gupta, P. Kerkar, R.K. Jain, R. Joshi, N. Chidambaram, D.S. Rao, Prevalence, treatments and outcomes of coronary artery disease in Indians: a systematic review, Indian Heart J. 67 (4) (2015) 302–310.

[12] C. Li, X. Hu, L. Zhang, The IoT-based heart disease monitoring system for pervasive healthcare service, Procedia Comput. Sci. 112 (2017) 2328–2334.

[13] A.K.M. Majumder, Y.A. ElSaadany, R. Young, D.R. Ucci, An energy efficient wearable smart IoT system to predict cardiac arrest, Adv. Human Comput. Interact. 2019 (2019).

[14] M. Rofouei, M. Sinclair, R. Bittner, T. Blank, N. Saw, G. DeJean, J. Heffron, A non-invasive wearable neck-cuff system for real-time sleep monitoring, in: 2011 International Conference on Body Sensor Networks, May, IEEE, 2011, pp. 156–161.

[15] W.Y. Lin, M.Y. Lee, W.C. Chou, The design and development of a wearable posture monitoring vest, in: 2014 IEEE International Conference on Consumer Electronics (ICCE), January, IEEE, 2014, pp. 329–330.

[16] E. Sardini, M. Serpelloni, V. Pasqui, Wireless wearable T-shirt for posture monitoring during rehabilitation exercises, IEEE Trans. Instrum. Meas. 64 (2) (2014) 439–448.

[17] R.S.B. Rosli, R.F. Olanrewaju, Mobile heart rate detection system (MoHeRDS) for early warning of potentially-fatal heart diseases, in: 2016 International Conference on Computer and Communication Engineering (ICCCE), July, IEEE, 2016, pp. 422–427.

[18] G. Wolgast, C. Ehrenborg, A. Israelsson, J. Helander, E. Johansson, H. Manefjord, Wireless body area network for heart attack detection [education corner], IEEE Antennas Propag. Mag. 58 (5) (2016) 84–92.

[19] A. Dewan, M. Sharma, Prediction of heart disease using a hybrid technique in data mining classification, in: 2015 2nd International Conference on Computing for Sustainable Global Development (INDIACom), March, IEEE, 2015, pp. 704–706.

[20] M. Koshti, S. Ganorkar, L. Chiari, IoT based health monitoring system by using raspberry pi and ECG signal, Int. J. Innov. Res. Sci. Eng. Technol. 5 (5) (2016).

[21] M. ElSaadany, A novel IoT-based wireless system to monitor heart rate: student research abstract, in: Proceedings of the Symposium on Applied Computing, April, ACM, 2017, pp. 512–515.

[22] D.S. Medhekar, M.P. Bote, S.D. Deshmukh, Heart disease prediction system using naive Bayes, Int. J. Enhanc. Res. Sci. Technol. Eng. 2 (3) (2013).

[23] P. Jambhulkar, V. Baporikar, Review on prediction of heart disease using data mining technique with wireless sensor network, Int. J. Comput. Sci. Appl. 8 (1) (2015) 55–59.

[24] M. Raihan, S. Mondal, A. More, M.O.F. Sagor, G. Sikder, M.A. Majumder, M.A. Al Manjur, K. Ghosh, Smartphone based ischemic heart disease (heart attack) risk prediction using clinical data and data mining approaches, a prototype design, in: 2016 19th International Conference on Computer and Information Technology (ICCIT), December, IEEE, 2016, pp. 299–303.

[25] A.L. Mamun, N. Ahmed, M. Alqahtani, O. Altwijri, M. Rahman, N.U. Ahamed, S.A.M.M. Rahman, R.B. Ahmad, K. Sundaraj, A microcontroller-based automatic heart rate counting system from fingertip, J. Theor. Appl. Inf. Technol. 62 (3) (2014).

[26] V.R. Parihar, A.Y. Tonge, P.D. Ganorkar, Heartbeat and temperature monitoring system for remote patients using Arduino, Int. J. Adv. Eng. Res. Sci. 4 (5) (2017).

[27] F. Sudhindra, S.J. Annarao, R.M. Vani, P.V. Hunagund, A GSM enabled real time simulated heart rate monitoring & control system, Int. J. Res. Eng. Technol. 3 (2014) 6–10.

[28] A. Kumar, R. Balamurugan, K.C. Deepak, K. Sathish, Heartbeat sensing and heart attack detection using Internet of Things: IoT, Int. J. Eng. Sci. Comput. 7 (4) (2017) 6662–6666.

[29] M.W. Alam, T. Sultana, M.S. Alam, A heartbeat and temperature measuring system for remote health monitoring using wireless body area network, Int. J. Bio-Sci. Bio-Technol. 8 (1) (2016) 171–190.

[30] V. Goel, S. Srivastava, D. Pandit, D. Tripathi, P. Goel, Heart rate monitoring system using finger tip through IoT, Heart 5 (3) (2018).

[31] N.M. Khoi, S. Saguna, K. Mitra, C. Åhlund, Irehmo: an efficient IoT-based remote health monitoring system for smart regions, in: 2015 17th International Conference on E-health Networking, Application & Services (HealthCom), October, IEEE, 2015, pp. 563–568.

A smart hand for VI: Resource-constrained assistive technology for visually impaired

Kanak Manjari, Madhushi Verma, and Gaurav Singal

Department of Computer Science Engineering, Bennett University, Greater Noida, India

1 Introduction

People with visual impairments can only use their sense of sight to a certain extent, or perhaps not at all. Hence, they require an assistive technology to carry out various daily activities. In our work, we have provided solutions to one indoor and outdoor problem that visually impaired people face to help them by providing them with independence and security. The aim is to develop an assistive technology for visually impaired to overcome what they face in their lives and their real-life issues [1]. One of the problems what they are facing in outdoor environment is protection from stray animals and it has always been considered as a primary requirement for assisted mobility. Another indoor solution in the form of extraction of text from any document, medical report, or medicines has also been considered and provided in this chapter. In the recent years, aid to individuals with visual impairments has come a long way. Until technology changed healthcare, magnifying lenses were the only choice for vision assistance. Today's technologies include improved magnification apps, software, and other items, rather than visual, that use tactile or audio. The term "visual impairment" is used to describe a vast array of disorders affecting the visual field and vision clarity. Technology can be of vital importance to visually impaired people, both as a medium for communication and learning and as a tool for providing visual stimulation. People with visually impaired access to basic facilities can be provided using a computer with suitable software and hardware. For example, speech synthesis enables a blind person to read a word-processed file without the need to have it translated into braille.

Computers can use voice recognition to read screen content and text that allows visually impaired and the blind users access. Adding voice to standard software can support visually impaired people. Those with severe conditions may require specialized software for voice screen reading. Early learning apps using voice and sound alongside a bright image will empower children with visual impairment to explore and communicate with their environment. Animal identification and recognition pose a difficult challenge at the moment and there is no clear method that offers a reliable and effective solution to all circumstances. Stray animal identification can improve mobility and the protection of visually impaired people, particularly in unfamiliar environments. The stray animals are initially identified and found, and then information about the animals is sent to the visually impaired people using different methods such as speech, sounds, etc.

IoT Based Data Analytics for the Healthcare Industry. https://doi.org/10.1016/B978-0-12-821472-5.00016-8

To better understand the role and shortcomings of animal detection and recognition, the following considerations must be done, since they can adversely affect the performance of the method of detection and recognition: (1) lighting and other image acquisition conditions—factors such as variations in lighting, distribution, and source or camera intensity features affecting animal image input and (2) occlusions, other objects and other animals that partially mask animal images.

This chapter aims to explore the method of creating a lightweight enriched assist with the ability to detect text, identify, and detect animals for visually impaired individuals to help them in leading an independent and normal life. Our contribution includes building a resource-constrained cane to guide visually impaired by using deep learning techniques. Lightweight and cost-effective hardware deployed with light software have been attached to the cane to perform text detection as well as recognition. Efficient and Accurate Scene Text (EAST) detector for CNN has been used for the purpose of image text detection and transfer learning using Faster RCNN and SSD models have been implemented for animal detection.

The overall chapter is arranged in the following structure. The related work done by other researchers is discussed in Section 2, which includes an explanation of the various techniques and models that they use. The methodologies adopted for carrying on this work are discussed in Section 3. Section 4 addresses the experimental results derived from this work followed by a conclusion in Section 5.

2 Related work

A wide variety of assistive solutions for visually impaired have been developed till now [2]. We have discussed some of the relevant solutions in this section. To support visually impaired people with text recognition and face detection, raspberry pi has been used with authors in [3], a camera-based assistive text reading to help people with visual impairments interpret the text on the captured image. The faces can also be recognized by controlling mode, when one person enters the frame. The approach proposed includes the extraction of text from the scanned image using Tesseract Optical Character Recognition (OCR) and the use of the e-Speak program to translate the text into expression.

In [4], a visually impaired or blind person based on a camera, consisting of a digital camera placed on the eyeglass or head of the person that takes photos on request. Nonreal-time image processing algorithms discern the attributes of the captured image by processing it for detection of edge patterns within the central region of the image. The results are categorized by artificial neural networks trained on a list of known artifacts, in a lookup table or by a threshold. A descriptive sentence of the object is created upon recognition of the pattern, and a computer-based voice synthesizer is used to verbally announce the descriptive phrase. The technology is used to assess an object's size or distance from another object, and may be used in combination with the IR-sensitive camera.

Text Scanner and Touch Reader for visually impaired users [5] is a very recent technology designed to give visually disabled people independence and privacy. This program provides the user with scenic explanations in audio format through the capture and further processing of the natural setting. For the analysis of the scene, a deep learning technique has been used. GuideCall is an android program that helps visually disabled people get video call assistance from friends and relatives when they are in need. In this application, even the position tracking feature is available. Smart Aid, a text to voice translator, allows visually impaired to retrieve audio-format information from eBooks. The performance can be heard in speakers or headphones using Arduino UNO, which makes it a low-cost solution.

A modern method, based on transforming feedback information into auditory or tactile sensory information, has been developed to help blind people [6]. Detection and text recognition explain the outside environment, so the blind person can understand the situation easily. This work is aimed at incorporating a framework for the identification and recognition of text in indoor scenes. The approach involves identifying the regions of interest that will include text using the segment connected to it. Instead, image analysis is used to identify text. It lets consumers get the most insightful feedback in the shortest possible period.

A further text detection and hand-held object recognition was introduced by authors in [7], which enabled visually impaired to read text patterns on hand-held objects in real time. This is the tool designed to help people with visual impairments interpret and transform text patterns into the audio output. The original software suggests the technique that helped the camera and object field identify the image to retrieve the text pattern from that object. Maximally stable texts are observed using the Maximally Stable External Regions (MSER) feature. A novel algorithm is evaluated on many scenes. The observed text is converted into voice production according to the example. The blind user earns the yield on expression. MSER shows this is a powerful algorithm for text detection. This chapter has done the study of identification and recognition of various objects with completely different text patterns. The cane prototype designed by us was shown in Fig. 1 using a microcontroller, an ultrasonic sensor, a camera, and a power bank connected to it.

FIG. 1

Prototype of cane.

Faster R-CNN and Yolo v1 were used in paper [8] to resolve the issue of zebra detection in real-world pictures. Scientists have encountered many challenges in identifying the animals from several overlapping zebras, such as complex zebras and occlusion perspectives. They created a labeled dataset of 2500 images manually, with 3541 plains zebras boundary boxes and 2672 Grévy zebras boundary boxes. In their research, YOLOv1 was the best plains zebras detector with 55.6% detection accuracy and 56.6% Grévy zebras. Snapshot Serengeti dataset is a massive public wildlife dataset that comprises 3.2 million annotated camera trap images and 48 different animal species [9]. Out of total 757,000 images, 707,000 were used by them for training and remaining 5000 images were used for testing purposes. Authors achieved 92% classification accuracy on their test set using a CNN architecture called ResNet-152 [10], improving the efficiency of previous approaches. This work shows promising results when classifying images of a single species, but it does not address multiple species identification difficulties.

A further CNN-based solution for object detection is proposed in [11]. The RGB-D architecture shown for object recognition consists of two independent CNN transmission sources, which are coupled with a late fusion network. ImageNet equipped CNNs [12]. Depth images are encoded as rendered RGB images, spreading the data in depth over all three RGB channels and then using a standard (pretrained) CNN to identify these. CNNs that are pretrained on ImageNet [13] are used due to the lack of large, labeled depth datasets. For enhancing detection in noisy real-world environments, a novel increase in data is suggested. Another approach to object detection, using deep CNN, was suggested in [14]. It also uses pretrained CNN to categorize images, and offers a rich collection of semisensitive features. In the representation of objects, the depth information is applied from a biblical viewpoint and the depth channel is defined by distance from the center of the object. Since we have trained the model on a larger and varied dataset, it can easily automatically classify up to six animal species in images under different conditions, providing more detail for accurate analysis and even identifying multiple animals in a single image.

3 Methodology

In this section, we discuss the used dataset for the stated problem and model design and implementation. As stated in Section 1, a different model was built for both the outdoor and indoor problems they face.

3.1 Dataset collection

To fulfill our objectives, we needed a proper dataset as there is no standard dataset available for serving our purpose of model development. So, we have created the dataset for both the objectives. The dataset needed for the extraction of information from medicines is a set of images of medicines. To generate good results, it is necessary to have a larger dataset. The dataset is prepared by creating videos of different types of medicinal items, particularly tablets, syrups, tubes, etc. To extract frames from a video, we have used VLC Media Player. VLC has the flexibility to extract the frames from a video. The recorded videos of the medicines are regenerated into frames at the speed of two frames per second. The desired range of pictures is often obtained during this method. The preprocessing of information ought to be done to get rid of noise data.

We have also created a dataset of animal images that includes dog, cat, horse, monkey, goat, and cow. To perform animal detection specifically for the Indian scenario, an appropriate dataset was not available. Therefore, this dataset can be used for solving other types of related problems as well which are specific to the Indian context. The dataset has been developed for the animal detection purpose also since there is no specific dataset suitable for this. The other dataset does not include or have only a very small amount of indigenous species. Due to these limitations, a custom dataset was needed to be created. The created custom dataset includes pictures of six animals—dog, cat, horse, monkey, goat, and cow. The dataset size is approximately 45,000 images, where approximately 5000 images for each class are annotated. Each class comprises nearly the same number of images so that one of the models is not biased to one or more animals. For all 30,000 images, the dataset was manually annotated using the labelImg tool. The images in the above-mentioned dataset contain only indigenous animals, taken from various video clips.

LabelImg is an image annotation tool that helps labeling objects present in the image. It is written in python and uses Qt for its graphical interface. We have used PascalVoc format for exporting XML files and annotation files are saved singly for each image in the supply folder. The generated pictures from the videos need to be annotated for different classes by creating a bounding box on the object. The annotated pictures are in .xml format by default. We have converted these .xml files to .txt files. The images taken were particularly indigenous so that the models become more accurate in detecting and classifying them, as they are trained on deployment environment scenarios. This helps in selecting the better model and also to fine-tune the hyperparameters for achieving higher precision rates.

3.2 Model selection

After dataset development and annotation, we can use a pretrained model, or we can create our model to perform the recognition function. We have used transfer learning in our research, which means using the weights of pretrained model and retraining it with our dataset. Transfer learning uses the knowledge gained from solving one issue and applying it to another problem, which is still related. For example, when learning to recognize cars, the information gained can be used to some degree to identify vehicles. If we train the network on a broad dataset (e.g., ImageNet), we train all of the neural network parameters and learn the model as such. Moving onto the GPU will take hours.

Text detection is a process in which text can be detected either from video, image, or in real time. The EAST text detector for OpenCV [15] is a deep learning model that supports a unique design and training pattern. It is capable of running on 720p quality images with 13 f and obtains gradual accuracy in text detection. The text region detected is translated into pyTesseract. Tesseract is an OCR engine designed for various operating systems. We prefer to use the Python-Tesseract, or just used the PyTesseract library, which is a wrapper for the Tesseract OCR framework used by Google. We need a pyTesseract v3 to perform text recognition that provides a highly accurate, deep learning-based model for text recognition.

The dataset that we have contains small texts, which are difficult to detect. So, we had to train our pretrained model dataset. A pretrained model is a computer that has been trained to solve a problem on a large dataset. We trained our dataset using Resnet50 [16] and Mobilenetv1. Comparatively, Resnet50 produced better results. Therefore, further implementation and experimentation were restricted to this model.

We have used the transfer learning approach to achieve the desired accuracy. The pretrained models were all trained on the COCO dataset to avoid bias due to the dataset. We take three different models to analyze the accuracy and loss of trade-off on the custom-created dataset. We train primarily on light-weight models as they need to be deployed on a mobile computing device. The hyperparameters of the pretrained models were not updated to extract the model's performance on its default parameters. The pretrained models that were taken for this work are SSD Inception 2 COCO, SSDLite Mobilenet V2 COCO, and Faster RCNN Resnet50 COCO. These models are selected to try a variety of detection and classification methods and to choose the best strategy for our approach.

The models were trained for 3000, 6000, 10,000, and 12,000 epochs. It is trained on different epochs to have an understanding of how good the model is performing. Occasionally, due to under-fitting, a model can underperform to avoid these discrepancies each model is trained four times and its output is observed.

3.3 Architecture used

We will discuss the architectures of EAST, SSD, and Faster RCNN models, we have used for fulfilling our objectives.

3.3.1 EAST

EAST algorithm is the pipeline for detecting scene text efficiently and accurately. The EAST pipeline can predict arbitrary word and text line orientations on 720p images, and can run on 13 FPS, too. More precisely, because the deep learning model is end to end, it is feasible to use side-by-side computationally expensive subalgorithms that are normally introduced by other text detectors, including candidate aggregation and word partitioning. The EAST method utilizes a novel, carefully built loss function to construct and train such a deep learning model. The architecture is expressed in Fig. 2.

3.3.2 SSD

In the Single Shot MultiBox Detector (SSD) model, a single forward pass of the network is used for the localization and classification of objects. The object detection is done using bounding boxes, and the object's classification is provided by the object's confidence scores if they are present on the given instance. It is also simple in relation to methods that involve object proposals, such as R-CNN and MultiBox, as it completely discards the proposal generation stage and encapsulates all the computation in a single network. This makes SSD easy to train and easy to implement in systems that require a part for detection. Fig. 3 displays the architecture. The SSD is based on VGG-16 architecture, except it discards the layers that are completely connected. Instead, we are adding a set of auxiliary convolutionary layers to enable us to extract features at multiple scales.

3.3.3 Faster R-CNN

Faster R-CNN [17] has two networks: a Regional Proposal Network (RPN) for regional proposals, and a network for the classification of objects using those proposals. The RPN is typically a simple network with three convolutionary layers. There is one unique layer that feeds into two layers—one for classification and the other for regression bounding. RPN aims to create a set of bounding boxes called region of interests, which are highly likely to contain those objects. The output from this network is bounding boxes marked with two pixel coordinates in diagonal corners, and a number or value.

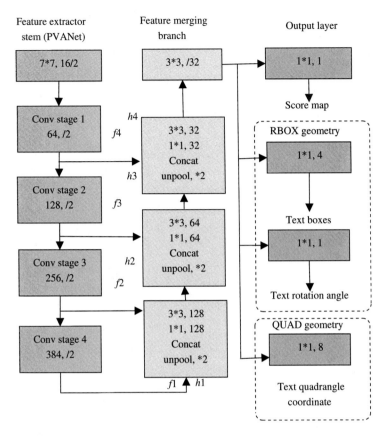

FIG. 2

Architecture of EAST.

RPN estimates the probability of an anchor being a backdrop or foreground, and maximizes the anchor to be more precise.

The detection network (also sometimes called the RCNN network) takes input from both the function network and RPN, creating the bounding box and the final class. Normally, it consists of four layers, which are completely connected or thick. Different layers are stacked by a layer of classification and a layer of bound box regression. The features are cropped to help it identify according to the boundary box.

4 Experimental results

We have tested the results on different edge devices such as Raspberry pi and Jetson NANO. We have tried to deploy and test the results achieved on the devices to make the solution portable and mobile. These devices can be deployed on cane or other aid for visually impaired to help them in their daily

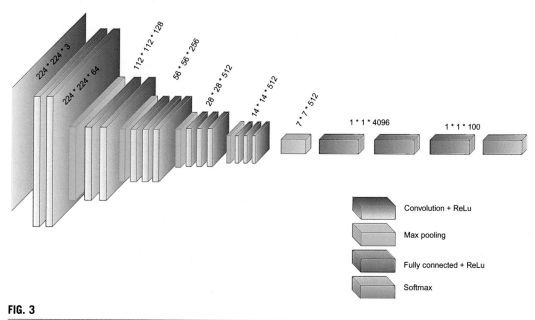

FIG. 3

Architecture of SSD model.

routine. This will help them in detecting the obstacles present around them to guide them while traveling. Text extraction will help visually impaired in reading any document such as newspaper, report, online pdfs, etc. for gaining the information. The IoT devices we have used here can help them in many ways by providing safety, independence, and mobility to them.

The results for text extraction from medicines include the detection of the text and recognition of the text from the medicines. We have used Resnet50 to train our model, which produced an accurate result up to some extent. The detection and recognition of the text is done and produced appropriate results. The original image is shown in Fig. 4A. The detected results for Mfg, Date, and Expiry are shown in Figs. 4B, 5A and B, respectively. Resnet, SSD Inception, and SSD Mobilenet models ran for 3000 epochs initially, the results obtained are shown in Figs. 6A and B and 7A, respectively.

The previous total loss comprises of localization and classification losses combined. It is clearly evident that Faster RCNN Resnet model performed better than the other models when compared on the basis of total loss. It was observed that total loss for SSD Mobilenet was too high. So, this model was discarded and the other models were trained for 6000 epochs and the result have been evaluated again and shown in Fig. 7B.

The momentum set into the SSD Inception model and the loss started hitting higher values. Further, the Faster RCNN Resnet losses came down further as the epochs were increased. Faster RCNN Resnet is trained further for 10,000 epochs and it was observed that the loss of model decreases after 8100. So, the final loss produced by the model is 0.21. The model when runs on the test dataset produced an accuracy of 0.73 on three classes. Few detected results for cat, cow, and dog are shown in Fig. 8A–C, respectively.

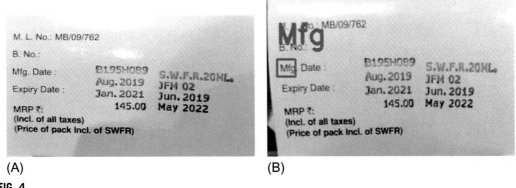

FIG. 4

Original image and image with manufacturing information. (A) Original image containing the required information; (B) text detected and recognized "Mfg."

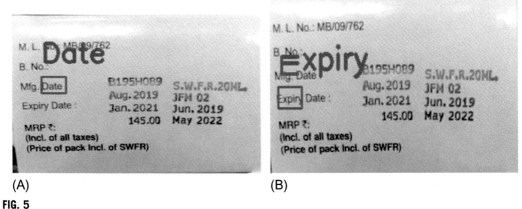

FIG. 5

Images with date and expiry information. (A) Text detected and recognized "Date"; (B) text detected and recognized "Expiry."

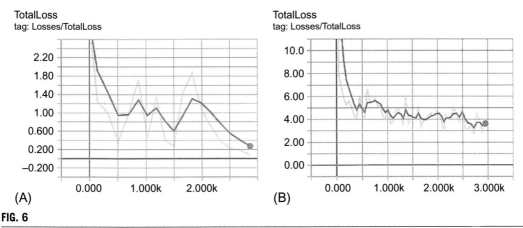

FIG. 6

Loss by resnet and SSD inception. (A) Loss by training resnet (3000 epochs); (B) loss by training SSD inception (3000 epochs).

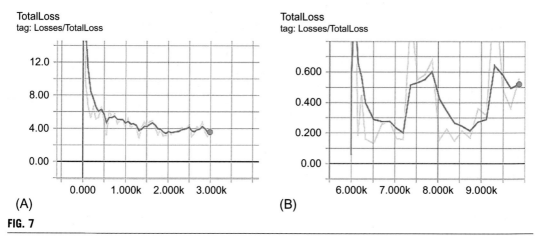

FIG. 7

Loss by SSD mobilenet and resnet. (A) Loss by training SSD mobilenet (3000 epochs); (B) loss by training resnet (6000–10,000 epochs).

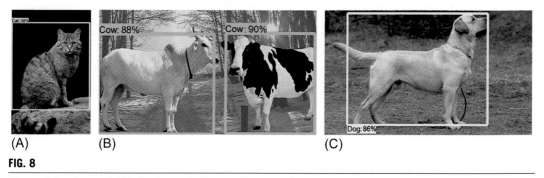

FIG. 8

Cow and dog detected. (A) Cat detected; (B) cow detected; (C) dog detected.

5 Conclusion

This work proposes a solution for visually impaired people to help them in text detection and recognition and to avoid stray animals while commuting. This aid can be attached to the user's cane with which they can get the required information from the medicine covers and get alerted from stray animals. It has been observed that our model for text detection and extraction could produce the most accurate results up to only a certain extent. For animal detection, the selected pretrained models were trained on the custom-created dataset and analysis was made on the models. However, we select the lighter models as it needs to be deployed on resource-constrained devices such as mobile. The model we trained needs more hyperparameter tuning. This model successfully classified three classes but for the other three classes, results were not accurate.

In future, we plan to improve the accuracy of model for detection of smaller texts also which can help the visually impaired to get accurate results. For animal detection, we plan to improve the performance of the model so that it can correctly classify all the classes for animal detection. We will also include experiments on other animal databases using this method.

References

[1] K. Manjari, M. Verma, G. Singal, CREATION: Computational ConstRained Travel Aid for Object Detection in Outdoor eNvironment. in: 2019 15th International Conference on Signal-Image Technology & Internet-Based Systems (SITIS), Sorrento, Italy, 2019, pp. 247–254, https://doi.org/10.1109/SITIS.2019.00049.

[2] K. Manjari, M. Verma, G. Singal, A survey on assistive technology for visually impaired, *Internet Things*, 11, 100188, (2020).

[3] M. Rajesh, B.K. Rajan, A. Roy, K.A. Thomas, A. Thomas, T.B. Tharakan, C. Dinesh, Text recognition and face detection aid for visually impaired person using raspberry PI, in: 2017 International Conference on Circuit, Power and Computing Technologies (ICCPCT), IEEE, 2017, pp. 1–5.

[4] R. Kumar, Assistive System for Visually Impaired Using Object Recognition (Ph.D. thesis), 2015.

[5] Z. Cui, Text scanner and touch reader for visually-impaired users, Technical Disclosure Commons, 2019.

[6] H. Jabnoun, F. Benzarti, H. Amiri, A new method for text detection and recognition in indoor scene for assisting blind people, in: Ninth International Conference on Machine Vision (ICMV 2016), vol. 10341, International Society for Optics and Photonics, 2017, p. 1034123.

[7] S. Deshpande, R. Shriram, Real time text detection and recognition on hand held objects to assist blind people, in: 2016 International Conference on Automatic Control and Dynamic Optimization Techniques (ICACDOT), IEEE, 2016, pp. 1020–1024.

[8] M.S. Norouzzadeh, A. Nguyen, M. Kosmala, A. Swanson, M.S. Palmer, C. Packer, J. Clune, Automatically identifying, counting, and describing wild animals in camera-trap images with deep learning, Proc. Natl. Acad. Sci. 115 (25) (2018) E5716–E5725.

[9] A. Swanson, M. Kosmala, C. Lintott, R. Simpson, A. Smith, C. Packer, Snapshot Serengeti, high-frequency annotated camera trap images of 40 mammalian species in an African savanna, Sci. Data 2 (2015) 150026.

[10] K. He, X. Zhang, S. Ren, J. Sun, Deep residual learning for image recognition, in: Proceedings of the IEEE Conference on Computer Vision and Pattern Recognition 2016, pp. 770–778.

[11] A. Eitel, J.T. Springenberg, L. Spinello, M. Riedmiller, W. Burgard, Multimodal deep learning for robust RGB-D object recognition, in: 2015 IEEE/RSJ International Conference on Intelligent Robots and Systems (IROS), IEEE, 2015, pp. 681–687.

[12] O. Russakovsky, J. Deng, H. Su, J. Krause, S. Satheesh, S. Ma, Z. Huang, A. Karpathy, A. Khosla, M. Bernstein, Imagenet large scale visual recognition challenge, Int. J. Comput. Vis. 115 (3) (2015) 211–252.

[13] A. Krizhevsky, I. Sutskever, G.E. Hinton, Imagenet classification with deep convolutional neural networks, in: Advances in Neural Information Processing Systems 2012, pp. 1097–1105.

[14] M. Schwarz, H. Schulz, S. Behnke, RGB-D object recognition and pose estimation based on pre-trained convolutional neural network features, in: 2015 IEEE International Conference on Robotics and Automation (ICRA), IEEE, 2015, pp. 1329–1335.

[15] X. Zhou, C. Yao, H. Wen, Y. Wang, S. Zhou, W. He, J. Liang, EAST: an efficient and accurate scene text detector, in: Proceedings of the IEEE Conference on Computer Vision and Pattern Recognition 2017, pp. 5551–5560.

[16] S. Targ, D. Almeida, K. Lyman, Resnet in resnet: generalizing residual architectures, arXiv preprint arXiv:1603.08029, (2016).

[17] S. Ren, K. He, R. Girshick, J. Sun, Faster R-CNN: towards real-time object detection with region proposal networks, in: Advances in Neural Information Processing Systems 2015, pp. 91–99.

MIoT: Medical Internet of Things in pain assessment

15

Sanjay Kumar Singh

Rajarshi School of Management and Technology, Varanasi, India

1 Introduction

Patients suffering from pain admitted in emergency medical care are a critical area of medicine whose outcomes are usually influenced by the time, accuracy, and accessibility of contextual information. However, data collection pertaining to pain is imprecise especially in the case of an intensive care unit where the information collection, processing, storage, and retrieval still remain manual and time consuming. Doctors are provided with the insufficient patient's health records obtained by the patient or their relatives. The approach is more patient centered instead of patient centric. Medical Internet of Things (IoT), computer vision, and wireless devices make sure the accurate collection of data. This chapter discusses the challenges of data collection and transportation, especially in developing countries. MIoT platform that is focused on patient centric is discussed. The results of the experiments conducted are presented and discussed.

A framework is proposed that uses MIoT to collect and disseminate data automatically in real time. It will prove to be quiet fruitful to mankind in developing a pain monitoring system that will provide more reliable and accurate patient-centric information and thus substantially will help in identifying the level of pain. The pain experience is reflected directly in the face of both humans and nonhuman mammals [1]. One of the major behavioral signs of pain is facial expressions, which may be used as an initial step for developing an automatic pain assessment tool meant for humans.

With the development of such a tool, collecting pain data using the self-report method could be substituted, generally for children lesser than 2 years or the patients admitted to the intensive care unit. In both cases the categories of patients are unable to report their level of pain.

The overall cost will reduce drastically upon the installation of remote health monitoring systems, either in home or hospital. Pain acts as a major monitoring index in disease diagnosis. In telehealth, various questionnaires have been developed that are answered by patients using the mobile application. The questionnaire reflects the pain intensity on a numerical rating scale, composed of time, location of pain, and medications for patients at home. The feedback received indicates that remote pain monitoring is effective and feasible in observing daily pain state. However, a self-report method has some limitations. Primarily, it is a human tendency that people express their unwillingness toward the manual entry of day-to-day information and generally in situations of long-term monitoring that is a major shortcoming of conventional approaches [2]. Another method does not comply with those groups of people who have restricted cognition and expressing abilities (patients in ICU and neonates). As a

IoT Based Data Analytics for the Healthcare Industry. https://doi.org/10.1016/B978-0-12-821472-5.00005-3

result, patients have to bear rigorous pain all day long, and their diagnosis and treatment lengthen. This is why it has become necessary for developing such MIoT devices that would automatically monitor the pain intensity level. Not much enhancement has been made for implementing an automatic pain assessment system, utilizing facial pain expression recognition looking at the videos of the face.

The following chapter is an attempt to explore the possibilities of incorporating MIoT with cloud technologies and human interaction so as to develop an automatic smart device that can be useful for remote pain monitoring.

2 Background

Various models for IoT in healthcare have been proposed by a number of researchers. This part covers the work done in these areas. It is a difficult task to assess and manage pain in a clinical setting [3]. The self-report provided by the patient is a widely used method for measuring pain, which does not count for technology and skills. It is composed of the questionnaire that asks the patient for their response, based on which the scores are calculated and analyzed. The other techniques include clinical interview and visual analogue scale (VAS) [4].

In the following chapter, we have tried to identify the work done by researchers using the concept of IoT for pain assessment, which are as follows:

Almotiri et al. [5] introduce a concept of mobile health (mhealth) that utilizes mobile phones for collecting real-time data of patients ultimately storing it on a server that is connected to the network, providing access to a limited number of clients only. Using wearable devices and body sensors, the data are extracted and collected, which are then used for the medical diagnosis of patients.

Ahn et al. [6] put forward a system for analyzing the physiological signals with ECG taken of the patient while they are sitting on the smart chair. The biosignals are monitored through a monitoring system that is developed by them proving the usefulness of IoT in healthcare.

Dwivedi et al. [7] proposed a framework that focused on the recording of clinical information that is required to be transmitted securely on the network for electronic patient record (EPR) systems. A multilayered system for healthcare to record information is presented that incorporates public key infrastructure, smart card, and biometric concept.

Gupta et al. [8] focused on the Intel Galileo development board responsible for the collection of data in the database. The doctor will use this for proper diagnosis every time the patient visits the hospital for any improvement in their health condition.

Barger et al. [9] have given the concept of a sensor network used in a smart house that records the movements of the patient. They have validated it by testing it on a prototype. The objective of their study is to check whether the system proposed if efficient enough so as to capture the behavioral patterns of humans.

Chiuchisan et al. [10] framed a model that focuses on preventing threats to those patients who are admitted to ICUs. If there is an anomaly in the body movements, room temperature, or in the health status of the patient, immediately a message will be conveyed both to the doctor and the patient relatives; the precautionary measures will be taken immediately. It is called a smart ICU.

Gupta et al. [11] proposed a model for measuring and recording ECG and various other health issues of the patient with the use of Raspberry Pi that would be of great help in hospitals to get data and effective diagnosis.

Lopes et al. [12] worked on an IoT framework centralized for disabled persons that would be fruitful in the healthcare community at large. Two use cases were considered for studying the effect of IoT technologies that could be used for the disabled segment of people.

Nagavelli et al. [13] proposed a method for predicting the rigorousness of the sickness based on the medical record of the patient through statistical approaches that is coined as the disease probability threshold. They have suggested an algorithm to support their goal.

Sahoo et al. [14] considered a huge amount of data received from patients generated by reports and studied the healthcare management system. They did an analysis of these parameters to predict patient health conditions likely to be in the future. They included the cloud-based big data analytics to achieve this with the use of probability.

Xu et al. [15] proposed a model for recording data using IoT. They designed a resource-based data access method for collecting and publishing it globally so that it could be assessed easily.

Tyagi et al. [16] focused on the effectiveness of IoT for healthcare, covering the technical aspects. They proposed a cloud-based model where the opportunities were recognized as to how the data could be transferred to the cloud and used by caregivers on having the patient's consent bringing together hospitals and laboratories together. The overall objective was to provide patients relief by reducing the expenses incurred during medical aid.

3 Methodological advances

We have undertaken this study due to the growing demand for MIoT in automatic pain assessment especially in hospitals and homes to improve the quality of diagnosis for patients suffering from pain, and it will also provide aid to the medical staff (Fig. 1).

In the pain assessment tool, the facial expressions represent the behavioral signs for those patients who cannot convey their pain, and hence, it is considered to be a unique aspect in the development of a distant pain monitoring system.

4 Distant pain monitoring system

The framework using MIoT and various other aspects including computational, medical, and psychological required for pain detection and intensity estimation is dealt with in detail in Fig. 2. It incorporates real-time central monitoring of patients in the hospital. It is also suggested that MIoT plays a significant part in capturing data, but without human involvement the results so obtained do not prove to be very efficient if pain intensity is considered [17, 18]. Hence, human-computer interaction is a must for detecting and estimating the pain intensity.

The system proposed includes sensors and a cloud acting as a back end, providing a supporting measure to patients who have access to the Internet. The cloud could be further assessed by doctors or patients who will provide aid to develop systems with advanced technologies incorporating data analytics and deep learning.

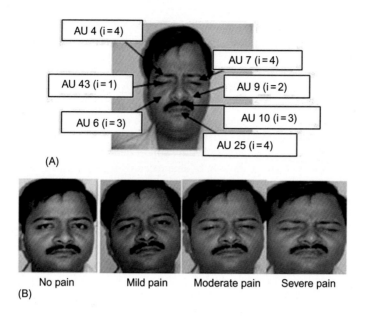

FIG. 1

(A) Face with severe pain with their action units (AUs) and their intensity levels. (B) Pain intensity represented by the FACS manual (i.e., AU4, AU6, AU7, AU9, AU10, AU25, and AU43). The onsets and offsets on the face are represented through zero and one reflecting whether the AUs are present or absent, respectively.

The framework composes of four steps:

(1) *Sensor node*

The sensing stage collects pain-related data. Human pain emotions are expressed and can be expressed in many ways, which include behavioral responses (e.g., pain, muscle contraction, and speech signals). The mobile devices are coming up with equipped physical sensors such as cameras, microphones, touch panels, and proximity sensors to see and feel the outer world. These prove to be suitable platforms for collecting emotional pain responses without the user's interruption.

(2) *Gateway*

It acts as an interface between the sensor and the cloud. It plays the role of a router, like a hot spot present in mobiles, a smart gateway having additional features like heterogeneity, reliability, and scalability. Smart gateways prove to be advantageous to the system where data are available from multidimensional techniques and communication path exists in the healthcare pain monitoring system.

(3) *Cloud server*

The data are transferred from the sensor nodes to the cloud server using TCP protocol, which is later forwarded to the data streaming channel that is further linked to another database server for storing data. This server is well suited to cloud computing models supporting platform as a service (PaaS) and software as a service (SaaS) that is efficient in signal processing and data mining.

System architecture

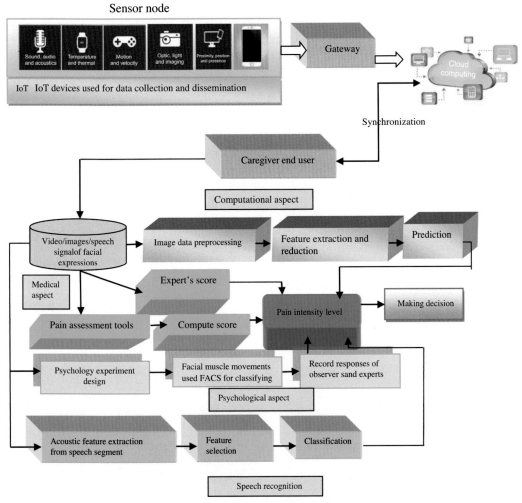

FIG. 2

IoT-based automatic pain detection and intensity estimation method using cloud and various other parameters required for an efficient pain monitoring system.

(4) *Caregiver end user*

This phase includes four different parameters for pain detection and intensity estimation.

(i) Computational aspect

In this aspect the data received from the cloud are in the form of facial images and speech signals. We have used two standard algorithms: scale-invariant feature transforms (SIFT) and speeded up robust feature (SURF) for feature extraction using the facial pain images.

Principal component analysis (PCA) has been used for dimensionality reduction and the support vector machine (SVM) for classifying the pain intensity level. The experimental results enhance the success rate of the automatic classification of the pain recognition system [19]. This will provide a much better diagnosis by the doctor as the pain intensity level is known to them and is free from any bias.

(ii) Medical aspect

The overall idea is to record and analyze the response of observers, experts (doctors), and the standard tools giving results using (e.g., visual analogue scale, numerical rating scale, and self-report) available for measuring the intensity level of pain. The facial pain expression recording was done using the MIoT device that includes webcam (Logitech) with a speed of 30 frames per second that are set before the subject that is linked to a computer (Fig. 3).

Further, comparison of pain intensity scores was performed from the three categories of respondents (observer, expert, and the patient) on each individual frame. The third categories here are the patients who made available their self-report on the VAS. It was observed that the participants were very comfortable with identifying normal and severe pain faces when presented; however, they had varied opinions when mild and moderate pain faces were displayed. It has been observed that lesser response time was recorded for normal and severe pain faces with comparison with mild and moderate ones.

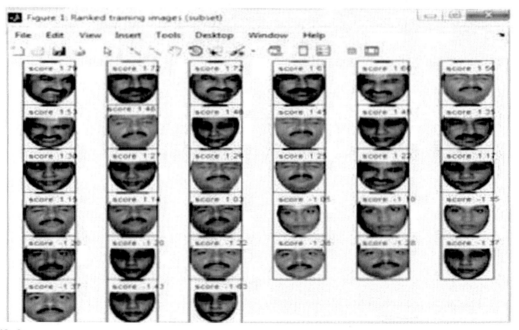

FIG. 3

Table representing severe to normal faces starting from the top based on the scores obtained on the database using computational approaches.

(iii) Psychological Aspect

These behavioral aspects of human beings are included. Facial action coding system (FACS) is used for predicting the pain intensity using facial expressions. The patients unable to report pain (e.g., patients in ICU or infant) are briefed by their observers to the experts (practitioner). The results obtained by both the observers and the practitioners are compared to reach some concrete results related to the pain intensity level.

(iv) Speech Recognition Aspect

Patient's voices were recorded who reported pain to formulate reports on various levels of pain. All the recordings were done at the medical center in the natural environment. From the recordings, voice samples were sliced to extract their audio features. Further the classification was made using machine learning techniques (e.g., SVM) to differentiate it between four different pain levels, that is, normal, mild pain, moderate pain, and severe pain. Performing classification using a large feature set gave better results. On analyzing the results, it conveys that there is a relationship among the measurable parameters of the human voice with the self-report of the patients regarding the pain intensity level. There are three key issues that need to be addressed for doing well with speech emotion recognition (SER) system: firstly, choice of a good pain emotion speech database; secondly, extracting effective features; and thirdly, designing reliable classifiers using machine learning algorithms. In fact the emotional pain feature extraction is a major issue in the SER system. Both techniques including feature extraction and selection improved learning performance leading to better generalizable models with less computational complexity and storage requirements.

5 Solutions and recommendations

I have proposed some challenges in automatic pain recognition and have given my suggestions that I believe is fruitful in overcoming those. I would like researchers to design effective MIoT devices and mobile apps to move a step forward toward more reliable and accurate pain assessment.
Few goals that need to be addressed are as follows:

(1) Designing more advanced MIoT systems for pain recognition to fulfill the requirements of doctors for effective diagnosis. This involves validating systems on patients suffering from pain, progressing the specificity and robustness in a real-time environment or at home conditions, and focusing on cost-effectiveness and advanced technology driven.

(2) More emphasis is required to access chronic pain rather than acute one as the majority of researchers have focused on acute nociceptive pain. Chronic pain does have a larger impact on society and therefore needs to be addressed. Further, finding the intensity of pain and suppression of pain where the medicine does not work along with pain quality and location are emphasized.

(3) Investigate the treatment provided and strive for better solutions.

5.1 Challenges

Using MIoT, much of the task pertaining to the data collection has been minimized, but for developing pain recognition systems, one has to have some additional knowledge about the following:

(1) The general physiology of pain and the responses measured. This prompts to the requirement of an automatic pain system composed of such feature sets.

(2) Awareness of the situations leading to pain will be helpful for automatic recognition.

(3) Data collected from clinical, psychological, and MIoT devices including the computational aspect will greatly enhance the proper diagnosis of the patient. As pain cannot be measured using one or two parameters. Therefore interaction of human–machine will be really helpful in automatic recognition.

(4) The main challenging problem with an automatic pain recognition system is the availability of a dataset. If we consider facial pain expressions, we know that the expression of pain is spontaneous and is available only for a few seconds [20]. It is therefore difficult to capture that expression. Another quality included in the dataset is that it has to be multimodal, annotated, and should compose of other states including pain to assess the specificity. Datasets should be shared and made available to take a leap in this field. In the present scenario, very few datasets are available for researches including BioVid and UNBC-McMaster shoulder pain databases that are validated.

In future, multiple datasets need to be validated across and checked for finding how effectively a system generalizes in other situations, pain types, and medical populations at large.

(5) Improved algorithms and MIoT devices are required for improving the performance in clinical settings.

6 Conclusions

From the patient end the distant pain monitoring system caters around the sensor node. The mobile web application and cloud assist in capturing the data automatically. The employed MIoT devices along with the cloud architecture at the back end help in online data storage functions, which provide caregivers to assess the level of pain intensity of the patient on analyzing it using various approaches discussed earlier in the framework (Fig. 2). We have involved human, that is, experts along with the machine to arrive at a more specific and accurate pain intensity score.

The users and the doctors receive reports through online processing in the form of graphical representation. The MIoT remote monitoring systems are scalable with respect to the functionality because of the architecture using a sensor gateway–cloud concept [19]. The cloud platform could be used for further performing online analysis of data and implementing the decision-making support algorithm required for the application of pain management.

The chapter presents the architecture to develop a prototype of a MIoT-based automatic pain recognition system along with the human interaction to provide more reliable results that involve various health parameters being constantly monitored so that predictions of any pain or disease could be made leading to proper precautionary measures being taken to provide relief to the patients suffering from pain and visiting hospitals for the cure. The system proposed has been validated and thus could be installed in hospitals that will ensure proper monitoring along with providing an ample amount of data that can be stored in an online database. An additional facility is that the results could be accessed through mobile. The concept for designing a system could be further enhanced by using augmented artificial intelligence to support both the doctors and the patients for improving their quality of life.

References

[1] D. Liu, D. Cheng, T.T. Houle, L. Chen, W. Zhang, H. Deng, Machine learning methods for automatic pain assessment using facial expression information: protocol for a systematic review and meta-analysis, Medicine 97 (49) (2018) e13421.

[2] R.S. Evans, Electronic health records: then, now, and in the future. Yearb. Med. Inform. (Suppl. 1) (2016) S48–S61, https://doi.org/10.15265/IYS-2016-s006.

[3] N. Wells, C. Pasero, M. McCaffery, Improving the Quality of Care Through Pain Assessment and Management, Agency for Healthcare Research and Quality (US), 2008.

[4] A.H. Gillian, M. Samra, K. Tetyana, F. Melissa, Measures of adult pain, Arthritis Care Res. 63 (S11) (2011) S240–S252.

[5] S.H. Almotiri, M.A. Khan, M.A. Alghamdi, Mobile health (m-health) system in the context of IoT, in: IEEE International Conference Fourth on Future Internet of Things and Cloud Workshops (FiCloudW), 2016, pp. 39–42.

[6] B.G. Ahn, Y.H. Noh, D.U. Jeong, Smart chair based on a multi heart rate detection system, IEEE Sensors (2015) 1–4.

[7] A. Dwivedi, R.K. Bali, M.A. Belsis, R.N.G. Naguib, P. Every, N.S. Nassar, Towards a practical healthcare information security model for healthcare institutions, in: International IEEE EMBS Fourth Special Topic on Conference on Information Technology Applications in Biomedicine, 2003, pp. 114–117.

[8] M.S.D. Gupta, V. Patchava, V. Menezes, Healthcare based on IoT using raspberry pi, in: International Conference on Green Computing and Internet of Things (ICGCIoT), 2015, pp. 796–799.

[9] T.S. Barger, D.E. Brown, M. Alwan, Health-status monitoring through analysis of behavioral patterns, IEEE Trans. Syst. Man Cybern. Syst. Hum. 5 (1) (2005) 22–27.

[10] I. Chiuchisan, H.N. Costin, O. Geman, Adopting the internet of things technologies in health care systems, in: 2014 International Conference and Exposition on Electrical and Power Engineering (EPE), 2014, pp. 532–535.

[11] P. Gupta, D. Agrawal, J. Chhabra, P.K. Dhir, IoT based smart healthcare kit, in: International Conference on Computational Techniques in Information and Communication Technologies (ICCTICT), 2016, pp. 237–242.

[12] N.V. Lopes, F. Pinto, P. Furtado, J. Silva, IoT architecture proposal for disabled people, in: IEEE 10th International Conference on Wireless and Mobile Computing, Networking, and Communications (WiMob), 2014, pp. 152–158.

[13] R. Nagavelli, C.V. Guru Rao, Degree of disease possibility (DDP): a mining based statistical measuring approach for disease prediction in health care data mining, in: International Conference on Recent Advances and Innovations in Engineering (ICRAIE-2014), 2014, pp. 1–6.

[14] P.K. Sahoo, S.K. Mohapatra, S.L. Wu, Analyzing healthcare big data with the prediction for a future health condition, IEEE Access 4 (2016) 9786–9799.

[15] B. Xu, L.D. Xu, H. Cai, C. Xie, J. Hu, F. Bu, Ubiquitous data accessing method in IoT-based information system for emergency medical services, IEEE Trans. Ind. Inform. 10 (2) (2014) 1578–1586.

[16] S. Tyagi, A. Agarwal, P. Maheshwari, A conceptual framework for IoT-based healthcare system using cloud computing, in: 2016 6th International Conference—Cloud System and Big Data Engineering (Confluence), 2016, pp. 503–507.

[17] G. Yang, M. Jiang, W. Ouyang, G. Ji, H. Xie, A.M. Rahmani, P. Liljeberg, H. Tenhunen, IoT-based remote pain monitoring system: from device to cloud platform, IEEE J. Biomed. Health Inform. 22 (6) (2018) 1711–1719.

[18] E. Jacob, J. Duran, J. Stinson, M.A. Lewis, L. Zeltzer, Remote monitoring of pain and symptoms using wireless technology in children and adolescents with sickle cell disease, J. Am. Acad. Nurse Pract. 25 (1) (2013) 42–54.

[19] P. Werner, D. Lopez-Martinez, S. Walter, A. Al-Hamadi, S. Gruss, R. Picard, Automatic recognition methods supporting pain assessment: a survey, IEEE Trans. Affect. Comput. 10 (V) (2019) 1–22.

[20] S.K. Singh, S. Tiwari, A.I. Abidi, A. Singh, Prediction of pain intensity using multimedia data, Multimed. Tools Appl. 76 (2017) 19317.

Applications of IoT for animals

Applications of Internet of Things in animal science

Sonal Saxena[a], Sameer Shrivastava[a], Abhinav Kumar[b], and Anshul Sharma[b]

Division of Veterinary Biotechnology, ICAR-Indian Veterinary Research Institute, Bareilly, Uttar Pradesh, India[a]
Department of Computer Science and Engineering, Indian Institute of Technology (BHU), Varanasi, Uttar Pradesh, India[b]

1 Introduction

The Internet of Things (IoT), first described by Kevin Ashton in 1999, refers to the ability of connected devices to sense and gather data, which are then shared through the Internet facility and finally used to achieve common goals. Interconnected digital tools can be remotely controlled and can be used to capture data from different locations with high efficiency and security. In the near future, billions of devices will have intent connectivity and will be accessible from anywhere in the world. Thus IoT can revolutionize the animal sector through a variety of IoT-enabled applications, such as smart animal health monitoring, smart animal farm, smart animal waste management, disease management, smart pet management, wildlife management, and environmental or pathogen monitoring devices. The livestock industry can benefit greatly from such sophisticated systems that can continuously monitor animal health, collect data, and report the results obtained to owners and regional animal centers. The Internet of Animal Health Things (IoAHT) therefore not only provides a framework for streamlining the collection of appropriate and usable data but also helps interpret meaningful information. Automated sensor-driven data collection in precision farming enables efficient monitoring of key production parameters, thus increasing profit margins without compromising animal health and welfare. Thus this chapters explore the various opportunities and challenges of adopting IoT in the animal sector.

2 Innovative IoT in management of smart farms and precision livestock farming

The growing human population has put the livestock sector under tremendous pressure. To cope up with the increasing demands for animal products, livestock farming has been intensified. Though this intensification increases productivity, it also affects animal health by making them susceptible to various diseases and outbreaks. Recently with the introduction of concepts of smart farming, IoT helps farmers in monitoring and management of livestock farms through data-driven insights. The IoT enables farmers to improve fodder production and support livestock health through the data-driven decision making and remote monitoring. Smart livestock farms make use of digital technologies

IoT Based Data Analytics for the Healthcare Industry. https://doi.org/10.1016/B978-0-12-821472-5.00001-6

for efficient production and management of farm operations. In such type of precision livestock farming, animal health, productivity, and waste management are controlled using IoT-based sensors and tools. Thus precision farming makes livestock farming more controlled and productive. In a smart system, remote monitoring and management of animal farm operations are made possible through the use of IoT-enabled control systems, sensors, autonomous vehicles, microcontrollers, robotics, automated hardware, and many more. In such a smart farm, surveillance of the farm can be done efficiently: Feed and water provided as, and when required, excess biogas produced by the animal's waste can be detected, animal health can be monitored, and moreover, any incidence of fire or theft can be detected in the farm. Such an intelligent system utilizes microcontrollers, IP cameras, and various types of sensor devices such as ultrasonic sensors, water level sensors, gas sensors, humidity, and temperature sensors, along with intranet connectivity with smart devices.

Now, let us discuss various ways in which IoTs help in managing animal farms.

2.1 Real-time monitoring of livestock health

Livestock owners are losing significant amounts of profit owing animal illnesses. IoT-supported livestock management solutions help farmers to promote better livestock health in a several ways. We have been using IoT devices in the form of wearables to track human activity and health, and now with technological advancements, IoT is making it possible to monitor animal health. Livestock wearable sensors allow livestock owners to monitor temperature, heart rate, respiratory rate, blood pressure, and other vital parameters. Data fed to the cloud directly from interconnected wearable devices help farmers identify and manage animal diseases and feed problems before they have a significant impact on the health of the entire herd. In addition, drones can be used to monitor weather, air quality, ventilation, etc. in livestock farms and facilities. An IoT-based intelligent animal health surveillance device was developed to monitor physiological parameters such as body temperature, heart rate, and rumination along with ambient temperature and humidity in real time [1]. In this system, Raspberry Pi3, which has built-in Wi-Fi, is used as a core controller for processing the data collected by the different sensors, and the information is forwarded to the cloud. The information can be accessed from anywhere via the Internet and the Android app. Fig. 1 shows how IoT can be used for real-time monitoring of cow's health through a model of a connected cow [2].

2.2 Monitoring livestock reproduction and fertility

IoT devices can be used to monitor the reproductive cycles and the calving of farm animals to ensure safe and successful outcomes. Livestock animals are fertile for a limited time period during which they are fertile, and this period can be even 8–9 h a month. IoT sensors help in the detection of this fertility period so that this fertility window is utilized maximally. Sensors within a range of 1000 m can collect the fertility information. Readings from different sensors are conflated together and help in deciding the fertility period for the entire herd, and thus this helps farmers to manage livestock efficiently. An IoT-based device has been developed by the Swiss start-up Anemon to detect cow's fertility [3,4]. The device is in the form of a sensor that is implanted in the cow's genitals for measuring body heat. Signals from this thermal sensor are then transmitted to another sensor on the animal's collar, which tracks body motion when the cow is fertile and then sends a text message to the farmer if the cow is in heat or in the estrus. The collar sensor is also equipped with a SIM card, and, when the cow is ready for

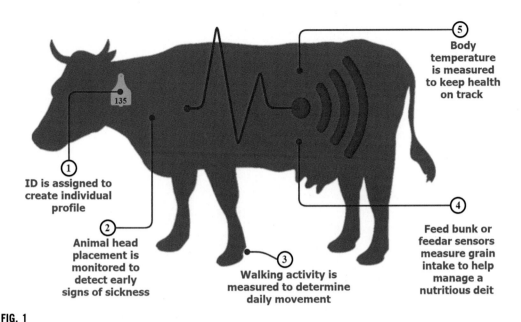

Body temperature is measured to keep health on track

ID is assigned to create individual profile

Animal head placement is monitored to detect early signs of sickness

Walking activity is measured to determine daily movement

Feed bunk or feedar sensors measure grain intake to help manage a nutritious deit

FIG. 1

IoT in the real-time monitoring of cow health.

reproduction, the farmer is alerted by SMS notifications. JMB, a North American company, is providing a system that alerts owners when the cows are pregnant and ready for calving. A model of livestock monitoring and estrus detection through IoT is displayed in Fig. 2.

2.3 Monitoring animal location and movement

Monitoring the movement of animals to locate their position and combating theft has been a serious necessity of the farmers for which solutions have come from the latest advancements in IoT technology. IoT is particularly useful in locating sick animals and establishing grazing patterns. A UK-based company has introduced IoT-based technology for pinpointing the cattle location. The technology is operational on any connected device, including smartphones, and thus helps in monitoring cattle roaming and theft. A robot named "SwagBot" has been designed for guiding cattle herds by the scientists at the Australian Centre for Field Robotics, University of Sydney [3,6]. Pilot studies have shown that the robot can herd cattle independently, clear weeds, and pull heavy loads also. An example of IoT-based system for detecting location of animal herd is shown in Fig. 3.

2.4 IoT in precision feeding and feed management

Intelligent IoT-based system helps to monitor optimum feeding and watering in animal farms. IoT-based sensors help in precision feeding each animal with the right amount of feed at the right time and in the right nutrient composition. A software named "fodjan smart feeding" has been developed

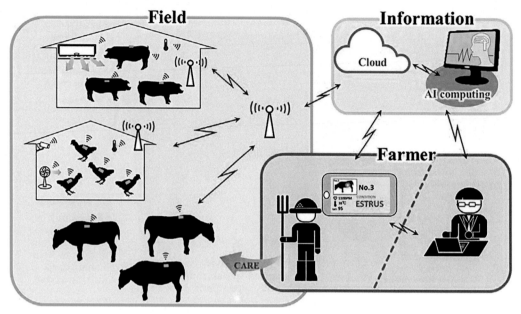

FIG. 2

Model of a smart livestock monitoring system based on IoT technology [5].

by fodjan, a Dresden-based company [8]. The system helps farmers to choose balanced food rations for the farm animals. The food rations are optimized, taking into account the cost of animal health and feed. The software also enables setting up animal feeding calendars and helps in monitoring food stocks. In this way, IoT-based systems help to feed animals cost-effectively balanced rations and reduce feed waste, thereby contributing to efficient farm management and increasing profitability. IoT-based precision feeding thus reduces feed wastage and feed intake per animal while simultaneously maximizing the individual growth rates.

2.5 Monitoring animal behavior and physiology

Remote monitoring of animal behavior and its physiology helps in better management of the animals. GPS collars and other devices allow researchers to monitor animal behavior and interactions with the environment. IoT-based devices have been developed for tracking activity of cows throughout the day [3]. Such devices can detect anomalies in their feeding and grazing patterns and can alert the farmers. The data can be accessed by the farmers with their smartphones for addressing issues related to animal behavior and physiology; thus they can focus on the management of their animals in a more authentic manner. Researchers have also developed a noninvasive wearable stress monitoring system (WSMS) for continuous monitoring of signs of stress in animals during transportation [9]. An example of such as stress monitoring system for sheep is shown in Fig. 4.

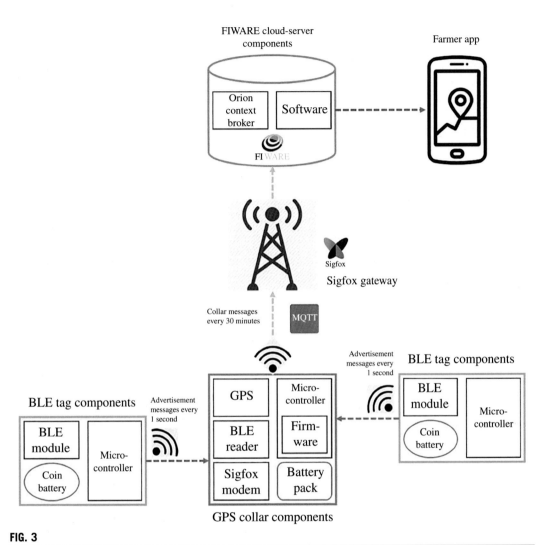

FIG. 3

IoT-based system to monitor the location of an animal herd [7].

2.6 Monitoring lactation and milk yields

Lactation is one of the most important part of revenue generation for livestock farms, and thus efficient monitoring of lactation patterns is very crucial. Therefore this is one area where IoT plays an important role. IoT can help in monitoring milking times and milk quality and can also alert the farmers when the cows are ready to milk or in cases of a sudden drop in milk yields. A smart dairy farm in New York utilizes Astronaut robots for milking more than a thousand cows [10]. The robots allow milking at optimal times when the cows prefer to be milked and thus increase the production of cows. Apart from this, robots and IoT-based devices help in collecting and monitoring milk yield records. In smart farms,

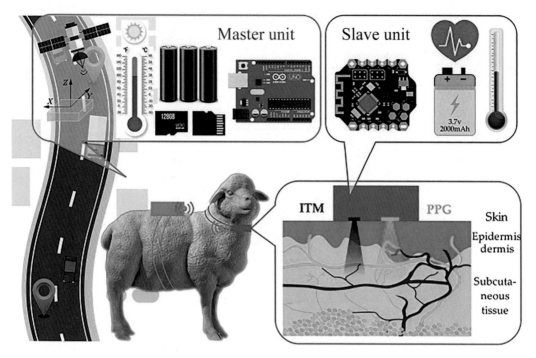

FIG. 4

Wearable stress monitoring system for sheep [9]: the system includes two interconnected modules: (1) a master unit fixed into an adjustable elastic band that is mounted on the back of the sheep and (2) a slave unit placed around the neck.

cows are equipped with IDs and transponders, which alert the farmers when they are ready to produce more milk. If a cow is producing more milk, dairy workers are accordingly directed to make changes in the dietary pattern to assist lactation. Fig. 5 shows IoT-based automated unit for collection of milk from cow.

2.7 Monitoring records

The accounting records of farms are essential for the efficient farm management as they monitor the productiveness of the farms. Maintaining animal data and records in large livestock farms is a cumbersome and time-consuming task. Besides managing farm animals, IoT-based devices are also very useful in collecting and maintaining farm records, which helps farmers to make important decisions. The use of sensor-operated devices automates data collection and reduces bias in the entry of data. Automated data collection is both time- and cost-efficient for the farmers as it allows the livestock owner to devote more time for the care and management of animals, thus maximizing returns.

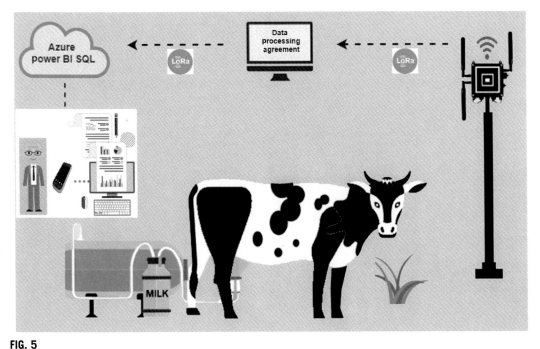

FIG. 5

IoT-based automated cow milking unit.

3 IOT-based smart management of pets

Monitoring pets is an area of concern to the pet owners as pets require special care and treatment. However, in the present scenario of an extremely busy lifestyle, this task is not as simple as it used to be. Nevertheless, pet owners can make use of a variety of IoT-based devices for smart pet management [11]. Several pet care devices are currently available in the market, and their use can also be great fun apart from making the pet management very easier. IoT-based smart pet management devices allow automated health monitoring and feeding of pets. A recent concept is smart pet houses with automated devices wherein pets can be left unattended by the owners.

3.1 Pet health and activity monitors

Pet's well-being can be tracked using various IoT-based sensors in the form of wearable tags and smart collars. These sensors help in tracking the health of pets by real-time monitoring of pet temperature, respiration, heart rate, pulse, feed and water intake, and other vital information. The devices may alert the owners if there is any deviation in animal health parameters. These sensors are also helpful in tracking the activity of pets during the day and can also monitor the duration and quality of their sleep. Fig. 6 shows neck-wearable and tail-wearable IoT devices for monitoring activities of pet dogs.

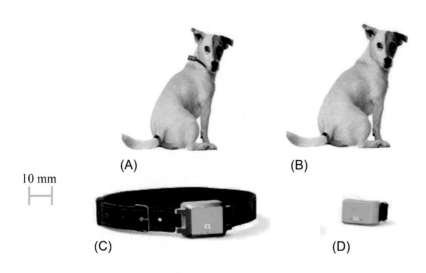

FIG. 6

Wearable sensors for dogs [12]: (A) Neck wearable and tail wearable sensor devices that are placed on the dog to detect activity. (B) The tail wearable sensor device is placed to detect emotions. (C) Neck wearable sensor device. (D) Tail wearable sensor device.

3.2 Smart pet feeders

Automated IoT-based pet feeder is one of the latest innovations for feeding pets and helps the owners to feed their pets when they are busy or not at home [13]. The IoT-based feeder automatically dispenses the required amount of food at the exact time, which can be controlled by the user by a wireless infrared remote control. Apart from scheduling the pet's feeding, these devices can be connected to the smartphones, and thus owners can get notifications about the time when the pet had lunch and also the amount of food eaten. Some pet feeders also have a bowl cover that opens and closes automatically and is operated by an infrared proximity sensor. With recent advancements, pet feeders can also easily integrate with smart home management systems or operated with digital assistants, such as Amazon's Alexa.

3.3 Pet monitors and interactive systems

IoT enabled pet monitors and interactive cameras are quite useful when the owners are not at home. In such cases they can still plugin and see how the pet is doing while they are away. Pet owners can also interact with their pets using their smartphones. Pet monitors are also useful in determining the position of pets. Pet collars with GPS tags are quite useful for determining the location of the pets when they go out and also help in preventing pet thefts. The GPS tag continuously transmits signals and helps to identify the whereabouts of a pet. Such type of wearable sensors continuously monitors the location of the pet and updates the user accordingly. This is particularly useful for the security of pets. Own et al. [11] have described an IoT-based pet monitoring system for monitoring pet location and activity of the pets.

3.4 Pet fitness apps

Advancements in Internet technology are a great boon to the animal fitness market. A number of apps have been developed to support the veterinarians and pet owners. PetDialog is an app that helps pet owners in monitoring vaccination schedules, exercises, feed intakes, and other pet activities and alerts the owners the routine activities and vaccination dates [14]. FitBark, Wag, Puppr, Chewy, Pet First Aid, Rover, BarkHappy, etc. are examples of other popular pet apps [15]. Such app-based digital systems make it easier to control animal fitness. Apps can send abnormal animal behavior alerts to veterinarians for faster medical aid.

3.5 IoT for treatment of pet animals

The use of IoT-based pet sensor devices helps veterinarians in better and effective treatment as the devices provide a detailed history and the data that would otherwise not possible in consultation with the owners. These devices help in monitoring recovery of pet's postsurgery or following medication. These devices also help in collecting and maintaining patient's data for further treatment and in monitoring trends and thus help in developing effective treatment strategies.

4 IoT in equine management

In the multibillion dollar equine industry, demand for connected devices for the IoT is growing rapidly. IoT-based sensor networks are now in place for remote monitoring and equine management.

　Various applications of IoT in the equine industry are listed in the succeeding text:

4.1 IoT in the horse racing industry

Horse racing has become a popular, lucrative sport that earns huge revenues and attracts visitors across the globe. With increasing global pressure the progress of the world's top horses has to be monitored similarly to top sports persons. Similar to athletes who use IoT-based wearable sensors for monitoring training progress, heart rates, and register work, IoT-based devices are becoming popular for racehorse training. Connected monitoring devices allow the end users to maximize the potential of their horses by allowing training of horses using the latest scientific training methods, apart from monitoring health problems. Gmax, a UK-based company, is providing IoT-based solutions for connecting equine technologies and provides huge amounts of data for effective training of equines. Gmax also facilitates multimedia presentations on horse racing and training, which is quite appealing for the new generation customers. Equimeter equipped with motion and performance sensors is IoT-based technology developed by France, which is designed especially for training and analysis of the performance of racehorses on a routine basis. Equimeter is a wearable girth-connected sensor for monitoring vital parameters and performance indicators using GPS data. Training sessions controlled by equimeter are displayed on the training pages of the mobile app and online web platform as well [16,17].

4.2 Smart IoT-based horse farms

IoT-monitored smart farm is the recent concept in management of equines. Spanish technology start-up company named EOIT has designed "Smart Horse," which is an integrated IoT-enabled system composed of wireless sensor networks for real-time monitoring the health of equines and managing the condition of stables and barns. "Smart Horse" is a component of "Smart Farm" platform that connects sensor devices and collects data, which is then analyzed along with other systems for effective management of horse farms. Smart Horse includes up to six calibrated sensors for measuring temperature, water flow, humidity, liquid levels, and open or close status of doors and windows. The data are collected and transmitted by the sensor nodes [16].

4.3 IoT for designing horseshoeing and horse prosthetic devices

IoT is providing the latest innovations in the design of horseshoes, orthopedic, and prosthetic devices. Australian firm, CSIRO, uses a 3-D horseshoe printing technology for designing horseshoes. The technology uses imaging software for close analysis of the hoof for providing horses with shoes having better ergonomic fitting. Such 3-D printing technology can also be utilized for creating splints, casts, and prosthetics for animals with injury or broken bones [16].

5 Wearable sensors for monitoring health

A number of companies are nowadays providing IoT-enabled wearable sensors for horses, which provide real-time monitoring into various health parameters and daily activities of the horses. The data gathered by the wearable sensors give valuable insight to the veterinarians about individual horses and the species as a whole, which helps in developing better therapeutic and diagnostic strategies for equine diseases. A France-based company has designed "Seaver," which is the first connected equipment of its kind for monitoring horse health and performance. The device is a wearable girth for recording heart rate, respiration rate, and horse movement. Another innovation by the France Voltaire group is the "smart saddle" having a sensor clip for collecting data about the gait of the horse, direction and time spent for each gait, number of jumps, etc. Protequus, Austin-based premier biomedical engineering firm, has developed yet another smart gadget for monitoring equine distress and wellness. The device is named "NIGHTWATCH Smart Halter" that utilizes sensors for 24-h monitoring of horses and generates automated alerts for the owners if the horse is in distress [18].

6 IoT for virtual veterinary training and imaging of horse diseases

The latest innovation in IT is the IoT-connected virtual classrooms, which puts forth the idea of live streaming and enhancing the quality of teaching by reaching out to more number of students. Rare and complicated surgeries involving equines and other veterinary patients could be practiced in a virtual reality classroom. This minimizes the risk to technicians, students, and teachers. Virtual learning is also cost-effective and time saving and also saves the animals from invasive procedures required for learning and training. IoT is also popular for equine thermography, which utilizes a camera for detecting infrared waves on the body surface of the horse. Veterinary Thermal Imaging, Ltd., uses similar technology to detect abnormalities in horse muscles, ligaments, tendons, nerves, and bones. Any abnormal signs can be detected weeks before the animal starts showing any symptoms allowing effective treatment of the horses [16].

7 IoT for equine data management

IoT technology is well known for its data management capabilities, and this potential of IoT is utilized in the equine industry. Blockchain technology is a secure global database technology used by horse owners and horse industry all over the world [16]. The database is utilized for storing valuable information related to horses, and any equine firm can access the database and register the horses and feed information relating to birth, vaccination, treatment, injections, surgery, height, and weight. Many equine federations are microchipping the horses for getting information about each horse, and the data are collected, stored, and shared using blockchain technology. The information can be accessed by the owners and the veterinarians.

8 Conclusions

IoT has the potential to revolutionize animal husbandry, pet ownership, and veterinary medicine. The market for IoT products in the animal sector, particularly for pets and the equine industry, is enormous. Application of IoT technology to hundreds of millions of farm animals, pets, and wildlife across the globe is a real challenge, and it will require many more years of research before IoT is exploited extensively in the animal sector. The development of microprocessors with increased speed of processing and miniaturization of sensors provides new insights for developing products to improve animal health and well-being.

References

[1] S. Kumari, S.K. Yadav, Development of IoT based smart animal health monitoring system using raspberry pi, Int. J. Adv. Stud. Sci. Res. 3 (8) (2018) Available from SSRN: https://ssrn.com/abstract=3315327.

[2] L. Bedord, Technology Helps Cattle Health Management, Retrieved 2 May 2020, from: https://www.agriculture.com/technology/livestock/technology-helps-cattle-health-management, 2018.

[3] Springwise, Sensor alerts farmers via SMS when a cow is in heat, Retrieved 10 April 2020, from: https://www.springwise.com/sensor-alerts-farmers-sms-cow-heat/, 2019.

[4] H. Reese, IoT for Cows: 4 Ways Farmers are Collecting and Analyzing Data from Cattle, Retrieved 10 April 2020, from: https://www.techrepublic.com/article/iot-for-cows-4-ways-farmers-are-collecting-and-analyzing-data-from-cattle/, 2016.

[5] W. Iwasaki, N. Morita, M.P.B. Nagata, IoT sensors for smart livestock management, in: Chemical, Gas, and Biosensors for Internet of Things and Related Applications, Elsevier, 2019, pp. 207–221.

[6] A. Klein, Cattle-Herding Robot Swagbot Makes Debut on Australian Farms, Retrieved 10 April 2020, from: https://www.newscientist.com/article/2097004-cattle-herding-robot-swagbot-makes-debut-on-australian-farms/, 2016.

[7] F. Maroto-Molina, J. Navarro-García, K. Príncipe-Aguirre, I. Gómez-Maqueda, J.E. Guerrero-Ginel, A. Garrido-Varo, D.C. Pérez-Marín, A low-cost IoT-based system to monitor the location of a whole herd, Sensors 19 (10) (2019) 2298.

[8] Fodjan, Fodjan Smart Feeding: Your Platform for Feeding Management, Retrieved 10 April 2020, from: https://fodjan.de/?locale=en_US, 2014.

[9] Y. Cui, M. Zhang, J. Li, H. Luo, X. Zhang, Z. Fu, WSMS: wearable stress monitoring system based on IoT multi-sensor platform for living sheep transportation, Electronics 8 (4) (2019) 441.

[10] Lely, Automatic Milking—Milking Robot—Astronaut A5, Retrieved 1 May 2020, from: https://www.lely.com/solutions/milking/astronaut-a5/, 2018.

[11] C.M. Own, H.Y. Shin, C.Y. Teng, The study and application of the IoT in pet systems. Adv. Internet Things 3 (1) (2013) 1–8, https://doi.org/10.4236/ait.2013.31001.

[12] S. Aich, S. Chakraborty, J.S. Sim, D.J. Jang, H.C. Kim, The design of an automated system for the analysis of the activity and emotional patterns of dogs with wearable sensors using machine learning, Appl. Sci. 9 (22) (2019) 4938.

[13] S.A. Yadav, S.S. Kulkarni, A.S. Jadhav, A.R. Jain, IOT based pet feeder system, Int. J. Adv. Res. Innov. Ideas Educ. 2395-43964 (2) (2018) 3656–3659.

[14] PetDialog from Zoetis, PETDIALOG—Your Local Vet Practice Mobile App, Retrieved 1 May 2020, from: https://www.zoetis.co.uk/conditions/petdialog.aspx, 2020.

[15] B. Gelbart, 9 Best Apps For Dog Lovers and Pet Owners in 2019, Retrieved 1 May 2020, from: https://fueled.com/blog/dog-lovers-apps/, 2019.

[16] K. Bohn, Top 8 IoT Disrupters in the Equine Industry, Retrieved from: https://www.arrow.com/ecs/na/channeladvisor/channel-advisor-articles/top-8-iot-disrupters-in-the-equine-industry/, 2020.

[17] Arioneo, EQUIMETRE VET: Heart Rate, ECG, Speed, Locomotion for Horse at Training Specially Design for Veterinarians, Retrieved 7 May 2020, from: https://www.arioneo.com/en/equimetre-vet-for-research-veterinarian/, 2020.

[18] P. Moorhead, Protequus Launches Nightwatch Edge-Computing Smart Halter: An Early Warning System For Horses, Retrieved 5 May 2020, from: https://www.forbes.com/sites/patrickmoorhead/2018/05/01/protequus-launches-nightwatch-edge-computing-smart-halter-an-early-warning-system-for-horses/#4536f9ba235f, 2018.

Internet of animal health things (IoAT): A new frontier in animal biometrics and data analytics research

17

Santosh Kumar[a], Sunil Kumar[c], Prerna Mishra[a], and Mithilesh K. Chaube[b]

Department of Computer Science and Engineering, Dr. SPM IIIT-Naya Raipur, Atal Nagar, Raipur, Chhattisgarh, India[a] Mathematical Sciences, Dr. SPM IIIT-Naya Raipur, Atal Nagar, Raipur, Chhattisgarh, India[b] Marine Engineering Research Institute (MERI), Kolkata, West Bengal, India[c]

1 Introduction

Recently the seamless combination of intelligent sensor technologies with traditional monitoring of livestock has driven to a miraculous transformation in smart farming and precision agriculture [1]. The primary motivation of all these disruptive technologies is to leverage a learning platform for interdisciplinary researchers, algorithm developers, scientists, and engineers to use these sensor technologies, global positioning system (GPS), guidance, management systems, unmanned vehicles (UAV), drones, and learning frameworks for ensuring better conditions for the health management of the animal [1, 2].

The basic fundamental for enabling smart sensing devices or systems is that the connected objects to be sensed and managed remotely using the network model. These sensed data are transferred from connected devices to the cloud server database system using WI-FI and other communication media [1, 2].

The integrated sensor frameworks enable to provide help in gathering sensed data in real-time data about the movement of animals, body posture, status about eating, body temperature, and surrounding conditions [2, 3]. The received data can be used in managing livestock activities by performing predictive analysis using various data science learning paradigms such as machine learning and deep learning techniques. It also integrates several multiple modalities to learn these data. These advanced multimodality-based learning paradigms reduce human errors by interpreting the massive amount of sensory data and facilitate smart sensing and processing capability about livestock health in the smart livestock-based framework.

Internet of Animal Health Things (IoAHT) is gaining more proliferated due to massive applications and achieved much attention for smart farming. It has reformed the research prospective for operation, connectivity, and reachability of smart livestock [1–5]. While the classical livestock management methods have been more intuitive, the growing enormous demands and supply of agricultural products had taken more time due to manual tracking and processing of data through the different thermal scanner of individual animals [6]. Therefore, the classical procedure becomes more time-consuming. It is not also a more cost-effective and optimal conditional-based predictive analysis for the health management of animals. The classical livestock frameworks have applied for the management of the small-size population of the animal. It becomes a more challenging problem with the increase in the farmhouse's size and scale [7].

The IoAHT-enabled learning framework is achieving intensification by technology improvements and its rapid utilization usage [7–9]. With usage, the data rate is also growing online. Conventional searching models lack in shoveling the knowledge needed by users from massive databases.

IoT is an integrated platform of connected devices, systems, and sensors, which have affected every individual's daily life than ever before. It has proffered a universal platform for combining intelligent devices and systems in the defined network model and enabled frameworks [1–11]. The application and uses of IoT framework have provided in several areas, such as connected industry, smart farming, smart city [12], smart home [13], smart energy, connected car [12], smart agriculture [13, 14], healthcare, and logistics [8]. It is supposed that the world population growing at a slower rate can reach up to 20.50 billion in 2050.

With this rise, food demands can also amplify. The world is now changing its trajectories toward highly reliable IoT-based applications, which are combined with data analytics techniques to meet the need for the requirement of smart livestock framework [10]. It is supposedly said that the incorporation of IoT devices and systems in the agriculture area can rise to more than 80 million in the upcoming decade. IoT systems can empower agriculture into smart agricultural practices offering improved functional efficacy and better harvest income [11–14]. IoT's primary objective is to increase the natural world and cyberspace, such as interconnected simulated environment using the network for interchanging the knowledge [15].

1.1 IoT-based livestock framework

The livestock framework uses several IoT paradigm services for enabling the smart sensing capability of animal health data. However, the integration of individual modalities and finding suitable feature representations of sensed data is an open-ended problem. The main central idea of this section is to illustrate the significant challenging problem in smart farmhouses. What are the basic processing units to process the massive amount of data from heterogeneous connected devices? How these modality or devices connected in an integrated framework. What are the primary learning techniques that leverage to find suitable data representation for IoT frameworks that provide health management solutions in a smart farming-based livestock management framework?

In this section, the IoT-based livestock framework is illustrated. IoT has increased its wings over multiple areas of science, medicines, business, and other societal spheres. Data are said to be the heart of IoT, making any ordinary entity a smart entity. This network produces a massive amount of data. Data management and data analysis are only possible if the facets and essence of data are adequately conceived [13–16]. Classical sensing techniques to comprehend this massive information might fail; these need to be excogitated and grow with the improved technology.

Storing all sensed data related to livestock farm and livestock animals are challenging for the farmers and other users (e.g., service provider, private organization, and government sectors) to do on the go.

Customers expect transparency in the means of food and drinks they receive from the farms. It is expected that dairy products should be of good quality retaining all the necessary nutrition. It is conceivable only when livestock is healthy, and their health is being monitored regularly either by giving them individual fodder or medicines if they are ill.

The systems, too, are changing around the globe for having the medicines and their usage [13, 16]. Due to the glassiness and documentation needed for livestock, monitoring makes the data management a necessary perspective.

The growing improvements in IoT and data analytics in the last decade have given the technology a new and vital dimension in regular activities [15]. The objectives of this chapter are to explore the opportunities and challenges of IoT and data analytics methods for the establishment of smart livestock health management for the animal health monitoring industry.

Moreover the Data-Driven Business Model (DDBM) Innovation Blueprint [16] portrays the ethical use of data collected from various sources for not only improving animal health and welfare but also increasing transparency of production processes. This leads to an efficient data-driven business model for livestock farming and monitoring. Additionally, the improvement of Internet connectivity has made all the latest information readily available. We had the Internet and our things in the form of animals, machines, and processes. Now, all these are unified under for larger platforms of health monitoring (as shown in Fig. 1).

The "Internet of Animal Health Things (IoAHT)" aids in streamlining the relevant data collected and making its interpretation easy and meaningful.

Farming and health monitoring are livestock framework that is now an automatic and sensor-driven data acquisition process that accumulates real-time, reliable, and precise sensed data by intelligent sensors. For the increase in profit rate and intense farming systems, farmers raise more

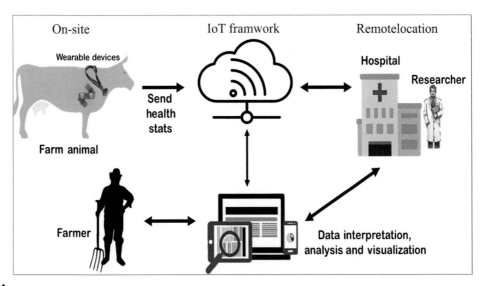

FIG. 1

IoAHT architecture for animal health monitoring.

animals over the smaller ground. With improving farm animals, it becomes difficult for the farmer to manage and monitor every animal. Additionally, due to the lack of trained workers, this problem has increased [17].

With IoT the process of farming and livestock management can develop against various monetary constrains, data acquisition, and health monitoring in constrained or unconstrained environments, for example, age parameters, food, health, veterinary care and medicines, and other requirements depending upon the environment. Although farmers solely are not benefitted by IoAHT, other investors have digital augmentation of their products and services [18].

Increasing the precision in intensive farming provides a chance to increase the efficiency in production and elevating the gaps in food supply between producers and consumers. With an increase in the number of farm animals, data management is now a crucial issue.

Using sensor-driven devices and systems acquired data can be automated, reducing the time required for data entry, which can be used in animal care.

Data acquisition and management are time effective and cost efficient.

The usage of sensor devices offers veterinarians a detailed and older history of animal health. Illness and improvement after disease both can be measured, enabling the traceability of animal growth and quicker follow-up when needed. Therefore a good standard of care is given [19]. IoT intelligence systems and veterinary interactions can promote customer recognition and customer loyalty.

Few start-up companies are implementing better solutions for livestock management and health monitoring using IoT, artificial intelligence, machine learning, and computer vision [18, 19].

These technologies also provide support to farmers and new owners of smart livestock monitoring systems. They can utilize learning frameworks and software packages for visualizing real-time health status, temperature, and other environmental changes, reproduction stats, geolocation, etc. through an android-based system and devices.

IoT in health monitoring and its promising societal benefits also equipped with a new room for interdisciplinary research and developments in telemedicine for health monitoring of livestock animals. The owners of smart livestock can get suggestions from different health experts through a telemedicine discussion forum.

At the same time, this new dimension of research has its challenges and opportunities, affecting not only the companies but also its investors, users, and all other relying on food production abilities. IoAHT has added a substantial potential supporting the management of individual animals. The role of animal biometrics, along with opportunity and significant challenges, is illustrated in the next section.

2 Role of animal biometrics: Opportunities and challenges

Animal biometrics is an upcoming research area in wildlife monitoring using computer vision. This emerging field helps in designing the methodology for health management and tracking of an individual animal or endangered species. For monitoring individual of animal, biometrics-enabled IoT has proffered with wide centralization and pervasive abilities. It allows remote monitoring, treatment via digital infrastructure. Medical facilities will enhance converting hospitals into smart hospitals [19].

2.1 Use cases

Following are the use cases where IoAHT is highly applicable.

2.1.1 Remote health monitoring

IoAHT learning framework utilizes the real-time health statistics for analysis of sensed data remotely. The heath of livestock animal can be easily monitored by any sensor device that can visualize sensed data through smartphones and smart systems. Great impact of health monitoring using devices has provided huge value to the market revenues and researches and treatment of diseases suffered mostly by animals. IoT network connects and tracks practically any sensors attached to the animal body. It can help in preventing seizures, Bovine respiratory disease complex, parainfluenza type 3, etc. by providing immediate medical help within time.

2.1.2 Mobile health (m-Health)

Mobile health is the easiest way of watching animal's location, behavior, fatigue rate, respiratory rate, or any other serious condition via smartphone apps. The m-Health has emerged as a technology that can hugely contribute to critical situations and routine treatment scenarios.

2.1.3 Enhanced chronic disease treatment

With IoT-powered wearable sensors or devices, analyzing data and real-time stats from m-Health helps veterinarians for battling reoccurring diseases. Chronic health issues can be monitored, and medicines can be given by analyzing data over time. Trends and many statistical health measures cater to the most effective health treatment of the animal.

2.2 Major challenges

Even with the improvements in IoT, still there exist few challenges that must overcome. With large integrated and interconnected systems, there are many technical complications and adaptation hardships. Sensors with extended battery life are vitally needed for accurate data.

IoAHT initiatives for battling chronic and fatal diseases need time to flourish. This technology must grow overall to provide end-to-end solutions proffering routinely improved results.

With time the software and hardware are needed to be updated on the latest hardware or software support required for better management. These updates are constant, demanding high efforts and facilities to overcome all the technical difficulties over the IoT network.

Health monitoring depends on intelligent sensors and devices connected to integrated IoAT frameworks. However, other factors such as wounds, scars on the animal body, bite, and other wounded parts on animal bodies can be turned into deadly infections if not treated on time. Moreover, these can be monitored by surveillance systems and smart cameras or any image-based intelligent devices to provide solutions to animals. However, at the same time, capturing and identifying this utilizing images is a difficult task due to unconstrained environmental issues such as lightening effects, partially or fully body covering obstacles.

Not all visibly prominent features are relevant. For example, thorns or spines on the animal's leg are not adequately visible [20]. It limits the IoT usage for monitoring complete health. The primary issues reside in acquiring the physical complexities of an animal's life cycle using wearable devices.

2.3 Advantages of IoAT in healthcare

Developers and research communities provide a better IoT system for monitoring livestock status. The animal health monitoring systems are used for detecting almost all the corporal factors such as body temperature; physical postures like sitting, standing, and eating; and other health stats such as heartbeat and breathing rate and environmental factors such as air temperature and relative humidity [21, 22].

Body temperature and surrounding temperature determine the propriety of weather for animals and what type of food should be given to them at these days. These signals and data are processed for removing other noise factors. These data can also be visualized and analyzed by breeders, farmers, and veterinarians on their devices by IoT networks and data processing platforms. These networks and platforms are always autonomous and customizable so that with time and as per requirement, new modules or hardware devices can be incorporated with existing systems.

In Fig. 2 the IoT-enabled system depicts the flow of information for analysis, interpretation, and visualization by doctors, farmers, and researchers. These parts of these systems can be connected on-site or at remote places via networks.

Acquired data are processed and are stored in a suitable database. The database has all the detailed information about each animal. On new health stats, new data are analyzed, while the older data of the same animal are studied to identify the pattern or cause. Depending upon medical history and new stats, treatment is provided to animals.

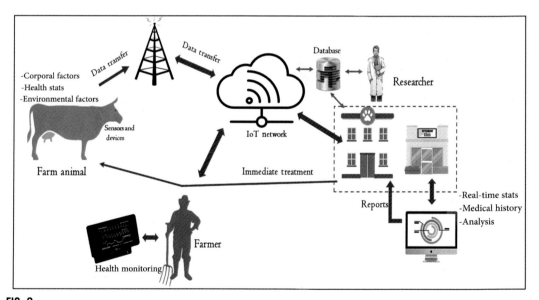

FIG. 2

Flow of actions in IoAHT.

3 Proposed IoAT-based livestock health monitoring system

In this section the proposed system is illustrated for the health monitoring of livestock animals (shown in Fig. 3). The system consists of several significant modules such as (1) data acquisition and intelligent sensor modules, (2) data preprocessing and communication modules step, (3) feature extraction and

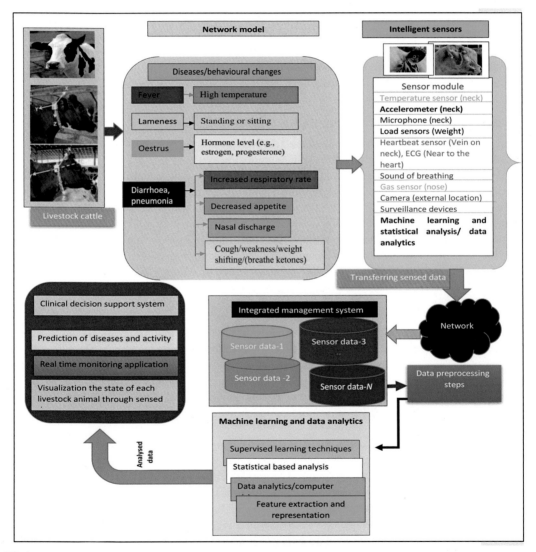

FIG. 3

Proposed learning model for health monitoring of livestock cattle.

representation using machine learning techniques, and (4) classification of disease based on analysis of sensed data using various statistical machine learning techniques and data analytics. The brief description of each module is depicted as follows:

3.1 Data acquisition and sensor module

The livestock animal can suffer from different types of diseases such as fever, lameness, oestrus, and other several health issues.

Each disease is associated with different kinds of behavior changes in the herd of animals. Intelligent sensors are embedded in the body or attached with body of animal of individual. Smart sensing enables the object to be sensed and controlled remotely using different network modules. The list of different sensors is depicted in the sensor module. The intelligent sensors provide help in collecting real-time health data of livestock animals [23].

3.2 Data processing and communication module

For preparing the integrated database, the sensed data are transferred from the client system to sever by using different data transferring protocols and commutation systems (e.g., WI-FI, Internet, and Zebee protocols). The transferred data are stored in an integrated database system in different hierarchical modules. Generally, data processing techniques include normalization of data, filtering of data, and cleaning the different artifacts from the stored data. The stored data are completed using different integrated modality [23].

These modalities sense the data from the various unconstrained environment. To get suitable data, there is a requirement for processing the data for better analysis purposes. Therefore the stored sensed data are preprocessed using data processing techniques such as min–max normalization methods.

3.3 Feature extraction, representation, and matching

Feature extraction is an essential step for raw sensory data to transform into statistical feature vectors. The discriminatory feature vectors are extracted using machine learning techniques for managing the activities of livestock animals by performing predictive analysis using various machine learning and deep learning techniques [23].

These IoT-enabled learning paradigms mitigate human errors by analyzing the massive amount of data sensory data from integrated modality and facilitate smart sensing in livestock health management architectures.

Representation of multiple sensed data in a suitable format is a major challenge in the smart sensing-based applications. On the other hand, surveillance/camera-based animal monitoring systems also suffer from major challenging problems. The primary challenge is how to capture structural variations, body changes exhibited by animals during vegetation using smart sensing devices, and learning models. Other challenges also include dynamic changes in shape and pose/posture of the animal body. The body surface of animals also reflects in different lighting conditions; therefore, animals are frequently appeared or not visual during data acquisition during vegetation.

3.4 **Classification module**

A classification model has been built using supervised machine learning techniques to perform animal disease prediction over sensory data (shown in Fig. 3).

The sensory database is prepared from the animal body equipped with sensor devices. The sensory database has been collected by intelligent sensors are such as the accelerometer sensors, gyroscope sensors, temperature sensors (equipped with animal's neck), load sensors (weight sensors), the microphones, heartbeat sensors, electrocardiogram (ECG) recording sensors, and gas sensor (attached with the nose (muzzle) of the animal).

Furthermore, to achieve better system performance, we have also captured multimedia databases (the animal's video and images (complete profile image, backside image, frontal image left side body images) using surveillance systems. The multimodal database provides better analysis based on collected sensory.

The sensory and multimedia databases are stored in the distributed integrated database system to train the learning models (shown in Fig. 3).

If the classification model failed to predict the correct health status (healthy or unhealthy) of animals using supervised machine learning techniques. The veterinary experts can then analyze the health status of animals quickly using their profile images and watching videos. The proposed model's solutions can be used for future applications such as real monitoring of animal health, clinical decision systems, tracking of animal, disease classification of animals.

It can also help establish a new avenue for smart farming using smart sensing models or disruptive technologies in precision agriculture frameworks.

4 **Current-state-of-the-art: IoAT-enabled smart livestock management framework**

IoAT-enabled framework is highly used for health monitoring of individual cattle in the smart livestock management. In these frameworks, several disruptive technologies are used to leverage services to smart sensing–based ecosystem using IoT and Fog computing–based resources. The overall architecture of the test bed is shown in Fig. 4.

The paradigms of different data analyses are also used to process the data from the sensing devices in the smart livestock management system.

An IoT-enabled smart system included IoT software packages and learning modality, edge computing-based paradigms (Fog computing), and hybrid computing models. The sensed data are processed and analyzed using machine learning techniques and deep learning modality [24–27].

The generated and sense data unit can be transferred to the receiver devices and sent to the corresponding receiver and transceiver in every 6 min. The range of the antennas attached to the receiver and transceiver is 2 km. Each gives enough coverage to collect data from livestock animal cows at all times, whether animals are grazing in the field. It also presents in their sheds (during adverse weather conditions) or being milked at the milking station (Fig. 5) [28–31].

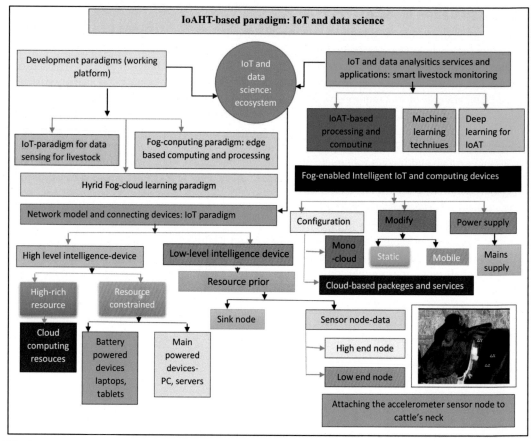

FIG. 4

The IoAT-enabled frameworks and learning techniques based on cloud architecture for smart animal health monitoring.

4.1 Deployment of the smart dairy farmhouse: Real-world application in rural development

In this section, we illustrate the working of a smart dairy farm. This working prototype model has been deployed in rural areas of several developing countries such as Israel, Ireland, India, and others. The working of a smart dairy farm was deployed on a local farm with a full dairy herd of 150 cattle in Waterford, Ireland. In this working model, Long-Range Pedometers (LRP) sensing devices are used to count the individual cow's number of steps/movement. It is designed explicitly for utilizing in dairy cattle where the Pedometer sensing device has been attached to the front leg of cattle cows, shown in Fig. 6.

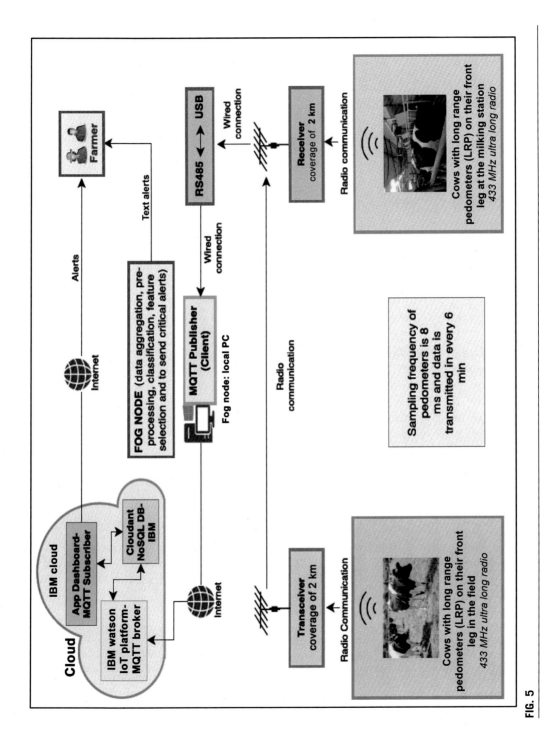

FIG. 5

Overall framework for livestock monitoring using IoT and cloud methods.

FIG. 6

The pedometer sensing device attached as a part of the experiment to the front leg of the cows.

The architecture of smart dairy farm

The overall architecture of the smart dairy farm is shown in Fig. 5. The enabled pedometer sensing device consists of an active system with a (backup) data holding capacity of up to 12 h. It evaluates livestock animals' activity such as walking, standing, lying, running, and others with a sampling frequency of 8 ms. The sensory data are generated, and the sensed data unit is sent to the corresponding receiver and transceiver systems every 6 min. The receiver and transceiver are equipped with an antenna, which provides coverage capability around 2 km. It provides better coverage to gather sensed data from dairy cows in real-time scenarios.

In real-time applications, dairy cows graze in the grass field, be available in their shelter or sheds during bad weather conditions, or be milked at the deployed milking station. As shown in Fig. 5, the receiver system is the master unit. It sends the received sensory data to the attached communication framework unit known as RS485 to USB. It is connected through a dedicated wired connection, which then transfers data to a connected working station or personal computer (PC) all the connected PC acts as a system controller and fog node. The used systems have configurations such as Intel Core 3rd Generation i7-3540M CPU @ 3.00GHz, 16.0 GB RAM, 500 GB Local Storage through wired connection via USB interface. The collected sensory data are then divided at the fog node using supervised machine learning programming-based approach [19] into three categories:

(1) *Latency insensitive data analysis paradigm*: The working model collects latency in-sensitive data of dairy cows. It does not need any immediate data analytics and decision making based on received data from the cattle behaviors' regular routine and logging of cattle data. The milking status and related data soil, water, and grass monitoring, etc.

(2) *Latency sensitive data analysis paradigm*: In this analysis, the working model collected sensory data with a critical value. Therefore, it needs immediate decision making to provide a better solution to cattle. For example, the model includes predicting pregnant cow activities before calving—routine check-ups such as calving alert and premature body variations for pregnant cows. The model needs better availability of streaming data related to pregnant cows in virtual fencing, where the decisions are based on continuous information feeds.

(3) *Latency tolerant data analysis paradigm*: The analysis paradigm includes sensory data, which is usually time insensitive. However, working models can get to the scale of sensitivity under certain intervals of time. The prosed model also provides regular analysis of periodic heat patterns (Estrous Cycle) owing to biological activities.

As a part of the experiment, commercially available long-range pedometers designed explicitly for use in dairy cattle were attached to the front leg of cows, as shown in Fig. 6.

The pedometer is a sensing device. It consists of an active system and also has backup storage [24–26].

It is used for data holding capacity of up to 12 h. It also measures the number of steps during the activity of cows in the smart livestock farming framework. It involves standing, lying, walking, etc. It has a sampling frequency of 8 ms.

5 Conclusion and future direction

The emerging field of Internet of Animal Health Things (IoAHT) is on the verge of providing powerful tools for smart farming, practitioners, and researchers used to collect and process collected information using these frameworks and tools in a standardized way and for abroad spectrum of applications.

Although existing livestock monitoring systems have shown that IoATH is feasible and useful to the setup innovation and smart farming in agriculture, numerous significant challenges lie ahead to develop the field into a widely accepted and applied subject. Bridging these gaps among industry, academia, and different disciplines involved remains the most significant challenge.

To achieve a significant impact, the applicability of IoAHT requires to be widened. A new and coherent set of ideas and fundamentals include automatic monitoring robotic systems (e.g., drones). These systems can actively seek data by traversing the agricultural fields and livestock framework sectors to increase both the quality and quantity of acquired data. These data can be processed and enable learning systems that adapt better to highly unpredictable environments. Those can continuously improve system capabilities.

Although IoAHT frameworks and learning systems are of use for a wide range of interdisciplinary disciplines, they can show their most significant potential in the emerging field of IoAHT for various applications. Following are the future directions in the field of IoAHT for diverse applications:

- There is a need to build a responsible multimodal machine learning model to meet the dual needs of privacy and on-demand personal data redaction.
- Early classification of diseases of animal in the massed sense data.
- Generation of integrated privacy models can be required for improving the animal medical outcomes that drive precision farming.
- There is a need to build a smart ubiquitous computing paradigm and protocol for standardizing IoAHT technology and its infrastructure to ensure proper privacy, security, and data management

while enabling the wider benefits of increased transparency and information sharing along the value chain.

- There is an urgent need to build the research communities and several research groups that work together to ensure data practices. The data can be collected and trusted by users and our society and across the world and set the benchmark and better standard for the responsible use of data.

References

[1] H. Wang, A.O. Fapojuwo, R.J. Davies, A wireless sensor network for feedlot animal health monitoring, IEEE Sensors J. 16 (16) (2016) 6433–6446.

[2] A. Kumar, G.P. Hancke, A zigbee-based animal health monitoring system, IEEE Sensors J. 15 (1) (2014) 610–617.

[3] R. Khan, S.U. Khan, R. Zaheer, S. Khan, Future internet: the internet of things architecture, possible applications and key challenges, in: 10th International Conference on Frontiers of Information Technology, December 17, 2012, IEEE, 2012, pp. 257–260.

[4] D. Smith, S. Lyle, A. Berry, N. Manning, M. Zaki, A. Neely, Internet of Animal Health Things (IoAHT) Opportunities and Challenges, University of Cambridge, 2015.

[5] J. Jin, J. Gubbi, S. Marusic, M. Palaniswami, An information framework for creating a smart city through internet of things, IEEE Internet Things J. 1 (2) (2014) 112–121.

[6] T.K. Hui, R.S. Sherratt, D.D. Sánchez, Major requirements for building smart homes in smart cities based on Internet of Things technologies, Futur. Gener. Comput. Syst. 76 (2017) 358–369.

[7] Y. Atif, J. Ding, M.A. Jeusfeld, Internet of things approach to cloud-based smart car parking, Procedia Comput. Sci. 98 (2016) 193–198.

[8] Y. He, P. Nie, F. Liu, Advancement and trend of internet of things in agriculture and sensing instrument, Nongye Jixie Xuebao = Trans. Chinese Soc. Agric. Mach. 44 (10) (2013) 216–226.

[9] M. Caria, J. Schudrowitz, A. Jukan, N. Kemper, Smart farm computing systems for animal welfare monitoring, in: 2017 40th International Convention on Information and Communication Technology, Electronics and Microelectronics (MIPRO), IEEE, 2017, pp. 152–157.

[10] H. Xiong, J. Tang, H. Xu, W. Zhang, Z. Du, A robust single GPS navigation and positioning algorithm based on strong tracking filtering, IEEE Sensors J. 18 (1) (2017) 290–298.

[11] M. Patil, V.D. Chaudhari, H.V. Dhande, H.T. Ingale, Hostel rooms power management and monitoring using Internet of Things, in: Computing, Communication and Signal Processing, Springer, Singapore, 2019, pp. 175–184.

[12] P.A. Laplante, N. Laplante, The internet of things in healthcare: potential applications and challenges, IT Prof. 18 (3) (2016) 2–4.

[13] A. Nettsträter, J.R. Nopper, C. Prasse, M. ten Hompel, The Internet of Things in logistics, in: European Workshop on Smart Objects: Systems, Technologies and Applications, June 15, 2010, VDE, 2010, pp. 1–8.

[14] A. Botta, W. De Donato, V. Persico, A. Pescapé, Integration of cloud computing and internet of things: a survey, Futur. Gener. Comput. Syst. 56 (2016) 684–700.

[15] Y. Sun, H. Song, A.J. Jara, R. Bie, Internet of things and big data analytics for smart and connected communities, IEEE Access 4 (2016) 766–773.

[16] A. Tzounis, N. Katsoulas, T. Bartzanas, C. Kittas, Internet of things in agriculture, recent advances and future challenges, Biosyst. Eng. 164 (2017) 31–48.

[17] P.P. Ray, Internet of things for smart agriculture: technologies, practices and future direction, J. Ambient Intell. Smart Environ. 9 (4) (2017) 395–420.

[18] DART, DART: German Antimicrobial Resistance Strategy, 2008, Published online: http://www.bmg.bund.de/fileadmin/dateien/Publikationen/Gesundheit/Sonstiges/DART_-_German_Antimicrobial_Resistance_Strategy.pdf.

[19] USDA, Global Agricultural Information Network: Stricter Control on Antibiotics in Animal Husbandry, 2011, Published online: http://gain.fas.usda.gov/Recent%20GAIN%20Publications/Stricter%20Control%20On%20Antibiotics%20In%20Animal%20Husbandry_Berlin_Germany_11-9-011.pdf.

[20] J. Brownlow, M. Zaki, A. Neely, F. Urmetzer, et al., Data and Analytics—Data-Driven Business Models: A Blueprint for Innovation, 2015, Published online: http://www.cambridgeservicealliance.org/uploads/downloadfiles/2015%20March%20Paper%20-%20The%20DDBM-Innovation%20Blueprint.pdf.

[21] Companies Proving Health Monitoring Apps, Published online: https://www.plugandplaytechcenter.com/resources/9-companies-revolutionizing-livestock-management/.

[22] S. Kumar, D. Datta, S.K. Singh, A.K. Sangaiah, An intelligent decision computing paradigm for crowd monitoring in the smart city, J. Parallel Distrib. Comput. 118 (2018) 344–358.

[23] S. Kumar, S.K. Singh, Monitoring of pet animal in smart cities using animal biometrics, Futur. Gener. Comput. Syst. 83 (2018) 553–563.

[24] H.S. Kühl, T. Burghardt, Animal biometrics: quantifying and detecting phenotypic appearance, Trends Ecol. Evol. 28 (7) (2013) 432–441.

[25] M. Taneja, J. Byabazaire, A. Davy, C. Olariu, Fog assisted application support for animal behaviour analysis and health monitoring in dairy farming, in: 2018 IEEE 4th World Forum on the Internet of Things (WF-IoT), IEEE, 2018, pp. 819–824.

[26] S. Kumar, S.K. Singh, Visual animal biometrics: survey, IET Biometrics 6 (3) (2016) 139–156.

[27] S. Kumar, S.K. Singh, A.K. Singh, S. Tiwari, R.S. Singh, Privacy preserving security using biometrics in cloud computing, Multimed. Tools Appl. 77 (9) (2018) 11017–11039.

[28] S. Kumar, S.K. Singh, R. Singh, A.K. Singh, Animal Biometrics: Techniques and Applications, Springer, 2018.

[29] J. Zhang, F. Kong, Z. Zhai, S. Han, J. Wu, M. Zhu, Design and development of IoT monitoring equipment for open livestock environment, Int. J. Simul. Syst. Sci. Technol. 17 (26) (2016) 2–7.

[30] G.J. Horng, M.X. Liu, C.C. Chen, The smart image recognition mechanism for crop harvesting system in intelligent agriculture, IEEE Sensors J. (2019).

[31] S. Kumar, S. Tiwari, S.K. Singh, Face recognition of cattle: can it be done? Proc. Natl. Acad. Sci., India, Sect. A: Phys. Sci. 86 (2) (2016) 137–148.

Internet of Things for control and prevention of infectious diseases

18

Sameer Shrivastava[a], Abhinav Kumar[b], Sonal Saxena[a], and Shrikant Tiwari[c]

Division of Veterinary Biotechnology, ICAR-Indian Veterinary Research Institute, Bareilly, Uttar Pradesh, India[a]
Department of Computer Science and Engineering, Indian Institute of Technology (BHU), Varanasi, Uttar Pradesh, India[b] Department of Computer Science and Engineering, Shri Shankaracharya Technical Campus, Bhilai, Chhattisgarh, India[c]

1 Introduction

The world is getting smaller every day with massive air travel. This has on one side helped in better connectivity among humans; on the other side, it has also increased the risk of spread of many infectious diseases across the globe, and the recent pandemic of SARS-CoV-2 is the best example; wherein the causative agent the corona virus has spread across the globe within a short period of time after its origin. Similarly, other infectious animal diseases, many of which are zoonotic in nature, also spread rapidly from one city to other, one country to another, and even from one continent to another, thereby infecting many animals and humans. To control any infectious disease, accurate diagnosis, as well as identification and detection of causative agent, is the first and most critical step. The diagnosis and prevention of infectious diseases require real-time information and analysis. Quick diagnosis and accurate information regarding spread of infectious diseases can have a great impact on the lives of susceptible population, both socially and economically.

Many researchers around the world are moving to the Internet of Things (IoT) for collecting such data in real time for many infectious diseases. The work includes tracking people, health systems, and environments and coming out with logical conclusions related to epidemiology of the disease, which may alert the policy makers to take quick decisions for disease management. The IoT is increasingly enabling the integration of devices capable of connecting to the Internet, providing information on the state of health of patients and providing information in real time to doctors who assist them. It is clear that chronic diseases such as heart disease, diabetes are a major problem at the global economic and social level [1].

The advent of IoT and related medical devices have strengthened various features of e-healthcare applications and resulted in generation of huge volume of data by IoT devices. Therefore cloud computing technology is used to manage these generated big data for secured storage and accessibility. Besides this, machine learning and other AI techniques also strengthened the medical data analysis by reducing the human intervention resulting in automated disease diagnostic system. Cancer is one of the leading causes of death, and thus, for early and accurate diagnosis, researchers have developed automated diagnosis model. Recently in Ref. [2], authors have proposed a fuzzy-based framework to automate cancer prediction. The fuzzy-based and neural network–based systems are also being proposed for healthcare monitoring [3]. Further, deep learning is another popular framework that was also investigated in cancer prediction [4]. Thus researchers are attracted toward the integration of AI and IoT for future medical data analysis and patient monitoring system. Fig. 1 depicts the multiple roles of IoT in prevention and control of infectious diseases.

2 Early warning system

IoT, a network of interconnected sensor nodes or system and advancements in data processing, AI, and ubiquitous global connectivity, can help in designing early warning system to minimize the transmission of communicable diseases. Detection and tracking of infectious agents and infected persons might be possible if national governments collectively construct a massive global sensor network integrated with facial recognition and location information, existing surveillance cameras for identification, tracking, and monitoring the people who may have contracted the causative agent. It would also be desirable to track every individual contacted by the infected person. The combination of IoT and AI may be the most logical way to develop early warning systems to prevent the rapid spread of highly infectious

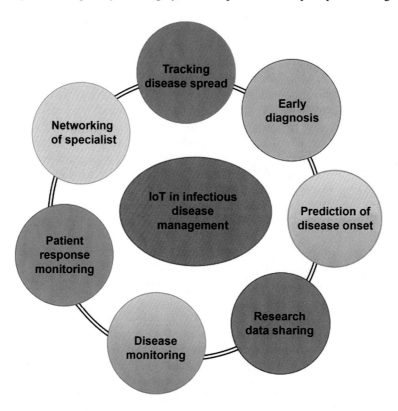

FIG. 1

Overview of IoT applications in infectious disease management.

diseases. This has been attempted for human and animal diseases by understanding disease biology and epidemiology along with application of different machine learning approaches to data collected through sensors. Use of such sensors will have massive benefits to many aspects of human and animal kind, especially in terms of infectious disease control. Investigators are using innovative technologies that combines different data format, use IoT applications, and advanced analysis for early disease detection in young cattle stock. This has helped in targeted use of antibiotics in the farm level thereby reducing the risks associated with overuse and misuse of antibiotics. Similarly, IoT techniques have shown overwhelming superiority in addressing the problem of cardiac diseases by triggering a health service based on the physical status of patients rather than on their feelings [5].

3 Early and quick diagnosis of diseases

The infectious diseases spread rapidly among susceptible population. In some infectious diseases the infected persons may remain asymptomatic and still act as spreaders of infection. In this situation the IoT networks and tools may facilitate early diagnosis. This is more true for diseases like tuberculosis wherein the confirmatory results based on isolation and culture of causative organism may take a long time, whereas systems that work with an IoT SIM card and global IoT communication platform may provide indicative diagnosis of tuberculosis within a period of 3 days only. Understanding the severity of disease often involves sending samples to diagnostic facilities for examinations. If an outbreak occurs in a remote area, the time taken to get those samples shipped and analyzed in a testing facility and receive guidance from experts could have devastating consequences on the spread of disease. Thus, while IoT could not help prevent the infectious diseases, it certainly has the potential to identify and enlighten the infected individuals. Thus infected individuals can be quarantined or provided with effective measures and stop spreading of diseases to other individuals.

4 Tracking the spread of diseases

To track the spread of diseases, a global network of IoT sensors can be used to monitor the diseased or "compromised" individuals at places like customs checkpoints at airports or when they crosses international borders. This will enable early warning systems for particular disease and will also help in targeted quarantine of selected persons. Using the IoT and big data analytics, automatic data collection may be possible. One of the most common examples is the use of smart thermometers and benchtop analyzers that are feeding data into global medical systems and instantly share data in real time with disease monitoring tools installed even miles away. The network dynamics, simulation, and disease monitoring tools may combine IoT data with population data, GIS data, land-use information, social data, and other sources to detect the emerging diseases of public health importance and threats like H1N1 and Zika. The data obtained from remote areas using IoT network of interconnected systems, machines, objects, and sensors can be hooked up to the wider network of global health data system and can be useful not only to track diseases in real time but also to apply predictive analytics to prevent their spread. In the case of a suspected outbreak, the health agencies can use a network of IoT devices to gather more focused data and pinpoint the source of an epidemic. Fig. 2 displays an IoT-based model

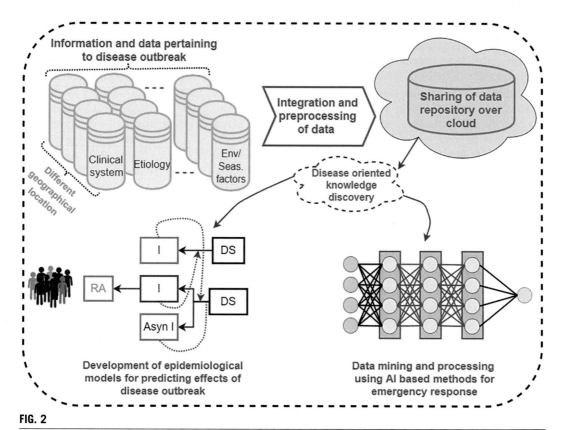

FIG. 2

IoT-based model for tracking, epidemiological monitoring, and development of architectures for disease surveillance.

for disease prediction, tracking, epidemiological monitoring, and development of architectures for surveillance of infectious diseases.

Similarly, in the event of an outbreak confirmed using diagnostic tools, the same network of IoT can be utilized to provide the necessary medical aids, supply of drugs, medical devices, and other diagnostic tools at the places of outbreak. A functional, affordable IoT system–based point-of-care instrument using fast, user-friendly, and miniaturized polymerase chain reaction has been demonstrated for its capability to amplify complementary deoxyribonucleic acid (cDNA) of the dengue fever virus. The resulting data were then automatically uploaded to an android-based smartphone via the Bluetooth interface and then wirelessly forwarded to the global network, making the test results available anywhere in the world [6]. Such systems can be useful for tracking the spread of infectious diseases.

5 Networking of disease domain experts

In case of epidemics and pandemics, the spread of causative agent is very rapid, and it becomes very difficult to control the disease. Several strategies are to be adopted to control such situation. It requires combined efforts of different persons, including government agencies and subject matter specialist/domain experts. Once the cause of disease is established and confirmed, the efforts are directed toward controlling the spread of the disease. But to achieve this, it becomes utmost important to understand the nature and scope of disease, its spreading pattern, and device strategies to prevent, manage, or treat infections. The situation becomes more challenging when the outbreak occurs in a very remote area, where it is highly unlikely to find domain experts, trained epidemiologists, pathologists, and clinicians on the ground. In such a situation the IoT-enabled digital imaging and pathology come into the picture. The symptoms and other raw clinical data may be analyzed from a distant place by the trained personnel, and immediate curative measures may be suggested to control further spread of infections. Thus IoT devices and tools bridge the gap between field level persons and trained pathologist/clinicians. Similarly, on a bigger network, the field to government approach may be established, wherein every single person who is important in the controlling the spread of disease can be linked up. These include the field level primary workers sitting in the remote locations, the city/district level officers, state level health personnel and the decision-making persons sitting at the center, and all those who can take part actively based on inputs provided through IoT tools. Thus a framework of committed and dedicated healthcare system may provide lot of insight for monitoring and controlling spread of diseases like COVID-19 [7].

6 IoT in monitoring patient's response

The IoT through its network of connected devices can help the clinicians and nurses to share the information of patients at the point of care in real time. The software-enabled interface can allow clinicians to view the medical histories, treatments, medications, and other clinical data of the patients, besides having an insight into their socioeconomic determinants. This becomes possible using a combination of visual graphics, images, and charts. This type of network also offers patients to easily access their own health data and monitor the progress of their health. Example of an IoT system for remote monitoring of patients at home is shown in Fig. 3.

The IoT-based devices like wireless stethoscopes are designed for anesthesiologists to listen to a patient's heart rate and breathing sounds during surgery. Wireless storage technology can digitally record physiological parameters such as heart rate, breathing rate, electrocardiogram, body temperature, and blood pressure. Further, with the development of technology, physiological parameter recorders based on wireless sensing technology are also introduced to the medical field. These include esophagus pH, pulse blood volume, CO_2 decompression, and wireless breathing frequency monitors during movement. A more recent advancement includes interdigitation of microelectronics and sensing technologies, and with the concept of instant "plug and play," modular installation of "sensor network" has become possible, and these can be easily applied for heart, lung function, and other joint monitoring. Similarly the RFID-based IoT medical models have been used as core technologies for item identification. The RFID applies a radio frequent spatial coupling signal to achieve wireless information transmission, for identification purposes, and also to monitor the progress of ill patients [9].

7 Predicting the seasonal onset and spread of diseases

There are certain diseases that have seasonal occurrence. This include flu, parasitic diseases, and some diseases caused by allergens. To manage such seasonally occurring diseases, the healthcare professionals around the world make lot of preparedness. Such efforts can give much better results if

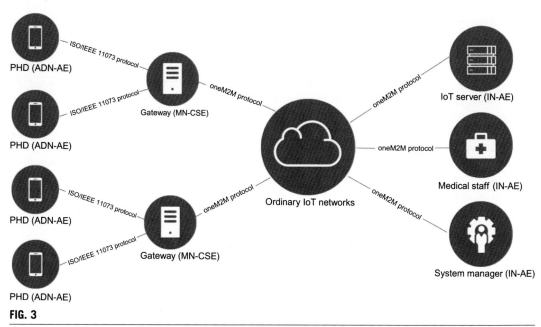

FIG. 3

Structure of the IoT system for monitoring patients at home [8].

integrated with IoT devices like smart thermometers, thermal imaging systems, and monitoring the behaviors and abnormal movements and activities. Different monitoring apps may be used to transfer relevant data, such as a person's daily temperature or symptoms to a panel of experts, who can suggest remedies. In addition, when such data are anonymized and aggregated, epidemiologists and public health professionals may find it useful to forecast about the occurrence of disease in different parts of the country. IoT-based curative data may provide more advanced reports of outbreaks and disease occurrence than traditional methods.

In another study, Verma and Sood [10] have proposed a cloud-assisted IoT-based m-healthcare monitoring disease diagnosing framework that predicts the potential disease with high level of severity. They suggested that big data generated by IoT devices in healthcare domain should be analyzed on the cloud instead of relying solely on limited storage and computing resources for handheld devices, thereby providing more efficient prediction systems for healthcare monitoring.

8 Role of IoT in sharing research data for devising better disease prevention mechanisms

The biology of infectious diseases is very complex. The problem-solving approach requires integration of several experts. In case of infectious diseases, some researchers work toward identifying the causative agents; some work toward gathering data related to spread of disease in susceptible population, while there may be groups working toward developing strategies for prevention and control of disease

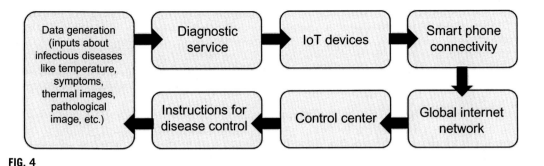

FIG. 4

Schematic representation of IoT network for controlling and prevention of infectious diseases.

in the susceptible population. All these groups may be working at different places in the globe and generating huge amount of data. In general, lack of readily available and relevant data needed to test a hypothesis has led to lack of evidence-based approaches in controlling infectious diseases. Health systems can easily overcome this challenge with the advent of IoT. Data can be collected from remote locations and fed into global health systems. Clinical researchers can be better equipped to perform an evidence-based analysis of an emerging outbreak. On the basis of this analysis, they can suggest better preventive measures or analyze them if the controlling measures suggested by experts and government agencies are being implemented properly. The real-time data of events at a farm or public places collected through IoT networks may be useful to study the impact of interventions in controlling disease spread. However, to be more effective, these tools may be used with proper care and taking into advice and guidance of several domain and subject matter specialists. Schematic representation of IoT workflow for controlling and prevention of infectious diseases is depicted in Fig. 4.

9 Conclusion

Prevention and control of infectious diseases is of utmost importance to save the man and animal kind, as well as environment, for the betterment of our life on the globe. With advances in the field of information and communication technology, it has become possible to interact with wider groups and discuss on a common platform. The IoT devices can make life easier for people who use them, while at the same time, they can also have significant benefits for communities or entire nations if we use them as a network to control or monitor the transmission of infectious diseases. It has proved useful to bring together different experts and healthcare staff, and as more and more experimentation advances and technologies become easily accessible, IoT will likely continue to astound people with its health-related capabilities.

References

[1] J. Gómez, B. Oviedo, E. Zhuma, Patient monitoring system based on Internet of Things, Procedia Comput. Sci. 83 (2016) 90–97.

[2] A. Kumar, S.K. Singh, S. Saxena, A.K. Singh, S. Shrivastava, K. Lakshmanan, N. Kumar, R. K. Singh, CoMHisP: A Novel Feature Extractor for Histopathological Image Classification Based on Fuzzy SVM With Within-Class Relative Density, IEEE Transactions on Fuzzy Systems (2020)https://doi.org/ 10.1109/TFUZZ.2020.2995968 Available from:https://ieeexplore.ieee.org/abstract/document/9097394.

[3] P.M. Kumar, S. Lokesh, R. Varatharajan, G.C. Babu, P. Parthasarathy, Cloud and IoT based disease prediction and diagnosis system for healthcare using fuzzy neural classifier, Futur. Gener. Comput. Syst. 86 (2018) 527–534.

[4] A. Kumar, S.K. Singh, S. Saxena, K. Lakshmanan, A.K. Sangaiah, H. Chauhan, S. Shrivastava, R. K. Singh, Deep feature learning for histopathological image classification of canine mammary tumors and human breast cancer, Inf. Sci. 508 (2020) 405–421.

[5] C. Li, X. Hu, L. Zhang, The IoT-based heart disease monitoring system for pervasive healthcare service, Procedia Comput. Sci. 112 (2017) 2328–2334.

[6] H. Zhu, P. Podesva, X. Liu, H. Zhang, T. Teply, Y. Xu, H. Chang, A. Qian, Y. Lei, L. Yu, A. Niculescu, C. Iliescu, P. Neuzil, IoT PCR for pandemic disease detection and its spread monitoring, Sensors Actuators B Chem. 303 (2020) 127098.

[7] M.S. Rahman, N.C. Peeri, N. Shrestha, R. Zaki, U. Haque, S.H.A. Hamid, Defending against the novel coronavirus (COVID-19) outbreak: how can the Internet of Things (IoT) help to save the world?. Health Policy Technol. 9 (2) (2020) 136–139, https://doi.org/10.1016/j.hlpt.2020.04.005.

[8] K. Park, J. Park, J. Lee, An IoT system for remote monitoring of patients at home, Appl. Sci. 7 (3) (2017) 260.

[9] Y. Song, J. Jiang, X. Wang, D. Yang, C. Bai, Prospect and application of Internet of Things technology for prevention of SARIs. Clin. eHealth 3 (2020) 1–4, https://doi.org/10.1016/j.ceh.2020.02.001.

[10] P. Verma, S.K. Sood, Cloud-centric IoT based disease diagnosis healthcare framework, J. Parallel Distrib. Comput. 116 (2018) 27–38.

Telemedicine system for animal using low bandwidth cellular communication post COVID-19

Jagdish Lal Raheja[a] and Ankit Chaudhary[b]

Control and Automation Unit, CSIR-CEERI, Pilani, Rajasthan, India[a] Department of Computer Science, The University of Missouri, St. Louis, MO, United states[b]

1. Introduction

The use of mobile phones and their capabilities has increased to a greater extent including real-time video conferencing and transmitting and receiving high-resolution data. Also the use of mobile phones in the area of healthcare has gained in recent years the attention of the research community a.k.a. telemedicine. To bring the benefit of the latest medical facilities and research to the people/animals who are not living close to the medical centers, high-end connectivity of the network is needed. The use of a remote monitoring system is increasing in almost all the expected applications. Also, after COVID-19, the use of online consultations and online checkups has grown exponentially. A cellular phone that is having a camera and some basic functionality can be used in many applications. The use of these mobile phones in the medical field has many possibilities to bring a new era of medicine, while in South Asia, many low-income people are having a phone with a low bandwidth connection.

The system discussed here is dealing with the monitoring of the animals for their healthcare using mobile communication. Mobile phones make it possible to transfer medical data including photos. In South Asia, domestic animals are cows, buffalos, horses, dogs, cats, camels (in the desert area), elephant, goat, sheep, hen, duck, etc. They are tamed and then trained for performing different works. Hence, these animals play an important role in the life of their owners, so the owner cares about them very well. Many diseases exist that can infect animals including the corona series virus. The spread of several diseases can be controlled by proper preventive and sanitary measures. An owner can easily see if an animal is sick by observing its normal posture, feeding habit, definite body temperature, pulse rates, and respiration rates [1]. Detection of sickness in animals at the early stages can aid the farmer in restricting the spread of disease in the flock and initiating treatment at the earliest. The owner can identify the symptoms of the disease in the animal and can take precautions, control, and treatment measures as soon as possible, but in the current scenario, he has to go to the hospital. So, there exists a need for a telemedicine to assist the owner of animals for diseases anywhere anytime [2].

2. System development phases

The systems development life cycle (SDLC) is the process of creating or altering systems and the models and methodologies that people use to develop these systems [3]. The SDLC model is shown in Fig. 1.

SDLC process used by a system analyst to develop the system, including requirements, validation, and maintenance. Any SDLC should result in a high-quality system and efficiently in the current and planned IT infrastructure and is inexpensive to maintain.

The SDLC framework provides a set of activities for project designers to follow. In this project, these stages of the SDLC are divided into nine steps from planning to the maintenance of the system as shown in Fig. 2.

2.1 Defining the problem

With the advancement in technology and the need for remote service, there exists a requirement of a remote telemedicine system for animal diseases. The goal of this project is to develop a reliable, efficient, and easily deployable system for animal diseases using mobile phones that can play an

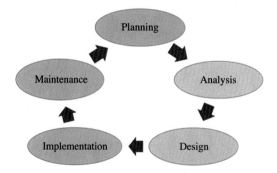

FIG. 1

SDLC model.

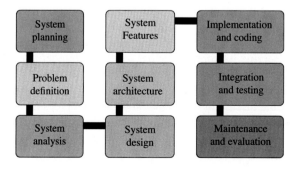

FIG. 2

Systems development life cycle phases.

important role in providing basic health services to the remotely located users at their doorstep. This system can enable expert doctors to monitor users in remote areas. In this project a mobile application will be available for the masters of the animals so that they can send the symptoms of the disease to the remote server. A desktop application at the server is operated by the veterinarian to analyze the symptoms and diagnose the disease. It uses DSS based on the fuzzy system for detecting the disease status. Finally a web application would be developed to monitor users online.

2.2 System analysis and design

Efforts have been made to define the requirements exhaustively and accurately [3]. SRS includes the general description of the system that defines product deliverables, product perspective, product activity, user characteristics, general constraints, assumptions taken, functional requirements, external interface requirements, hardware and software requirements, and various design constraints. For this project the user requirements have been categorized into various modules. Accordingly the system has been divided. The modular diagram is shown in Fig. 3. In this system, there exist two types of

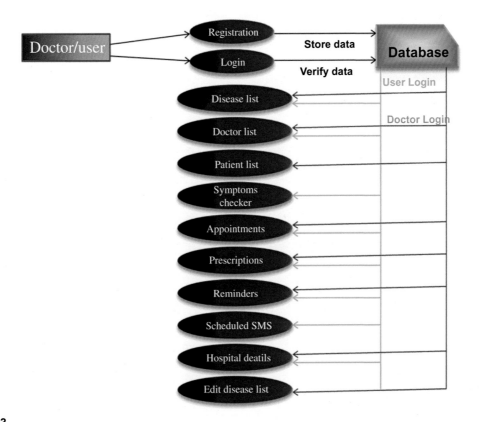

FIG. 3

The system modularization.

users—user and doctor. Each user has different operations to perform, and hence the diagram shows the various operations of the system.

This phase of system development includes the description of the project and its various modules. A procedural design tool is used to translate a narrative into "graphs and network diagrams." The network diagram of various modules of the system is shown in Fig. 4.

SDS helps to establish that all requirements are specified by the software design and indicates which components are critical to the implementation of specific requirements. Design constraints such as

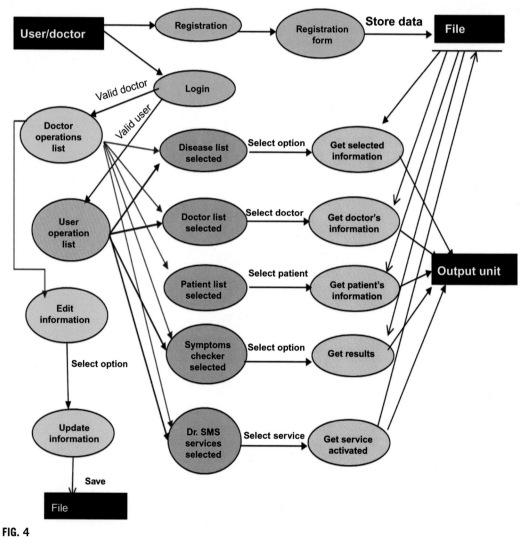

FIG. 4

The system network diagram.

physical memory limitations or the necessity for a specialized external interface may dictate special requirements for assembling or packaging the software. Supplementary data such as algorithm description, alternative procedures, tabular data, excerpts from other documents, and other relevant information are also presented.

2.3 System architecture

The flow of the system is shown in Fig. 5. The system consists of mainly four parts: The first is the Java-enabled and GPRS-enabled mobile phone to send the input to the server. The second is the remote server that communicates with the database, fuzzy system, and the mobile phone. Its major function is to generate a report of animal data including information about its disease. It also stores and manages the data in the information system, that is, database. The third is the fuzzy-based DSS, which is used to analyze the animal's symptoms and calculate the results to detect the disease and its status. Finally the information system that stores and manages the information related to the system and its services [4].

Concretely, three functions are developed. The first is storing the animal's information in the database. The second is the communication between the server and the mobile phone. The third is the analysis of the symptoms using the fuzzy system.

2.3.1 The Java-enabled mobile phone

The mobile phone that is used in this study is a GPRS-enabled, Java-enabled Nokia mobile phone. It provides a comprehensive application development platform [5] for creating networked products and applications for the consumer. In addition, GPRS enabling Internet service using the mobile phone is provided by the reliance connection. It is used to give a static IP address to the system. The overall mobile communication flow is shown in Fig. 6 [6]. The system contains three steps. First, take the input from the mobile and send it to the central server. Second the central server analyzes the data and sends it to the user. If required, it communicates with the database server also. Third the server sends the result back to the mobile in the form of alerts or the form of a message.

2.3.2 Remote information server

A desktop application is developed as a server in the hospital. The desktop application is used to get information from the MySQL server running on a Windows XP operating system. To get information dynamically the PHP is used. PHP is a general-purpose scripting language that is especially suited for dynamic application development. Its main advantage in comparison with other scripting languages such as Perl is that it can be incorporated with the application itself, which brings an opportunity to

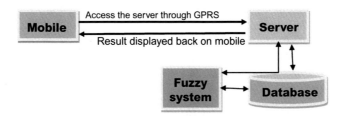

FIG. 5

The basic flow diagram.

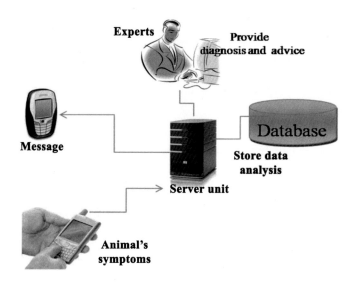

FIG. 6

The mobile communication flow.

the server itself to generate a part of the source code [7, 8]. On the user's side, Doctor's PDA/laptop is used to access the server system. This server is accessed through a Java application developed in eclipse. There are two functions: (1) to link between the hospital's network and the Internet gateway and (2) to generate a report of animal data including information on its disease. It also stores and manages the data in the information system, that is, database. The server system is shown in Fig. 7 [9].

2.3.3 Decision support system: Fuzzy system

The DSS for medical diagnosing was designed using the fuzzy logic. The symptoms send from the mobile are send to the fuzzy system to analyze symptoms and detect the disease [10]. In DSS, for every disease, there is a separate logic for fuzzy systems. For example, for anthrax, we are maintaining the anthrax fuzzy, which will analyze the anthrax disease. Similarly a fuzzy DSS is created for the rest of the diseases. FL incorporates a simple, rule-based approach to solving problems rather than attempting to model a system mathematically [11].

Fuzzy symptoms table: The symptom in Table 1 contains the details of the symptoms that can be used to find the disease. The following are the details of the fuzzy system that we have developed in our DSS to find out the anthrax disease by using the following symptom table.

Fuzzy rules: Fuzzy rules are the rules that are used to find the disease.

Rule statement: The fuzzy rule statements use rule-based approach [12]. For example, rule 2 from Table 2 is stated as,

IF s1 IS low AND s2 IS low AND s3 IS mid AND s4 IS mid AND s5 IS low THEN result IS low;

2.4 System features

There exists the requirement of the development of an affordable and reliable solution to the problem of the provision of expert healthcare to people in India at their doorstep. The system should enable expert

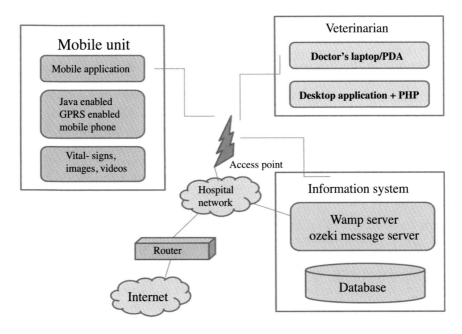

FIG. 7

The remote server architecture.

Table 1 Fuzzy table for symptoms of anthrax disease.	
Fuzzy variable	**Symptom**
S1	Fast breathing
S2	High temperature (splenic fever)
S3	Uncoordinated movements, weakness
S4	Bloody discharge from natural body openings
S5	Tiredness, loss of appetite, and nausea

Table 2 Fuzzy rules for symptoms of anthrax disease.							
Rule	**S1**	**S2**	**S3**	**S4**	**S5**	**Result**	**Stage**
1	L	L	L	L	M	L	Low
2	L	L	M	M	L	L	Low
3	L	L	L	M	M	M	Medium
4	M	M	H	H	L	M	Medium
5	H	H	M	M	M	H	High
6	M	M	H	H	H	H	High

doctors to monitor people in remote areas. As a result the user no longer needs to travel long distances to reach the nearest basic health units and then just be examined by. This project includes the following features.

Three models concept: The remote monitoring system for animal diseases is available in three different models. The mobile version is in the form of a mobile application to be installed on mobile phones. The desktop version is in the form of a desktop application to be installed on the desktop system. The web version in the form of a website is to be available online on the Internet for all the remote users.

The mobile application: It is the mobile version of the monitoring system as shown in Fig. 8. It is to be installed on mobile phones. It helps in monitoring users via mobile phones, and communication is between the server and the mobile. It can access all the remote monitoring system services provided in this project.

The server application: It is the desktop view of the remote monitoring system as shown in Fig. 9. It is to be installed on the desktop systems. It can access all the remote monitoring system services provided in this project. It is also used by the doctor for analyzing the animal's symptoms for detecting disease and its level using fuzzy-based DSS.

The web application: It is the web application of the remote monitoring system as shown in Fig. 10. It is in the form of a website. It is to be available on the Internet for all the remote users. It can also access all the remote monitoring system services provided in this project.

The information system: It is an MYSQL database, which stores the information of all the diseases, their symptoms, and the measures of prevention, control, and treatment, covered in this project. It also

FIG. 8

Mobile application.

FIG. 9

Server application.

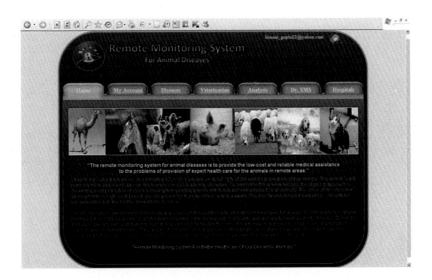

FIG. 10

The web application.

includes information about all the registered doctors, users, and infected animals with their diseases, symptoms, status, and the level of infection. It also maintains information related to Dr. SMS services and other remote monitoring system services.

Fuzzy-based DSS: The disease diagnosis, its status, and its level are calculated using fuzzy-based DSS with the help of various fuzzy logic rules.

Flexibility: The system is available in three major versions. Users can use any of them based on their convenience. All these versions provide a high level of flexibility.

Platform independent: The system is developed using platform-independent languages like Java.

Interfaces: Easily accessible options are provided. User-friendly interfaces are developed.

Help: It is available for each module for all three versions.

2.5 Major system services

This project provides the remote user with various services for medical assistance.

2.5.1 Symptoms checker

The architecture is shown in Fig. 11. This is the most vital service provided for remote users. It includes receiving and storing the symptoms of the disease of an infected animal in the database and the further processing of these symptoms to detect the disease and its level for remote user assistance [10]. In this project the DSS is designed for diagnosing animal diseases by using the symptoms information given by the user by mobile phones. The symptoms are shown in Fig. 12, and symptom analysis is shown in Fig. 13.

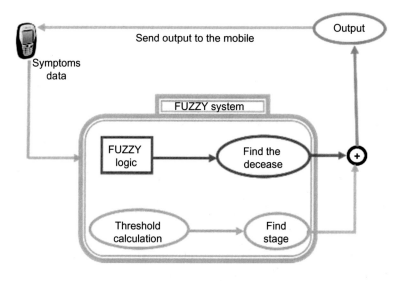

FIG. 11

The architecture of symptoms checker.

FIG. 12

The symptoms checker showing symptoms and their level of the animal.

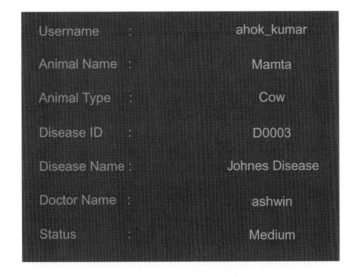

FIG. 13

The symptoms checker showing symptom analysis and the results.

2.5.2 Dr. SMS

The architecture is shown in Fig. 14. This is another important feature provided by the remote monitoring system for healthcare. This feature includes various types of message facilities to assist remote users [13].

Here, four types of message services are provided.

First is the *scheduled SMS*. Using this the user can get the scheduled SMS for talking the medicines on time for the proper treatment. The message includes the medical prescriptions with the timing to take

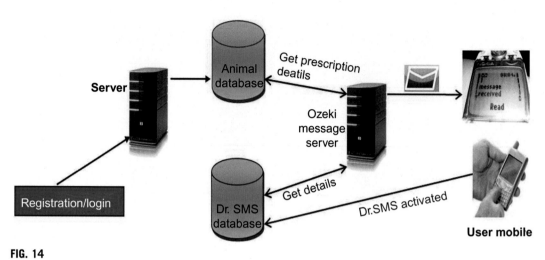

FIG. 14

The architecture of Dr. SMS feature.

the medicine. The message will be sent by the *Ozeki* message server at the prescribed timings at the user's mobile phone to inform about the intake of medicines.

Second is the *appointment SMS*. Using this the user can fix the appointment with the doctor by sending an SMS. The user can select the doctor and can mention the date and the time of the appointment. The system will send the message to the selected doctor regarding the appointment. If the doctor accepts the appointment, then the user will receive a confirmation message; otherwise a rejection message will be sent. The user can refix the appointment. The *Ozeki* message server will also remind the user about the appointment before the appointment timings through a message so that the user can meet the doctor at the fixed timings.

The third is the *alert SMS*. Using this the user can set various alert reminder SMSs for their better healthcare. It includes alert for new disease in a city named "new disease alert." It also includes alerts for new hospitals and medical shops in the specified city in a particular area. It allows the user to activate/deactivate scheduled message alert and appointment message alert [14].

Fourth is the *search SMS*. Using this the user can find the details of the hospitals, medical shops, doctors, etc. by just mentioning the city and the area in the SMS. The user will receive a message that will contain the details about the hospital/medical shop including the roadmap details and the contact details. This is provided to help users in the way that they don't know much about the city and need urgent information for medical aid.

All these message services make our remote monitoring system different from others. This project deals almost all the aspects of medical assistance with reliability and efficiency.

The functioning of Dr. SMS is as follows. User has to do registration/login to provide their details into the system. The animal details are sent to the server, and the server stores the information into the database. If the user activates the Dr. SMS facility, then he/she is getting assisted with all the services related to it. The *Ozeki* message server accesses the database. The messages are sent to the users according to the data and the services on their mobile phones [15].

2.6 Implementation

This phase of system development includes the actual implementation and coding of the system. It includes defining the methodology adopted for system development and other implementation issues.

Implementation of the system is done by using the system requirements. This project is implemented using three basic models. The mobile application is developed using NetBeans software in J2ME, while desktop server application is developed using Eclipse software in J2ME. The database used is MySQL. The scripting language used is PHP. The information server used is Wamp Server, and the message server is Ozeki server. The project is developed such that it acquires all the specified features with efficiency and reliability to achieve the objectives of the system.

2.7 Integration and testing

The basic function of testing is to detect errors in the software and quality control. During requirement analysis and design, the output is a document, while after the code, computer programs are available that can be used for white and black box testing. The goal of testing is to uncover requirements, design, and coding errors in the programs. Consequently, different levels of testing are used.

Testing has two phases, unit testing and integration testing. Each test report contains the set of test cases and the corresponding results. The generated error report describes the error encountered and the action taken to remove the errors. This project is tested using two basic approaches: functional testing and structural testing. If the program fails to show the required behaviors, then debugging and correction are required.

The evaluation framework will evaluate the system deployed in the controlled conditions based on these indicators. These performance indicators include an increase in the fuzzy logic rules to determine the disease, number of diseases considered, the messaging services for timely preventive measures, the amount of correct information available, and ultimately the reliable monitoring of animal diseases.

The field of telemedicine has seen tremendous technical development in developing countries. Though a lot of work has been done in different avenues of telemedicine, like wireless transmission, video conferencing between patients and doctors, however, little research is done in developing a low-cost remote monitoring system that can be of significant value in providing basic health services to the users at their doorsteps.

3. Conclusions

The field of telemedicine has seen tremendous development and social acceptance during a few months in the pandemic. Though a lot of work has been done in different avenues of telemedicine, however, little research is done in developing a low-cost remote system that can be of significant value in providing basic health services to the users at their doorsteps [16–22]. Users can get any information about the disease using the mobile phone itself. The user can also be monitored by the messages to take the medicines on time. So by using the aforementioned services, the user can get help from the doctors from the remote locations at any time.

In this chapter a remote monitoring system using a Java-enabled mobile phone was developed. We developed functions that are gathering animal characteristics in the server, creating an application for the remote information server, a Java application for a mobile phone, providing system features, and confirming its performance. In the future, we will improve the system on the mobile phone for seamless browsing information and increase useful functions in the system. Moreover, security measures are expected to be introduced in the future. We are also proceeding to include image processing and analyzing for medical diagnosing.

References

[1] G. Harris, Telemedicine's great new frontier, IEEE Spectr. 39 (2002) 16–17.
[2] J. Jiehui, Z. Jing, Remote patient monitoring system for China, IEEE Potentials 26 (2007) 26–29.
[3] I. Sommerville, Software Engineering, tenth ed., Pearson Education Limited, London, 2016.
[4] A.K. Sangalb, S. Bhatia, L.S. Satyamurthy, A. Bhaskarnarayana, Communication Satellite-Based network for Telemedicine in India, in: Proceedings of 6th International Workshop on Enterprise Networking and Computing in Healthcare Industry, 2004.
[5] T. Tsumura, H. Ohtani, The Mobile Phone's Application Program Textbook, September 2004.
[6] NTT DoCoMo, Appli Contents Development Guide for Doja-4.0, November 2004, Available: http://www.nttdocomo.co.jp/p_s/imode/java/pdf/jguideforDoja4_041117.pdf.
[7] G. Hojtsy, The Manual of PHP, 2004, Available: http://www.php.net/manual/jp.
[8] R. Hirokawa, J. Kuwamura, T. Koyama, The Full Capture of PHP4, Practical Version, July 2003.
[9] K. Hung, Y. Zhang, Implementation of a WAP-based telemedicine system for patient monitoring, IEEE Trans. Inf. Technol. Biomed. 7 (2) (2003) 101–107.
[10] C.W. Holsapple, A.B. Whinston, Decision Support Systems: A Knowledge-Based Approach, West Publishing, St. Paul, 1996, ISBN: 0-324-03578-0.
[11] L.A. Zadeh, et al., Fuzzy Sets, Fuzzy Logic, Fuzzy Systems, World Scientific Press, 1996, ISBN: 9810224214.
[12] E. Cox, The Fuzzy Systems Handbook: A Practitioner's Guide to Building, Using, Maintaining Fuzzy Systems, AP Professional, Boston, 1994, ISBN: 0-12-194270-8.
[13] S. Bhatia, Development of telemedicine technology in India: "Sanjeevani"—An integrated telemedicine application, J. Postgrad. Med. 51 (4) (2005).
[14] J.S. Bhatia, M.K. Randhawa, H. Khurana, S. Sharma, India's Tryst with Telemedicine, e-Health, May 2007.
[15] P.N. Mechael, Mobile Phones for Mother and Child Care (Case Study of Egypt), London School of Hygiene and Tropical Medicine, UK, 2005.
[16] V. Gay, P. Leijdekkers, E. Barin, A mobile rehabilitation application for the remote monitoring of cardiac patients after a heart attack or a coronary bypass surgery, in: PETRA '09, June 2009 Article No.: 21, 2009, pp. 1–7.
[17] Z. Jin, Y. Sun, A. Cheng, Predicting cardiovascular disease from real-time electrocardiographic monitoring: an adaptive machine learning approach on a cell phone, in: Conference Proceedings—IEEE Engineering in Medicine and Biology Society, 2009, pp. 6889–6892.
[18] J.L. Raheja, Dhiraj, D. Gopinath, A. Chaudhary, GUI system for elder/patients in intensive care, in: Proceedings of the 4th IEEE International Conference on International Technology Management Conference, Chicago, USA, 12–15 June, 2014, pp. 1–5.
[19] A. Chaudhary, J.L. Raheja, ABHIVYAKTI: a vision based intelligent system for elder and sick persons, in: Proceedings of the 3rd IEEE International Conference on Machine Vision, Hong Kong, 28–30 Dec, 2010, pp. 361–364.

[20] Y. Lin, I. Jan, et al., A wireless PDA-based physiological monitoring system for patient transport, IEEE Trans. Inf. Technol. Biomed. 8 (4) (2004) 439–447.

[21] A. Palvictor, W.A. Mbarika, F. Cobb-Payton, P. Datta, S. McCoy, Telemedicine diffusion in a developing country: the case of India, IEEE Trans. Inf. Technol. Biomed. 9 (1) (2005).

[22] J.L. Raheja, H. Gupta, A. Chaudhary, Monitoring animal diseases in remote area, Pattern Recognit. Image Anal. 28 (1) (2018) 133–141.

Internet of things and other emerging technologies in digital pathology

Abhinav Kumar[a], Sonal Saxena[b], Sameer Shrivastava[b], Vandana Bharti[a], and Sanjay Kumar Singh[a]

Department of Computer Science and Engineering, Indian Institute of Technology (BHU), Varanasi, Uttar Pradesh, India[a] Division of Veterinary Biotechnology, ICAR-Indian Veterinary Research Institute, Bareilly, Uttar Pradesh, India[b]

1 Introduction

Pathology is the most crucial field in the area of healthcare. Various surveys and research demonstrate that the demand for pathology services is continuously growing owing to the increased prevalence of cancer and other diseases. Due to the significant growth in the pathology market, there is increased demand for trained and experienced pathologists. Recent technologies allow a pathologist to provide diagnosis through quantitative analysis of images on computer screens, and an electronic digital record is an emerging concept to provide a diagnosis. Digital pathology supports the development of optimized pathology workstations, whole-slide imaging, image analysis using artificial intelligence (AI)-based approaches, efficient data management, including data handling and storage, and many more. Proper deployment of digital pathology solutions can streamline the workflow of pathology services, and can potentially increase the efficiency and decrease the time for diagnosis. Digitization of pathology makes it possible to generate and store vast amounts of data. As a result, new methods, technologies, and computing resources are accessible via the cloud at a low cost. It is indeed a challenging task to make effective use of new and emerging digital pathology modalities to process and model all data and information contained in whole-slide imaging (WSI). Recent advancements in the field of information technology, including cloud computing, coupled with Internet of things (IoT) and AI are revolutionizing the entire healthcare paradigm. An overview of IoT-based digital healthcare is presented in Fig. 1. Machine learning and deep learning are AI techniques that use algorithms to recognize patterns and make connections between different items in ways similar to that of the human mind—but at a much faster rate and in a more comprehensive manner than any medical expert.

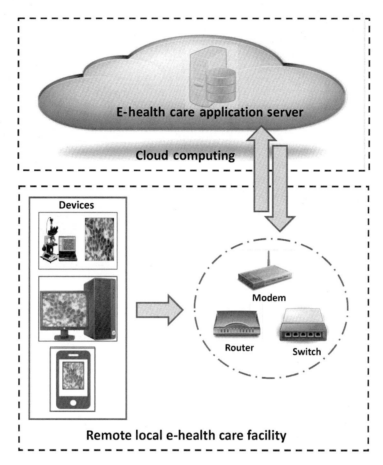

FIG. 1

Overview of IoT-based e-healthcare [1].

2 Recent trends in digital pathology

2.1 Mobile diagnostics using smartphones

Since smartphones are nowadays having high-resolution cameras and processing power, accelerometer, along with unlimited access to internet data, smartphones can now be used to diagnose health problems [2]. Smartphones have a massive reach today, and with emerging IoT technology, it can double up as a sort of laboratory. The advancement of mobile technology adds a degree of flexibility in providing a diagnosis in the remote areas. Many times due to the limited availability of pathologists and experts in remote areas, many patients are not able to get timely medical advice. A workflow for rapid IoT-based diagnostic framework for automated diagnosis for blood smears using smartphones is shown in Fig. 2.

Thus, in quest of developing rapid, cost-effective systems for providing a diagnosis in remote areas, Auguste developed a technology for accurately providing medical diagnosis using just a smartphone

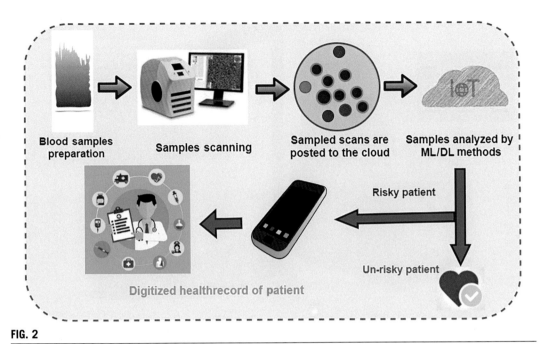

Blood samples preparation   **Samples scanning**   **Sampled scans are posted to the cloud**   **Samples analyzed by ML/DL methods**

Risky patient

Un-risky patient

Digitized healthrecord of patient

FIG. 2

IoT-based diagnostic framework for automated diagnosis for blood smears using smartphones.

[3]. An emerging technology that has potential to revolutionize the histopathological diagnosis is the fourth generation of virtual slide telepathology system. The technology is based on the concept of virtual microscopy and WSI. This innovative imaging technique involves high-resolution scanning of tissue slides, image digitization, and automated computer-based processing of whole-slide microscopic images. Automated WSI is helpful in the extraction of disease-specific features and quantification of the selected features for supporting diagnoses and disease management. Another software application of the start-up mWSI (mobile Whole Slide Imaging) helps stitch together individual images captured under the moving slide stage into a composite whole. Together with the mWSI live feature, it allows users to view specimens live over the Internet, transforming it into a "Skype for microscopes." Such sort of telepathology systems has great potential to avoid delay in diagnosis of life-threatening illnesses by easy access to experts in remote locations [3].

2.2 Portable digital microscope platform coupled with IoT technology

Since decades, optical microscopes have been the key tool to assist pathologists in the diagnosis process. Portable hand-held digital microscopes are now commercially available to produce fast, accurate, and reliable results for a wide range of applications. The integration of these portable systems with IoT and cloud can be very helpful in providing on-site diagnostics. Researchers have recently developed portable digital microscope platform utilizing IoT technology [4]. This system is a portable platform for capturing high-resolution digital images that are transferred directly to the computer via Wi-Fi. In addition, the system has the ability to record and display, allowing users to capture, store, and process digital images of the object being viewed. As a result, users do not have to manually transfer the images

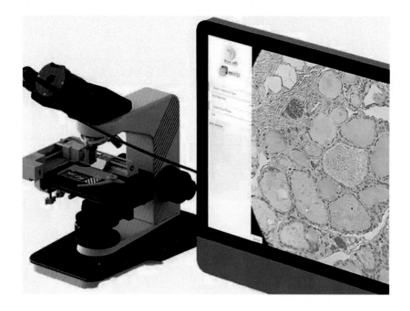

FIG. 3

IoT device for converting microscope into a virtual imaging machine [3].

to perform the analysis. Such an integrated portable system is expected to deliver a great deal of benefit to people by providing faster analysis and accurate results.

2.3 Lab in a pocket: IoT device for virtual imaging machine

Alexapath, a Bangalore-based company, has introduced a solution to make the traditional microscope smarter, allowing fast sharing of images and video feeds to specialists working remotely [3]. Alexapath makes microscope smarter and faster to provide an affordable diagnosis of diseases. The hardware component of Alexapath is Auto Diagnostic Assistant (ADA) that can be easily attached to the stage of any microscope and allows auto capturing of the slide images from the camera mounted on the eyepiece. ADA is IoT robotic slide holder that converts a conventional microscope into a virtual imaging device (Fig. 3), which is then linked with a smartphone. This enables slides to be automatically captured and the Wi-Fi-enabled ADA can automatically share slide images or videos in real time to experts working remotely. Thus, it is a great tool to take second opinions and collective consultations on complex cases.

3 Importance of IoT and AI in the field of cancer diagnosis

Cancer has a significant impact on the society across the world, owing to high morbidity and mortality rates. Proper treatment and detection at the early stages can significantly reduce cancer-related deaths. Various imaging techniques are used for assisting in diagnosis, including mammography, magnetic resonance imaging (MRI), ultrasonography, and thermography. However, histopathological analyses of H&E-stained tissue sections by an experienced pathologist remain the gold standard for most reliable diagnoses. Histopathological images are used for the diagnosis of all types of cancers, including breast cancer. In cancer diagnosis, differentiation between benign and malignant lesions is essential to determine the appropriate therapeutic regime. However, the accuracy of a manual analysis of histopathological slides varies from 65% to 98% on the basis of the experience and knowledge of the pathologist [5]. Human error may lead to inappropriate diagnoses, with literature reporting 25%–26% discordance between pathologists in differentiating between malignant and benign neoplasms [6]. Manual analysis is also laborious and time consuming, and often, the time involved in manual diagnosis may delay the diagnosis, eventually resulting in fatalities. In the era of image digitization, computer-aided diagnosis (CAD) has become a reality and has the potential to transform the field of cancer diagnosis. Automation overcomes the limitations of manual histopathological image analysis and can reduce the workload of pathologists by facilitating accurate and rapid diagnosis possible. Various studies have shown that CAD-based approaches can provide a rapid diagnosis in complex cases also. CAD systems have shown to enhance diagnostic quality and reduce analysis time in the detection of breast cancer [7–9], cervical cancer, prostate cancer, and lung cancer [10]. The new developments in AI-based approaches make this process more consistent, reliable, and cost effective than conventional methods. For automated image analysis, various AI-based algorithms like artificial neural networks, support vector machine (SVM), random forest, Naïve Bayesian, decision trees, etc. have been explored. Deep learning has emerged as one of the most powerful tool in the healthcare industry for image analysis.

Recently, the IoT and powerful infrastructure like the cloud are fast-growing phenomena in the field of e-healthcare that supports interconnecting devices and gathering data so as to provide efficient and robust services to users. These remotely connected devices are useful for exchanging patient-related information among pathologists, physicians, and healthcare experts. The term Internet of Medical Things (IoMT) is used to describe the exchange of medical information through devices and software applications, which are remotely connected through the Internet to the cloud platform. IoMT has the potential to improve the accuracy, productivity, and reliability of diagnosis. IoMT has numerous applications such as patient monitoring, wearable healthcare devices, tracking medical orders, sending feedback to users, and many more. Through IoMT, various pathological reports are made directly accessible to surgeons and experts on their smartphones, which greatly helps them to make timely decisions during complicated surgeries. Most of the IoMT devices, like smartphones and tablets, generally have low computation power and limited memory. Thus, most of the applications of IoMT are executed through the cloud. IoMT devices collect health information from patients and the information is accessible to doctors for diagnosis through cloud data centers. There is a huge demand for IoMT devices and IoMT trade is expected to touch 136.8 billion per year up to the year 2021 globally [11]. In the remote areas, with a shortage of cancer pathologists, the IoMT services are of great help. The patient data in the form of cytological or histological images can be fed through the mobile devices to the e-health care expert system, which can accurately detect and classify cancer cells.

The cloud-based e-Health care server can effectively perform data analysis for providing rapid cancer diagnosis. Processing vast amounts of data generated by IoMT devices is not possible using conventional data processing algorithms. Therefore, intelligent deep learning and machine learning-based algorithms are required for efficient data processing and analyses. Recently, nature-inspired algorithms (NIA) are widely used in deep learning model to find the best architecture for specific application as well as in image analysis through deep learning model and resulted in new evolution of intelligent deep

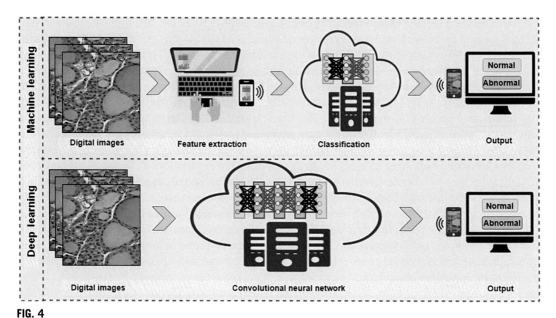

FIG. 4

Cloud-assisted machine learning and deep learning-based approaches for automated analysis of digital histopathological images.

learning [12]. Thus, machine learning and deep learning-based techniques are rapidly evolving along with IoMT technology for the early detection and classification of cancer. Fig. 4 shows how machine learning and deep learning-based architecture can be used for automated cancer diagnosis in cloud-assisted framework. Further, healthcare industry also looking for secure health data transmission for IoMT, which is essential for these cloud-based applications [13]. Recently, medical image encryption using NIA is presented in [14]. Subsequently, a secure and intelligent deep learning can be seen in near future for the early prediction of cancer.

4 IoT in the detection of breast cancer

The early and accurate detection of breast cancer is vital for recovery and poses a significant challenge for pathologists. Biopsy analysis, mammography, and MRI are conventionally used for breast cancer diagnosis. Various serological assays and gene expression-based assays have also been explored for cancer diagnosis [15–17]; however, histopathology of H&E-stained tumor tissues remains the gold standard. However, histopathological analysis is laborious, time consuming, and involve complex analysis by experienced pathologists. Various machine learning and deep learning-based approaches have been explored to automate breast cancer diagnosis. A technique employing three classifiers, namely probabilistic neural network, radial basis function, and multilayer perceptron, has been described by Azar and El-Said for the diagnosis of breast cancer [18]. Recently, Hasan et al. [19] have reported automated detection of breast cancer images using symbolic regression of multigene genetic programming with 99.28% accuracy. A breast classification algorithm using fuzzy-GA technique with 97.36% accuracy has also been proposed [20]. F-score method and SVM-based detection method for breast cancer features selection have also been proposed with good results [21]. Researchers have also used k-means algorithm along with SVM selection and extraction of features for binary breast cancer classification [22]. Hybridized principal component analysis (PCA), in combination with different classifiers has also shown promise in breast cancer diagnosis [23]. A memetic pareto artificial neural network-based breast cancer classification has also been explored [24] in early 2000. Later, artificial meta-plasticity multilayer perceptron-based algorithm has been explored for breast cancer with classification performance accuracy of around 99.2% [25]. Liu et al. [26] have utilized decision tree-based breast cancer prediction strategy along with under-sampling method for balancing of the training data. Onan [27] utilized a novel fuzzy-rough nearest neighbor algorithm for automated breast cancer detection. Researchers have also developed particle swarm optimization-based architecture utilizing a nonparametric kernel density-based method for accurate prediction of breast cancer [28]. Framework based on mixed ensemble of convolutional neural networks (CNN) with accuracy of 96.39% has also been developed by Rasti et al. [29]. A fused VGGNet-16-based framework has also been described with binary classification of breast cancer with accuracy of around 97% [7]. Researchers have also investigated VGGNet-19 for IDC histopathology breast cancer analysis [9]. Now researchers are also exploring for new efficient feature extraction technique for cancer diagnosis. Very recently, a novel CoMHisP framework based on a fuzzy SVM with within-class density information (FSVM-WD) has also been reported for the classification of breast cancer [30]. Thus so far, various deep learning and machine learning-based architectures have shown promise for the detection of breast cancer. The integration of such promising algorithms with the IoT technology can lead to developments of novel platforms for on-site diagnosis of breast cancers. An IoT-based tool for disease diagnosis and monitoring of patients has been developed by Ani et al. [31]. The system utilizes ensemble classifier and can provide diagnosis with an accuracy of 93%. An IoT-cloud-integrated wearable ECG monitoring system has been proposed recently by Yang et al. [32]. The system proves the efficacy of smart healthcare devices in the diagnosis of diseases. A machine learning-based IOT predictive model has been successfully tested for differentiating benign and malignant breast cancer patients [33]. The method adopted recursive feature selection-based approach for selection of features, which improves the performance of the SVM classifier [33]. From these studies, it can be concluded that deep learning and machine learning-based algorithms combined with IoT-enhanced diagnoses accuracy, and reduces the time and labor involved in the manual diagnosis of breast cancer. The application of these systems is very reliable in all aspects of IoT healthcare for breast cancer.

5 IoT in lung cancer diagnosis

Lung cancer is the most frequently reported cancer in men and the third most common cancer among women. Survival rate can be greatly increased if precancerous lung lesions can be detected in early stages. Analysis of the lung nodule using computed tomography (CT) is used for diagnosing and monitoring the disease severity. Using the current methodologies, diagnosis is usually made in the middle or advanced stages of cancer. CAD models have been found to be effective in the early detection of lung abnormalities. Deep learning and machine learning-based models have been applied for the detection of lung cancer using CT images. Both supervised and unsupervised models have been tried for image classification. Recently, a model for diagnosis of IoT coupled with cloud using SVM has been developed by Valluru et al. [34]. The researchers in this study have designed an optimal SVM for classifying lung cancer images. For the optimization of SVM parameters and feature selection, a modified gray wolf optimization algorithm is integrated with genetic algorithm (GWO-GA). The improved gray wolf optimization (IGWO) procedure extracts the significant features, and thereafter, SVM is applied for the classification of image data. The IoT-based model proposed in the study method achieved an average classification accuracy of 93.5% for lung cancer. Detailed analysis revealed that the model allows real-time data analysis and monitoring in hospitals. The efficacy of the model can be further enhanced by the incorporation of deep learning-based algorithms. In yet another study [35], IoT-based predictive algorithm for lung cancer has been designed employing fuzzy clustering-based augmentation by using the affected CT images. The model has the potential for lung cancer prediction and can even improve healthcare by providing real-time instructions. The model is based on fuzzy c-means clustering algorithm for categorizing transition region features for the effective image segmentation, along with Otsu thresholding method for the extraction of features from lung cancer images. Furthermore, morphological thinning and right edge images are also utilized for effective segmentation. Moreover, a new incremental classification algorithm effectively combining the existing association rule mining, decision tree with temporal features, and the CNNs have been utilized. The experimentation is done on the images collected from standard databases and the health data collected from patients through IoT devices. Thus, various studies have successfully demonstrated the application of IoT in the diagnosis of lung cancer; however, the challenge is the large-scale application of such systems in the healthcare sector.

6 IoT in the diagnosis of cervical cancer

Cervical cancer is a cancer caused by the cervix that generally affects women over 30 years of age. Representing 6.6% of all cancers, this is the fourth most common cancer among women. The diagnosis is usually made through screening of cervical smears for the detection of cancerous lesions. However, for accurate diagnosis, quality control is essential in cytological screening. For reducing the incidence of false negatives, the cervical smears need to be analyzed twice or thrice by at least two pathologists. This takes a lot of time and efforts of trained pathologists, resulting in a delay in diagnoses. Thus, several systems have been tried for automated cervical cancer screening so as to reduce the workload and efficiency of pathologists. The first automated pap-smear screening device was a US-developed cytoanalyzer, based on the concept of nuclear size and optical density [36]. Extensive experiments carried out using the device, however, revealed that the device had produced too many false results. Another automated application CYBEST, developed in Japan, was based on nuclear characteristics such as area, density, and nuclear to cytoplasmic ratio [37]. Though the technology showed promising results, it was never commercialized. Another cervical screening image analysis project was BioPEPR, which also used nuclear features for image identification [38], but due to a lack of in-depth studies the product could not reach the market. Yet, another technology was FAZYTAN [39], which relied on TV-image pickup and parallel processing. The system was efficient but could not be commercialized due to lack of cost effectiveness. In 2004, BD Focal Point Slide Profiler imaging system was developed, and a new liquid-based specimen preparation technique called Sure Path was added later to further enhance the efficiency of pap-smear slide analyzes [40]. Although these automated cervical cancer screening systems are available, the high costs associated with them limit their use in low- to middle-income countries. Automated cost-effective IoT-based devices are now coming to speed up the diagnosis process. Different instruments are available today: computerized microscopes (Ac Cell Series 2000 Pathfinder System), semiautomated or interactive systems (PAPNET, ACCESS, AutoCyte, or CytoRich), and automated systems (AutoPap 300), which are available in the healthcare sector and provide possible automation. These devices utilize algorithmic image analysis, with the exception of computerized microscopes, which values single cells using the morphological feature. The PAPNET system, an automated interactive instrument for examining traditional cervical smears (Papanicolaou), was the first to incorporate connectivity into an automated screening [41]. The technique incorporates the use of algorithmic image analysis with neural network, and the system performance is in the form of ranking of the cells detected for abnormality. The abnormal images are stored on a magnetic tape, and then displayed to the cytotech at a review station, where the decision is made about the abnormality.

Another device developed to minimize errors in interpreting cervical smears is the AutoPap 300 QC system [42]. The system is an automated device that has been approved by the US Food and Drug Administration to analyze conventionally prepared cervical cytology slides. The use of the AutoPap 300 QC System can significantly improve the number of false-negative cases. The device assigns a score based on the probability of a slide being abnormal. In remote instances, even in the absence of pathologists, AutoPap can be used to analyze visual image data acquired by IoT connected devices and detect regions of abnormal cytology. Images of such abnormal fields can then be shared with a pathologist through a smartphone. Very recently, pap-smear analysis tool (PAT) employing Trainable Weka Segmentation classifier has been described for detection of cervical cancer [43]. The study used a sequential elimination approach for debris rejection, and for the selection of features, simulated annealing integrated with a wrapper filter was used. Finally, a fuzzy c-means algorithm was used to achieve classification. Studies have shown that the system significantly reduces the time compared to manual analysis, and is a low-cost system that could be of considerable use in the developing economies.

7 Open challenges and future directions in digital pathology

The technical foundation of future healthcare lies in intelligent IoT along with cloud computing and fog computing. However, IoT-enabled healthcare gives a new edge and takes a step forward in modern automated intelligent healthcare applications. Further it facilitates users to access remote service in spite of different geographical location but it also faces the challenges in practical healthcare implementation in real time. Among these, major challenges in robust IoT are spatial diversity of the entities involved in communications, due to the limitations of the propagation media, and due to the varying temporal characteristics of the environment. Besides this, health data are mostly uploaded to a cloud server for further analysis and, therefore, due to different transmission modes, noise is introduced in data and thus makes data vulnerable for cyberattacks. Since these critical applications, such as remote surgery and cancer diagnosis, require reliable networking solutions that are robust for disturbances and disruptions, including high mobility, cyberattacks, disasters, high density, infrastructure failures, etc. Therefore, an efficient secure technology needs to be developed.

Data privacy, scalability, data storage, interoperability, private prediction, etc. are the crucial points as well as open problems that still need to be explored for practical implementation of IOMT.

Further, medical pathology data belong to high-dimensional data that need to be uploaded over cloud for further analysis; however, medical devices and cloud are usually resource constrained and not well efficient to handle big data is another issue that can be solved by designing an efficient data management technique for IoMT in near future. Even more reducing computation cost as well as communication cost among device and server is also an interesting future direction. Distributed data modeling for pathology images and its deployment over cloud for private prediction can be a new future of digital pathology.

8 Conclusion

The IoT plays a key role in the field of digital pathology. AI including machine learning and deep learning, cloud computing expands the IoT powers even further. Integrated with these technologies, IoT can be nothing short of a revolution in the field, but there are still many difficulties, peculiarities, and technological barriers to overcome before the widespread use of technology in the healthcare sector.

This chapter, therefore, discusses the recent work on cancer diagnosis in order to provide a deeper insight and clarity into emerging technologies in the field of digital pathology. We also set out in detail the obstacles and challenges to the practical implementation of IoMT along with its future directions. In addition, we also pointed out different AI techniques used to monitor and diagnose different types of cancer. Finally, the need for a secure intelligent deep model as well as a secure AI for the distributed environment for future IoMT is mentioned.

References

[1] S.U. Khan, N. Islam, Z. Jan, I.U. Din, A. Khan, Y. Faheem, An e-health care services framework for the detection and classification of breast cancer in breast cytology images as an IoMT application, Futur. Gener. Comput. Syst. 98 (2019) 286–296.

[2] I. Hernández-Neuta, F. Neumann, J. Brightmeyer, T. Ba Tis, N. Madaboosi, Q. Wei, A. Ozcan, M. Nilsson, Smartphone-based clinical diagnostics: towards democratization of evidence-based health care, J. Inter. Med. 285 (1) (2019) 19–39.

[3] A. Thomas, Lab in a pocket: this IoT device turns a standard microscope into a virtual imaging machine, 2018, https://economictimes.indiatimes.com/small-biz/startups/features/lab-in-a-pocket-this-iot-device-turns-a-standard-microscope-into-a-virtual-imaging-machine-alexapath/articleshow/62454623.cms?from=mdr.

[4] N.S. Zamani, M.N. Mohammed, S. Al-Zubaidi, E. Yusuf, Design and development of portable digital microscope platform using IoT technology, in: 2020 16th IEEE International Colloquium on Signal Processing & Its Applications (CSPA), IEEE, 2020, pp. 80–83.

[5] A. Muratli, N. Erdogan, S. Sevim, I. Unal, S. Akyuz, Diagnostic efficacy and importance of fine-needle aspiration cytology of thyroid nodules, J. Cytol. 31 (2) (2014) 73.

[6] D.S. Gomes, S.S. Porto, D. Balabram, H. Gobbi, Inter-observer variability between general pathologists and a specialist in breast pathology in the diagnosis of lobular neoplasia, columnar cell lesions, atypical ductal hyperplasia and ductal carcinoma in situ of the breast, Diagn. Pathol. 9 (1) (2014) 121.

[7] A. Kumar, S.K. Singh, S. Saxena, K. Lakshmanan, A.K. Sangaiah, H. Chauhan, S. Shrivastava, R.K. Singh, Deep feature learning for histopathological image classification of canine mammary tumors and human breast cancer, Inf. Sci. 508 (2020) 405–421.

[8] J. Xie, R. Liu, J. Luttrell IV, C. Zhang, Deep learning based analysis of histopathological images of breast cancer, Front. Genet. 10 (2019) 80.

[9] R. Singh, T. Ahmed, A. Kumar, A.K. Singh, A.K. Pandey, S.K. Singh, Imbalanced breast cancer classification using transfer learning. IEEE/ACM Trans. Comput. Biol. Bioinf. (2020). https://doi.org/10.1109/TCBB.2020.2980831.

[10] F. Bray, J. Ferlay, I. Soerjomataram, R.L. Siegel, L.A. Torre, A. Jemal, Global Cancer Statistics 2018: GLOBOCAN estimates of incidence and mortality worldwide for 36 cancers in 185 countries, CA Cancer J. Clin. 68 (6) (2018) 394–424.

[11] G.J. Joyia, R.M. Liaqat, A. Farooq, S. Rehman, Internet of medical things (IOMT): applications, benefits and future challenges in healthcare domain, J. Commun. 12 (4) (2017) 240–247.

[12] V. Bharti, B. Biswas, K.K. Shukla, Recent trends in nature inspired computation with applications to deep learning, in: 2020 10th International Conference on Cloud Computing, Data Science & Engineering (Confluence), IEEE, 2020, pp. 294–299.

[13] H. Qiu, M. Qiu, G. Memmi, M. Liu, Secure health data sharing for medical cyber-physical systems for the healthcare 4.0. IEEE J. Biomed. Health Inf. (2020), https://doi.org/10.1109/JBHI.2020.2973467.

[14] V. Bharti, B. Biswas, K.K. Shukla, A novel multiobjective GDWCN-PSO algorithm and its application to medical data security, ACM Trans. Internet Technol. (2020), https://doi.org/10.1145/3397679.

[15] S. Hussain, S. Saxena, S. Shrivastava, R. Arora, R.J. Singh, S.C. Jena, N. Kumar, A.K. Sharma, M. Sahoo, A.K. Tiwari, et al., Multiplexed autoantibody signature for serological detection of canine mammary tumours, Sci. Rep. 8 (1) (2018) 1–14.

[16] S.C. Jena, S. Shrivastava, S. Saxena, N. Kumar, S.K. Maiti, B.P. Mishra, R.K. Singh, Surface Plasmon resonance immunosensor for label-free detection of BIRC5 biomarker in spontaneously occurring canine mammary tumours, Sci. Rep. 9 (1) (2019) 1–12.

[17] S. Hussain, S. Saxena, S. Shrivastava, A.K. Mohanty, S. Kumar, R.J. Singh, A. Kumar, S.A. Wani, R.K. Gandham, N. Kumar, Gene expression profiling of spontaneously occurring canine mammary tumours: insight into gene networks and pathways linked to cancer pathogenesis. PLoS ONE 13 (12) (2018). https://doi.org/10.1371/journal.pone.0208656. e0208656.

[18] A.T. Azar, S.A. El-Said, Probabilistic neural network for breast cancer classification, Neural Comput. Appl. 23 (6) (2013) 1737–1751.

[19] M.K. Hasan, M.M. Islam, M.M.A. Hashem, Mathematical model development to detect breast cancer using multigene genetic programming, in: 2016 5th International Conference on Informatics, Electronics and Vision (ICIEV), IEEE, 2016, pp. 574–579.

[20] C.A. Pena-Reyes, M. Sipper, A fuzzy-genetic approach to breast cancer diagnosis, Artif. Intell. Med. 17 (2) (1999) 131–155.

[21] M.F. Akay, Support vector machines combined with feature selection for breast cancer diagnosis, Expert Syst. Appl. 36 (2) (2009) 3240–3247.

[22] B. Zheng, S.W. Yoon, S.S. Lam, Breast cancer diagnosis based on feature extraction using a hybrid of K-means and support vector machine algorithms, Expert Syst. Appl. 41 (4) (2014) 1476–1482.

[23] G.N. Ramadevi, K.U. Rani, D. Lavanya, Importance of feature extraction for classification of breast cancer datasets—a study, Int. J. Sci. Innov. Math. Res. 3 (2) (2015). 763-368.

[24] H.A. Abbass, An evolutionary artificial neural networks approach for breast cancer diagnosis, Artif. Intell. Med. 25 (3) (2002) 265–281.

[25] A. Marcano-Cedeño, J. Quintanilla-Domínguez, D. Andina, WBCD breast cancer database classification applying artificial metaplasticity neural network, Expert Syst. Appl. 38 (8) (2011) 9573–9579.

[26] Y.-Q. Liu, C. Wang, L. Zhang, Decision tree based predictive models for breast cancer survivability on imbalanced data, in: 2009 3rd International Conference on Bioinformatics and Biomedical Engineering, IEEE, 2009, pp. 1–4.

[27] A. Onan, A fuzzy-rough nearest neighbor classifier combined with consistency-based subset evaluation and instance selection for automated diagnosis of breast cancer, Expert Syst. Appl. 42 (20) (2015) 6844–6852.

[28] R. Sheikhpour, M.A. Sarram, R. Sheikhpour, Particle swarm optimization for bandwidth determination and feature selection of kernel density estimation based classifiers in diagnosis of breast cancer, Appl. Soft Comput. 40 (2016) 113–131.

[29] R. Rasti, M. Teshnehlab, S.L. Phung, Breast cancer diagnosis in DCE-MRI using mixture ensemble of convolutional neural networks, Pattern Recogn. 72 (2017) 381–390.

[30] A. Kumar, S.K. Singh, S. Saxena, A.K. Singh, S. Shrivastava, K. Lakshmanan, N. Kumar, R. K. Singh, CoMHisP: a novel feature extractor for histopathological image classification based on fuzzy SVM with within-class relative density, IEEE Trans. Fuzzy Syst. (2020) 1–14.

[31] R. Ani, S. Krishna, N. Anju, M.S. Aslam, O.S. Deepa, IoT based patient monitoring and diagnostic prediction tool using ensemble classifier, in: 2017 International Conference on Advances in Computing, Communications and Informatics (ICACCI), IEEE, 2017, pp. 1588–1593.

[32] Z. Yang, Q. Zhou, L. Lei, K. Zheng, W. Xiang, An IoT-cloud based wearable ECG monitoring system for smart healthcare, J. Med. Syst. 40 (12) (2016) 286.

[33] M.H. Memon, J.P. Li, A.U. Haq, M.H. Memon, W. Zhou, Breast cancer detection in the IOT health environment using modified recursive feature selection, Wireless Commun. Mobile Comput. 2019 (2019). 1–19.

[34] D. Valluru, I.J.S. Jeya, IoT with cloud based lung cancer diagnosis model using optimal support vector machine, Health Care Manag. Sci. (2019). https://doi.org/10.1007/s10729-019-09489-x.

[35] D. Palani, K. Venkatalakshmi, An IoT based predictive modelling for predicting lung cancer using fuzzy cluster based segmentation and classification, J. Med. Syst. 43 (2) (2019) 21.

[36] E.G. Diacumakos, E. Day, M.J. Kopac, Exfoliated cell studies and the cytoanalyzer, Ann. N. Y. Acad. Sci. 97 (2) (1962) 498–513.

[37] N. Tanaka, T. Ueno, H. Ikeda, A. Ishikawa, K. Yamauchi, Y. Okamoto, S. Hosoi, CYBEST model 4. Automated cytologic screening system for uterine cancer utilizing image analysis processing, Anal. Quant. Cytol. Histol. 9 (5) (1987) 449–454.

[38] D.J. Zahniser, P.S. Oud, M.C.T. Raaijmakers, G.P. Vooys, R.T. Van de Walle, Field test results using the BioPEPR cervical smear prescreening system, Cytometry 1 (3) (1980) 200–203.

[39] R. Erhardt, E.R. Reinhardt, W. Schlipf, W.H. Bloss, FAZYTAN: a system for fast automated cell segmentation, cell image analysis and feature extraction based on TV-image pickup and parallel processing, Anal. Quant. Cytol. 2 (1) (1980) 25–40.

[40] T.F. Kardos, The FocalPoint system: FocalPoint slide profiler and FocalPoint GS, Cancer Cytopathol. 102 (6) (2004) 334–339.

[41] M. Cenci, C. Nagar, M.R. Giovagnoli, A. Vecchione, The PAPNET system in cytological rescreening of cervical smears, Minerva Ginecol. 49 (4) (1997) 139–145.

[42] S.F. Patten Jr, J.S.J. Lee, D.C. Wilbur, T.A. Bonfiglio, T.J. Colgan, R.M. Richart, H. Cramer, S. Moinuddin, The AutoPap 300 QC system multicenter clinical trials for use in quality control rescreening of cervical smears: II. Prospective and archival sensitivity studies, Cancer Cytopathol. 81 (6) (1997) 343–347.

[43] W. William, A. Ware, A.H. Basaza-Ejiri, J. Obungoloch, A pap-smear analysis tool (PAT) for detection of cervical cancer from pap-smear images, Biomed. Eng. Online 18 (1) (2019) 16.

Index

Note: Page numbers followed by *f* indicate figures, and *t* indicate tables.